Aquatic Oligochaeta

Developments in Hydrobiology 24

Series editor
H. J. Dumont

Aquatic Oligochaeta

Proceedings of the Second International Symposium on Aquatic Oligochaete Biology, held in Pallanza, Italy, September 21–24, 1982

Edited by
G. Bonomi and C. Erséus

Reprinted from Hydrobiologia, vol. 115 (1984)

1984 **DR W. JUNK PUBLISHERS**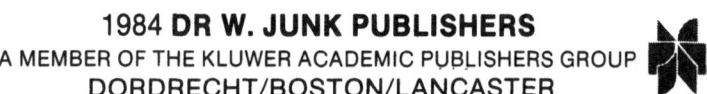
A MEMBER OF THE KLUWER ACADEMIC PUBLISHERS GROUP
DORDRECHT/BOSTON/LANCASTER

Distributors

---

for the United States and Canada: Kluwer Boston, Inc.,
190 Old Derby Street, Hingham, MA 02043, USA
for the UK and Ireland: Kluwer Academic Publishers, MTP
Press Limited, Falcon House, Queen Square, Lancaster
LA1 1RN, UK
for all other countries: Kluwer Academic Publishers Group,
Distribution Center, P.O. Box 322, 3300 AH Dordrecht,
The Netherlands

Library of Congress Cataloging in Publication Data

```
International Symposium on Aquatic Oligochaete
   Biology (2nd : 1982 : Pallanza, Italy)
   Aquatic Oligochaeta.

   (Developments in hydrobiology ; 24)
   "Reprinted from Hydrobiologia, vol. 115."
   Includes indexes.
   1. Oligochaeta--Congresses.  2. Aquatic invertebrates--
Congresses.  I. Bonomi, G. (Giuliano)  II. Erséus, C.
(Christer)  III. Hydrobiologia.  IV. Title.  V. Series.
QL391.A6I57  1982          595.1'46'0916          84-4361
```

ISBN-13: 978-94-009-6565-2   e-ISBN-13: 978-94-009-6563-8
DOI: 10.1007/978-94-009-6563-8

Cover design: Max Velthuijs

*This volume is dedicated
to the memory of H. R. Baker*

Preface

At the First International Symposium on Aquatic Oligochaete Biology in Sidney, B.C., the suggestion was made to hold the second in Pallanza, on Lake Maggiore, at the C.N.R.-Istituto Italiano di Idrobiologia. At the same time it was decided that there should be a symposium every third year.

The organization of the symposium was made considerably easier by the Senior Editor's having been in personal contact with several Russian colleagues (courtesy of a kind invitation from the U.S.S.R. Academy of Sciences), the 'Hamburg Group' and the Junior Editor. Correspondence with various students of Oligochaeta also furnished many useful suggestions.

The Second International Symposium on Aquatic Oligochaete Biology was therefore held in Pallanza in late September 1982 and was attended by 53 scientists from 16 countries in Europe, North America, Asia and Australia. No review papers were formally invited as this had already been done at the first symposium: all the papers were accordingly allotted the same time for oral presentation. These proceedings contain all papers presented at the symposium, including those by Hrabĕ and Slepukhina, which were given *in absentia*, as it was impossible for the authors to join the group. The main topics were taxonomy and evolution of Oligochaeta, life-cyle and population studies, the role of Oligochaeta in assessing water pollution, physiological studies, community and distribution studies. Special taxonomic workshops took place in the evenings.

After an informal 'al fresco' reception on the evening of September 20th, the opening session was held the following morning at the C.N.R.-Istituto Italiano di Idrobiologia. Participants were formally welcomed by Prof. Ettore Grimaldi, Director of the Institute, and by Prof. Livia Tonolli, the S.I.L. National Representative. In the afternoon the group moved to the hotel, in whose conference hall the remaining sessions took place. During the symposium we took some breaks – a visit to the beautiful botanic gardens of Villa Taranto (thanks to the kind generosity of Livia Tonolli) and a tour of the Istituto Italiano di Idrobiologia laboratories. The official dinner was a very pleasant occasion in the fairy-tale setting of Orta on the lake of the same name 30 km from Pallanza.

Mrs. Adreani and Mrs. Nobili very much appreciated the many expressions of thanks which arrived after the end of the symposium, in recognition of their valuable efforts in running the secretariat; Mr. Arca, Mr. Piazza and Miss Sabadini were of particular help in the conference rooms, but we acknowledge the contribution of the entire staff of the Istituto to the success of the meeting. Miss Sandra Spence revised almost all of the English texts and generously assisted in the symposium activities.

We are especially grateful to our colleague Dr. Carla Bonacina whose help and support were inestimable at every stage of our work, including the editing of the papers.

The symposium was sponsored and entirely financed by the National Research Council of Italy, to whom we give our grateful thanks. S.I.L. sponsorship was also given.

The same atmosphere of friendly cooperation which was noticeable in Sidney was also in evidence in Pallanza, and we are certain that the Hamburg Symposium of 1985 will carry on this happy tradition.

G. Bonomi
C. Erséus

Contents

PART THREE: LIFE CYCLE, PRODUCTION AND POPULATION DYNAMICS

PART FOUR: POLLUTION AND COMMUNITY STRUCTURE

PART FIVE: DISTRIBUTION

List of participants and contributors

Luciana Adreani, C.N.R.-Istituto Italiano di Idrobiologia, Largo V. Tonolli 50/52, I-28048 Pallanza NO, Italy

Randy Baker, Department of Biology, University of Victoria, P.O. Box 1700, Victoria, British Columbia, Canada V8W 2Y2

Carla Bonacina, C.N.R.-Istituto Italiano di Idrobiologia, Largo V. Tonolli 50/52, I-28048 Pallanza NO, Italy

Giuliano Bonomi, C.N.R.-Istituto Italiano di Idrobiologia, Largo V. Tonolli 50/52, I-28048 Pallanza NO, Italy

Ralph O. Brinkhurst, Institute of Ocean Sciences, P.O. Box 6000, Sidney, British Columbia, Canada V8L 4B2

Sandra Casellato, Istituto di Biologia Animale, Università di Padova, Via Loredan 10, I-35100 Padova, Italy

Christina Centurioni, Museo Civico di Storia Naturale, Lungadige Porta Vittoria 9, I-37129 Verona, Italy

Peter M. Chapman, E.V.S. Consultants Ltd., 195 Pemberton Avenue, North Vancouver, British Columbia, Canada V7P 2R4

Bent Christensen, Institute of Population Biology, Universitetsparken 15, DK-2100 Copenhagen, Denmark

Kathryn Coates, Institute of Ocean Sciences, P.O. Box 6000, Sidney, British Columbia, Canada V8L 4B2

Robert J. Diaz, Department of Estuarine and Coastal Ecology, Virginia Institute of Marine Science and School of Marine Science, The College of William and Mary, Gloucester Point, VA 23062, U.S.A.

Michael Dzwillo, Zoologisches Institut und Museum der Universität Hamburg, Martin-Luther-King-Platz 3, D-2000 Hamburg 13, Federal Republic of Germany

Christer Erséus, Department of Zoology, University of Gothenburg, Box 25059, S-400 31 Göteborg, Sweden

Marco Ferraguti, Dipartimento di Biología, Sezione di Zoologia e Citologia, Università di Milano, Via Celoria 26, I-20133 Milano, Italy

Nonna P. Finogenova, Zoological Institute, U.S.S.R. Academy of Sciences, University Embankment, Leningrad 199034, U.S.S.R.

Stuart R. Gelder, Biology Section, Division of Math/Science, University of Maine at Presque Isle, Presque Isle, ME 04769, U.S.A.

Narcisse Giani, Laboratoire d'Hydrobiologie, Université P. Sabatier, 118 Route de Narbonne, F-31062 Toulouse Cédex, France

Olav Giere, Zoologisches Institut und Museum der Universität Hamburg, Martin-Luther-King-Platz 3, D-2000 Hamburg 13, Federal Republic of Germany

Antanas Grigelis, Institute of Zoology and Parasitology, Lithuanian Academy of Sciences, 232021 Vilnius, Lithuanian S.S.R., U.S.S.R.

Walter J. Harman, Department of Zoology, Louisiana State University, Baton Rouge, LA 70803, U.S.A.

Brenda Healy, Department of Zoology, University College, Belfield, Stillorgan Road, Dublin 4, Ireland

Sergěj Hrabě, Stojanova 9, CS-602 00 Brno, Czechoslovakia

Marice Hullé, École Normale Supérieure, Laboratoire de Zoologie, 46, Rue d'Ulm, F-75230 Paris Cédex 05, France

Barrie G. M. Jamieson, Department of Zoology, University of Queensland, St. Lucia, Brisbane, Queensland, Australia 4067

Petur M. Jónasson, Freshwater Biological Laboratory, University of Copenhagen, 51 Elsingørsgade, DK-3400 Hillerød, Denmark

Jacques Juget, Université Claude Bernard, Lyon I, Dpt. Biologie animale et Ecologie 'L.A.C.N.R.S. 367, 43, Boulevard du 11 Novembre 1918, F-69622 Villeurbanne Cédex, France

Krzysztof Kasprzak, Provincial Administration, Department of Environment Protection, Stalingradzka 16/18, PL-60-967 Poznań, Poland

Deedee Kathman, Institute of Ocean Sciences, P.O. Box 6000, Sidney, British Columbia V8L 4B2, Canada

Klaus J. Kossmagk-Stephan, II. Zoologisches Institut und Museum der Universität Göttingen, Berliner Str. 28, D-3400 Göttingen, Federal Republic of Germany

Michael Ladle, Freshwater Biological Association, River Laboratory, East Stoke, Wareham, Dorset BH20 6BB, U.K.

Michel Lafont, Laboratoire d'Hydrobiologie du CEMAGREF: 3, Quai Chauveau, F-69009 Lyon, France

Barbara Lang, Conservation de la Faune, Ch. du Marquisat 1, CH-1025 St-Sulpice, Switzerland

Claude Lang, Conservation de la Faune, Ch. du Marquisat 1, CH-1025 St-Sulpice, Switzerland

Michael S. Loden, Environmental and Development Control Department, Compliance Section, 3600 Jefferson Highway, Jefferson, LA 70121, U.S.A.

Enrique Martinez-Ansemil, Colegio Universitario de Orense, Universidad de Santiago, C. General France 35, Orense, Spain

Göran Milbrink, Institute of Zoology, University of Uppsala, P.O. Box 561, S-751 22 Uppsala, Sweden

Carlo Monti, C.N.R.-Istituto Italiano di Idrobiologia, Largo V. Tonolli 50/52, I-28048 Pallanza NO, Italy

Anna Maria Nocentini, C.N.R.-Istituto Italiano di Idrobiologia, Largo V. Tonolli 50/52, I-28048 Pallanza NO, Italy

Pietro Omodeo, Istituto di Biologia Animale, Università di Padova, Via Loredan 10, I-35100 Padova, Italy

Andreina Paoletti, Dipartimento di Biología, Università di Milano, Via Celoria 26, I-20133 Milano, Italy

Celia Pascar Gluzman, Istituto de Embriología, Biología e Histología, Facultad de Ciencias Médicas, Universidad Nacional de La Plata, La Plata 1900, Argentina

Olaf Pfannkuche, Institut für Hydrobiologie und Fischereiwissenschaft, Universität Hamburg, Zeiseweg 9, D-2000 Hamburg 50, Federal Republic of Germany

Tamara L. Poddubnaya, Institute of Biology of Inland Waters, U.S.S.R. Academy of Sciences, Laboratory of Zoobenthos, Borok, Nekouf, Jaroslavl 152742, U.S.S.R.

Narcis Prat, Departamento de Ecología, Facultad de Biología, Universidad de Barcelona, Av. Diagonal 637–645, Barcelona-28, Spain

Ludwig Probst, Institut für Seeforschung, D-7994 Langenargen (Bodensee), Federal Republic of Germany

Beatrice Sambugar, Museo Civico di Storia Naturale, Lungadige Porta Vittoria 9, I-37129 Verona, Italy

Tatyana D. Slepukhina, Institute of Lake Research of the Academy of Sciences of the U.S.S.R., Sevastyanova 9, 196199 Leningrad, U.S.S.R.

Valerie Standen, Department of Zoology, University of Durham, South Road, Durham DH1 3LE, U.K.

Rosmarie Steinlechner, Zoologisches Institut, Universitätsstr. 4, A-6020 Innsbruck, Austria

Tarmo Timm, Võrtsjärv Limnological Station of the Institute of Zoology and Botany of the Estonian S.S.R. Academy of Sciences, Rannu 202454, Tartu District, Estonian S.S.R., U.S.S.R.

Piet F. M. Verdonschot, Provincial Water Authority, Department of Watershed Management, P.O. Box 73, NL-8000 AB Zwolle, The Netherlands

Benno Wagner, Voralberger Umweltschutzanstalt, Montfortstrasse 4, A-6901 Bregenz, Austria

Mark J. Wetzel, Illinois Natural History Survey, 172 Natural Resources Bldg., 607 E. Peabody, Champaign, IL 61820, U.S.A.

Rudolf Zahner, Institut für Seeforschung, D-7994 Langenargen (Bodensee), Federal Republic of Germany

Howard R. Baker (1955–1983) in memoriam

Howard Randall Baker, known to all his many friends and colleagues as Randy, was a big man in every sense of the word. Having completed his B.Sc. at the University of Victoria in 1977, he first joined my laboratory as a technician in February, 1978. We were immediately impressed by those characteristics that many came to know and admire, his basic decency, generosity of spirit and warm good humour. He also proved to be so adept at handling marine oligochaetes and so responsible with valuable material that he was admitted to a special course of study in oligochaete microstructure and eventually as an M.Sc. candidate in July, 1979. His work was supported by the Government of British Columbia and first Dobrocky Seatech Ltd. and then E.V.S. Consultants Ltd. of Vancouver. After a year of study he decided to devote his time to taxonomy rather than to the more applied aspects of the work, and obtained funding from the National Scientific and Engineering Research Council from June 1981 to June 1983. As early as April, 1981, the supervisory committee proposed that Randy had progressed so fast that his enrollment be upgraded to allow him to proceed directly to the

Hydrobiologia 115, 1–2 (1984).
© Dr W. Junk Publishers, Dordrecht.

Ph.D. degree, with the proviso that he gain experience through travel and work at centres outside Victoria. Accordingly, Randy planned and executed a tour of Europe that took him to laboratories from Ireland and England through Norway and Sweden. In 1982 he acted as station manager of the Smithsonian Institution field station at Carrie Bow Cay, Belize, and worked with the visiting scientists and on his own collections. He attended the first and second International Aquatic Oligochaete Symposia, extending his European travel experience. This travel was twice supported by University grants.

Randy was a member of the Canadian Society of Zoologists, the International Association of Meiobenthologists, the Biological Society of Washington, and the Society of Systematic Zoology. He was a Voluntary Research Associate of the British Columbia Provincial Museum, and was elected Vice President, University of Victoria Graduate Student Society in April, 1983.

Randy combined scholarship and dedication with infectious charm, and all the shining promise of youth that has found its focus in life. His pioneering work is his monument. His place will not be filled, he will be truly missed.

Ralph O. Brinkhurst

Publications

Baker, H. R. & C. Erséus, 1979. *Peosidrilus biprostatus* n.g., n.sp., a marine tubificid (Oligochaeta) from the Eastern United States. Proc. biol. Soc. Wash. 92: 505–509.

Brinkhurst, R. O. & H. R. Baker, 1979. A review of the marine Tubificidae (Oligochaeta) of North America. Can. J. Zool. 57: 1553–1569.

Baker, H. R., 1980. A redescription of Tubificoides pseudogaster (Dahl) (Tubificidae: Oligochaeta). Trans. am. microsc. Soc. 99: 337–342.

Baker, H. R. & R. O. Brinkhurst, 1981. A revision of the genus *Monopylephorus* and redefinition of the subfamilies Rhyacodrilinae and Branchiurinae (Tubificidae: Oligochaeta). Can. J. Zool. 59: 939–954.

Baker, H. R., 1981a. A redescription of *Tubificoides heterochaetus* (Michaelsen) (Tubificidae: Oligochaeta). Proc. biol. Soc. Wash. 94: 564–568.

Baker, H. R., 1981b. *Phallodrilus tempestatis* n.sp., a new marine tubificid (Annelida: Oligochaeta) from British Columbia. Can. J. Zool. 59: 1475–1478.

Baker, H. R. & C. Erséus, 1982. A new species of *Bacescuella* Hrabě (Oligochaeta, Tubificidae) from the Pacific coast of Canada. Can. J. Zool. 60: 1951–1954.

Erséus, C. & H. R. Baker, 1982. New species of the gutless marine genus *Inanidrilus* (Oligochaeta, Tubificidae) from the Gulf of Mexico and Barbados. Can. J. Zool. 60: 3063–3067.

Baker, H. R., 1982a. A note on the genitalia of *Potamothrix hammoniensis* (Oligochaeta: Tubificidae). Proc. biol. Soc. Wash. 95: 563–566.

Baker, H. R., 1982b. Two new Phallodriline genera of marine Oligochaeta (Annelida: Tubificidae) from the Pacific northeast. Can. J. Zool. 60: 2487–2500.

Baker, H. R., 1982c. *Vadicola aprostatus* nov. gen., nov. sp., a marine oligochaete (Tubificidae: Rhyacodrilinae) from British Columbia. Can. J. Zool. 60: 3232–3236.

Baker, H. R., 1984. Diversity and zoogeography of marine Tubificidae (Annelida, Oligochaeta) with notes on variation in widespread species. In G. Bonomi & C. Erséus (eds.), Aquatic Oligochaeta. Proceedings of the Second International Symposium on Aquatic Oligochaete Biology. Developments in Hydrobiology (this volume).

Baker, H. R. & K. A. Coates, in press. Key to the marine Oligochaeta of Puget Sound and British Columbia. In E. N. Kozloff (ed.), Keys to the Marine Invertebrates of Puget Sound, the San Juan Archipelago, and Adjacent regions. Second edition.

Baker, H. R., 1983. New species of *Tubificoides* Lastockin (Annelida, Oligochaeta) from the Pacific Northeast and the Arctic. Can. J. Zool. 61: 1270–1283.

Baker, H. R., 1983. New species of *Bathydrilus* Cook (Oligochaeta: Tubificidae) from British Columbia. Can. J. Zool. 61: 2162–2167.

A phenetic and cladistic study of spermatozoal ultrastructure in the Oligochaeta (Annelida)

B. G. M. Jamieson
Department of Zoology, University of Queensland, St. Lucia, Brisbane, Queensland, Australia 4067

Keywords: aquatic Oligochaeta, oligochaete spermatozoa, acrosome, electron microscopy, phenetics, phylogeny

Abstract

Spermatozoal ultrastructure of nine oligochaete families has been examined for congruence with phylogenetic and taxonomic systems for the Oligochaeta based on general morphology, particularly the holomorphological hennigian analysis of Jamieson (1978a, 1980, 1983). Estimation of congruence has been made following phenetic and cladistic (phylogenetic) analysis. Correspondence, in phenograms and phylograms, of sperm types with taxonomic and phylogenetic groupings previously recognized is generally good. Departure from this rule in the similarity of the phreodrilid sperm to that of the Lumbricina suggests a corresponding alteration of fertilization biology in the phreodrilids. The results indicate that the Haplotaxidae lie at the base of the opisthopores though they do not unequivocally contraindicate acceptance of a *Haplotaxis* like form as a stem form of the Haplotaxida (opisthopores and Haplotaxidae) and Tubificida. An even more basal position for prosopores, now represented by the Lumbriculida, cannot yet be dismissed.

Introduction

The subclass Oligochaeta was divided into three orders by Brinkhurst & Jamieson (1971): the Lumbricida, Moniligastrida and Haplotaxida. Jamieson (1977) showed that the Moniligastrida were in fact transitionally opisthoporous and should be placed in the Haplotaxida, a view with which Brinkhurst & Fulton (1981) concurred. In a detailed computer analysis of the opisthoporous families and the Haplotaxidae, employing the principles of Hennig (1966), Jamieson (1978a, 1980) concluded by extrapolation that the Tubificina, of which major families are the Tubificidae, Naididae, Phreodrilidae and Enchytraeidae, should be removed from the Haplotaxida to constitute the separate order Tubificida, thus giving three orders with the Lumbriculida and Haplotaxida.

Independently, Timm (1981) recognized a similarly constituted order Tubificida but as the sole order of a superorder Naidimorpha separate from the superorder Lumbricomorpha (both superorders *sensu* Chekanovskaya, 1962). Brinkhurst (1982b, 1984) has come to accept a separate order for the Tubificida (despite adverse criticism of this step in another recent work, Brinkhurst, 1982a) but has further advanced the contention of Brinkhurst & Jamieson (1971) that the Haplotaxidae are representatives of very basal (plesiomorph) oligochaetes to the extent of deriving all oligochaetes from a haplotaxid-like stem form. He emphasizes the significance of the Haplotaxidae by restricting the order Haplotaxida to this family and consequently erects an order Lumbricida for the opisthoporous oligochaetes while recognizing also the orders Lumbriculida and Tubificida.

Derivation of Tubificida, modern Haplotaxidae and the opisthopores from an octogonadial, plesioporous haplotaxid-like form accords with the views of Brinkhurst & Jamieson (1971) and is not contrary to the phylogenetic suggestions of Jamieson (1978a, 1981a), but the author has differed from

Hydrobiologia 115, 3-13 (1984).

Brinkhurst's current system in regarding the proso-porous, multigonadial condition, which is retained in the Lumbriculidae, as precursory to the octogo-nadial, plesioporous haplotaxid condition.

The author (Jamieson, 1983) has examined, by intuitive means, congruence of spermatozoal ul-trastructure with his earlier, holomorphological phylogeny (Jamieson, 1978a) and has found the two studies to be, on the whole, consistent although the position of the Lumbriculidae cannot be consid-ered settled.

The present study was designed to investigate to what extent a computerized phenetic and cladistic classification of spermatozoal ultrastructure for all oligochaete families for which this is comprehen-sively known might illuminate these and other ob-scure areas in our understanding of oligochaete phylogeny and taxonomy. Most particularly, it is concerned with investigating congruence of sperm ultrastructure with the holomorphological phylog-eny proposed by the author (Jamieson, 1978a, 1980).

Material and Methods

Electron micrographs of the sperm of species listed in Table 1, from the sources indicated, were

Table 1. Species and sources from which spermatozoal data were obtained.

1. *Bythonomus lemani;* Ferraguti & Lanzavecchia 1977 (Lum-briculidae). (Probably *Stylodrilus lemani* (Grube).
2. *Lumbricillus rivalis;* Webster & Richards 1977 (Enchytraei-dae).
3. *Rhyacodrilus arthingtonae;* Jamieson, Daddow & Bennett 1978, and original (Tubificidae). (Now *Rhizodrilus arthing-tonae;* Baker & Brinkhurst 1981).
4. *Limnodriloides winckelmanni;* Jamieson, Daddow & Benett 1978; Jamieson & Daddow 1979 (Tubificidae). (Now *L. aus-tralis* Erséus, 1982).
5. *Phreodrilus* sp. Jamieson 1981b, and original (Phreodrili-dae).
6. *Haplotaxis ornamentus;* Jamieson 1982 (Haplotaxidae).
7. *Sparganophilus tamesis;* Richards, Fleming & Jamieson 1981; Jamieson, Fleming & Richards 1982; and original (Sparganophilidae).
8. *Hormogaster redii;* Ferraguti & Jamieson 1984 (Hormo-gastridae).
9. *Lumbricus rubellus;* Jamieson, Richards, Fleming & Erséus 1983 (Lumbricidae).
10. *Fletcherodrilus unicus;* Jamieson 1978b (Megascolecidae).
11. *Amynthas gracilis,* original (Megascolecidae).

used to assemble the data for attributes listed in Table 2. Phenograms of spermatozoal similarity were prepared by group-average and nearest neigh-bour sorting from similarity matrices computed with the Gower metric from one of which an ordi-nation was prepared by the principal coordinates procedure. A divisive classification was also con-structed. Ordered values were coded as cumulative binary states and disordered multistate characters as binary states; a hierarchy was then produced using a binary divisive technique based on a infor-mation statistic (DMIS). Computations and den-drograms were prepared on the CSIRONET sys-tem of the Commonwealth Scientific and Industrial Research Organization.

Results

The analyses are based on data for the attributes shown in Table 2 obtained from the micrographs of sperm listed in Table 1. The cladistic study has required recognition of plesiomorph states detailed in Jamieson (1983). Phenetic and cladistic nearest neighbours for each species are listed in Table 3.

Phenetic analyses

The attributes used for computing affinities, whether phenetic or cladistic, and in the latter case their deduced polarity with plesiomorph and apo-morph states, are listed in Table 2.

Phenetic dendrograms (Figs. 1–5) and the ordi-nation (Fig. 10) show close congruence whether resulting from group-average (Figs. 1–3) or nearest neighbour sorting (Figs. 4 and 5) and show good agreement with relationships and taxonomic group-ings proposed from general morphology by Jamie-son (1978a, 1980, 1981a, 1983) (Fig. 11). Thus *Ha-plotaxis* and *Sparganophilus* lie at the base of the opisthoporous oligochaetes; ignoring *Phreodrilus, Hormogaster* links with *Lumbricus* and these lum-bricoids link with the Megascolecidae (Megascole-coidea) represented by *Fletcherodrilus* and *Meta-phire.* Furthermore, as proposed in the previous work, the Tubificida, represented by the tubificids *Limnodriloides* and *Rhyacodrilus* and by the en-chytraeid *Lumbricillus,* form the nearest-affinity group of the Haplotaxida (Haplotaxidae and fami-lies Sparganophilidae through Megascolecidae).

Table 2. Sperm attributes computed. Q1-4, qualitative (binary); D1-3, disordered multistate; N1-19, numeric attributes; 1-78 binary equivalents of these used only for the divisive analysis.

		Attribute	Plesiomorph state	Apomorph state
Q1	1	Connectives	Absent	Present
Q2	2	Connectives	Anterior	Posterior
Q3	3	Nuclear tip	Domed	Flat
Q4	4	Nuclear pad	Without	With central boss
D1	5- 6	Secondary tube	Straight	(1) divergent (2) nodelike
D2	7-10	Limen	Absent	(1) conjoined (2) pointed (3) bulbous (4) flat
D3	11-12	Nucleus	Simple	(1) spiral (2) spirally flanged
N1	13-15	Acrosome length	Short	Long (3 states)
N2	16-18	Acrosome tube length	Short	Long (4 states)
N3	19-22	PAV emergence:length	Large	Small (4 states)
N4	23-25	PAV withdrawal:tube 1	Small	Large (3 states)
N5	26-29	Axial rod projection:tube length	Large	Small (4 states)
N6	30-32	Axial rod projection:rod length	Large	Small (3 states)
N7	33-36	Distance axial rod to nucleus:tube length	Large	Small (4 states)
N8	37-41	Axial rod length:tube length	Large	Small (5 states)
N9	42-43	Capitulum	Absent	(1) rudiment, (2) full
N10	44-46	Axial rod width:tube width	Large	Small (3 states)
N11	47-52	Axial rod width:length	Large	Small (6 states)
N12	53-58	Axial rod fraction post PAV	Large	Small (6 states)
N13	59-61	Sec. invagination:acrosome length	Small	Large (3 states)
N14	62-65	Sec. invagination:PAV length	Large	Small (4 states)
N15*	66-68	Sec. acrosome tube:acrosome tube length	Large*	Small (3 states)
N16	-	Projection axial rod behind sec. tube:rod length	-	-
N17*	69-73	Midpiece length:width	Large*	Small (5 states)
N18	74-75	No. of mitochondria	4	(1) 6, (2) 8
N19* or	76-78	No. of midpiece gyres	15*	(1) 6, (2) 3, (3) 0
N19	-	Mitochondria (1) straight, (2) intermediate, (3) spiral		
N20	-	No. midpiece gyres if spiral		

*Polarity reversed for Fig. 8.

An exception to this correspondence with previous phylogenetic taxonomy based on non-spermatological data, is the close similarity of the sperm of *Phreodrilus* (also Tubificida) to those of *Hormogaster* and *Lumbricus,* a similarity already noted from intuitive considerations (Jamieson, 1982).

Insufficient data are available for lumbriculid (*Bythonomus*) sperm to allow satisfactory determination of their similarities. The significance of Figs. 2 and 3 in which the *Bythonomus* sperm groups within the Tubificida, a position not supporting the very distinct and basal position of the Lumbriculidae previously proposed (Brinkhurst & Jamieson,

1971; Jamieson, 1978a, 1981a) is therefore uncertain. *Bythonomus* was eliminated from later analyses because of the inadequacy of its data set.

Cladistic analyses

From a survey of the ultrastructure of the spermatozoa of eight oligochaete families (those for which we have an extensive knowledge for one or more species) and from a comparison with related groups (leeches and Branchiobdellida), Jamieson (1983) recognized plesiomorph states which were attributed to a hypothetical sperm type in the ances-

Table 3. Nearest neighbours of the 10 genera from matrices of sperm similarity. Phenetic data correspond with Figs. 4 and 5*. Phylogenetic data correspond with Fig. 7* and 8**.
Similarity values shown (for Figs. 4 and 7) are on an arbitrary scale but serve to show degree of resemblance, lowest values being highest similarities.

Genus	Phenetic nearest neighbour		Phylogenetic nearest neighbour*	
1. *Bythonomus*	–		–	
2. *Lumbricillus*	*Rhyacodrilus*	.24	*Rhyacodrilus*	.37
3. *Rhyacodrilus*	*Limnodriloides*	.19	*Limnodriloides*	.18
4. *Limnodriloides*	*Rhyacodrilus*	.19	*Rhyacodrilus*	.18
5. *Phreodrilus*	*Hormogaster*	.20	*Hormogaster*	.23
6. *Haplotaxis*	*Sparganophilus*	.24	*Sparganophilus*	.295
7. *Sparganophilus* *Sparganophilus*	*Haplotaxis*	.24	*Metaphire* **Haplotaxis*	.28
8. *Hormogaster* *Hormogaster*	*Lumbricus* *Phreodrilus*	.20	*Phreodrilus* *Phreodrilus*	.23
9. *Lumbricus*	*Hormogaster*	.20	*Metaphire*	.21
10. *Fletcherodrilus*	*Metaphire*	.13	*Metaphire*	.197
11. *Metaphire*	*Fletcherodrilus*	.13	*Fletcherodrilus*	.197

  * Numeric 16 eliminated.
 ** N15 reversed: short secondary acrosome tube plesiomorph.
 ** N17 & N19 reversed: short, uncoiled midpiece plesiomorph.

tral stock of the Oligochaeta. The attributes, with their plesiomorph states and apomorphic transformation series are listed in Table 2.

Deduced plesiomorph features not used in the computer analysis are the condition, constant in clitellates but rare among other taxonomic groups, of failure of the axoneme to penetrate the midpiece mitochondria (the sole, distal centriole or basal body lying behind the mitochondria) and a 9 + 2 axoneme differing from that of leeches and branchiobdellids in lacking a sheath around the pair of central microtubules (a notable synapomorphy of those two groups). The axoneme differs from the classical condition, however, in the addition of two supernumerary fibrils in a plane at right angles to that of the two singlets to give a tetragon configuration, as shown for many oligochaetes (see Jamieson, 1981c). *Pheretima s. lat.* (pers. observ. of *Amynthas gracilis* and *Metaphire californica*) is exceptional in having a 9 + 1 axoneme with a central core lacking evidence of singlets, though two free singlets have been demonstrated distal to the glycogen region in *A. gracilis.*

Construction of phylograms. Although the phenograms (Figs. 1–5 and the ordination, Fig. 10) validly represent overall similarly of the spermatozoa of the constituent taxa according to the similarity measure and sorting strategies used, and although, despite frequent assertions to the contrary, phylogenetic interpretation of phenograms is not unreasonable, the objection may be levelled at them from the view-point of hennigian phylogenetic systematics that association of taxa is partly based on what is considered an invalid matching of plesiomorph states (symplesiomorphies). To obviate this problem, in the cladistic analyses, the deduced plesiomorph states have been suppressed in preparation of the matrix of coefficients of resemblance while using the same (Gower) similarity measure and the same (group-average and nearest neighbour) sorting strategies.

Comparison of phylogenies. Suppression of plesiomorphies with group-average sorting (Fig. 6) gave very similar results to the phenetic analysis (Fig. 5) with no suppression, including continued linkage of *Haplotaxis* and *Sparganophilus*, but these linked with the tubificids and the enchytraeid occupied an even more isolated position.

Suppression of plesiomorphies with nearest neighbour sorting, which is the strict hennigian approach, gave a phylogram (Fig. 7) closely resembling the phenograms and the group average phylogram in many respects but with some major differences despite a general agreement of sperm

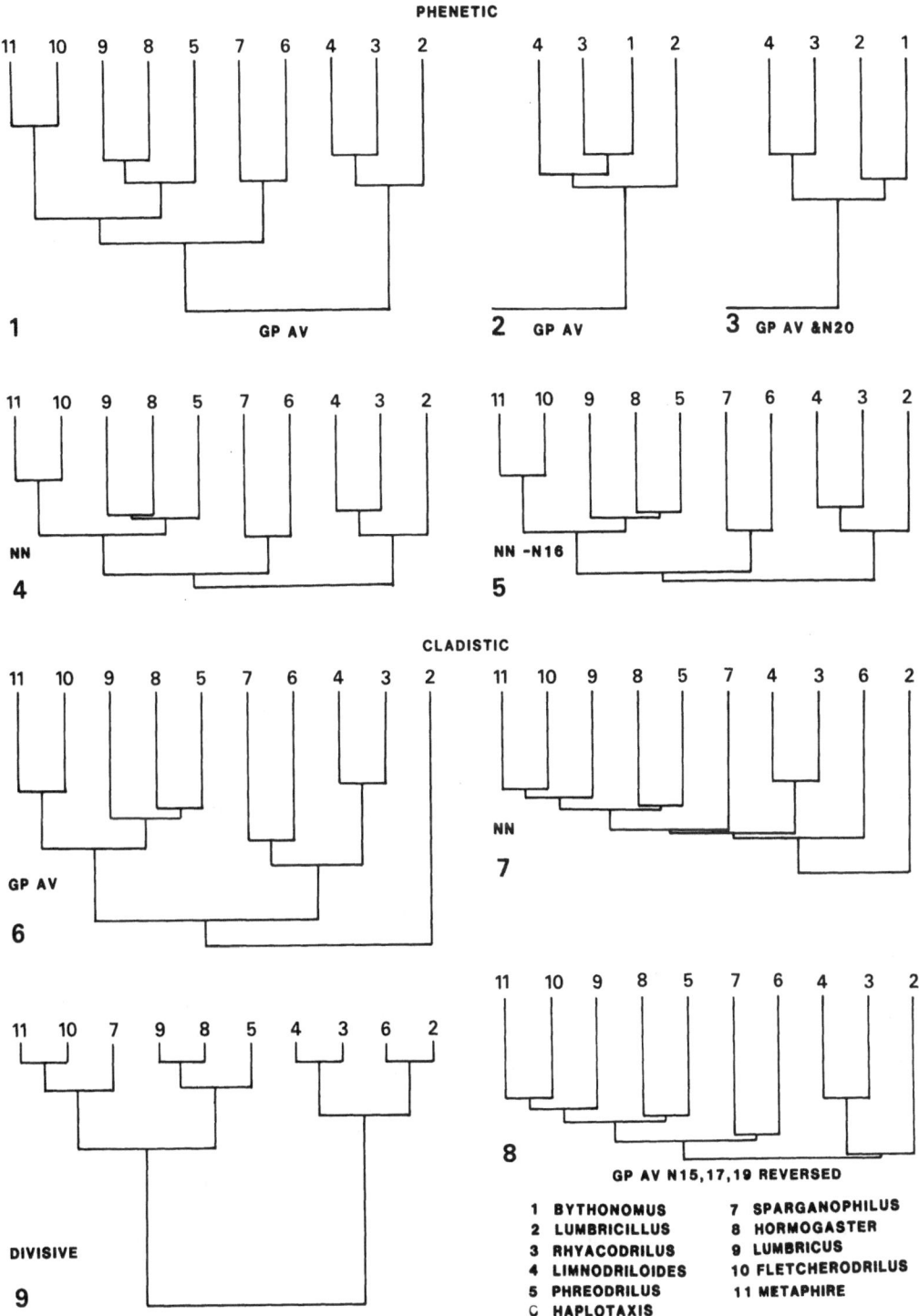

Figs. 1-9. Phenetic and phylogenetic dendrograms of sperm ultrastructure all except 9 based on similarity matrices computed with the Gower metric, 9 being divisive. Numeric attributes 19 and 20 separate in Fig. 3, fused in all others (see Table 2). 1-3. Phenetic, group-average sorting: (1) for 10 species with numeric attributes 19 and 20 fused; (2-3) identical groupings obtained but with addition of the lumbriculid, numerics 19 and 20 fused and separate respectively. 4-5. Phenetic with nearest neighbour sorting, (4) with, (5) without, numeric 16. 6-8. Cladistic, with plesiomorph states and numeric 16 suppressed: (6) group-average; (7) nearest neighbour sorting; (8) as 7 but with short, uncoiled midpiece and short secondary tube regarded as plesiomorph. 9. Divisive, with numeric 16 suppressed and plesiomorph states contributing but coded as zero in cumulative scoring.

8

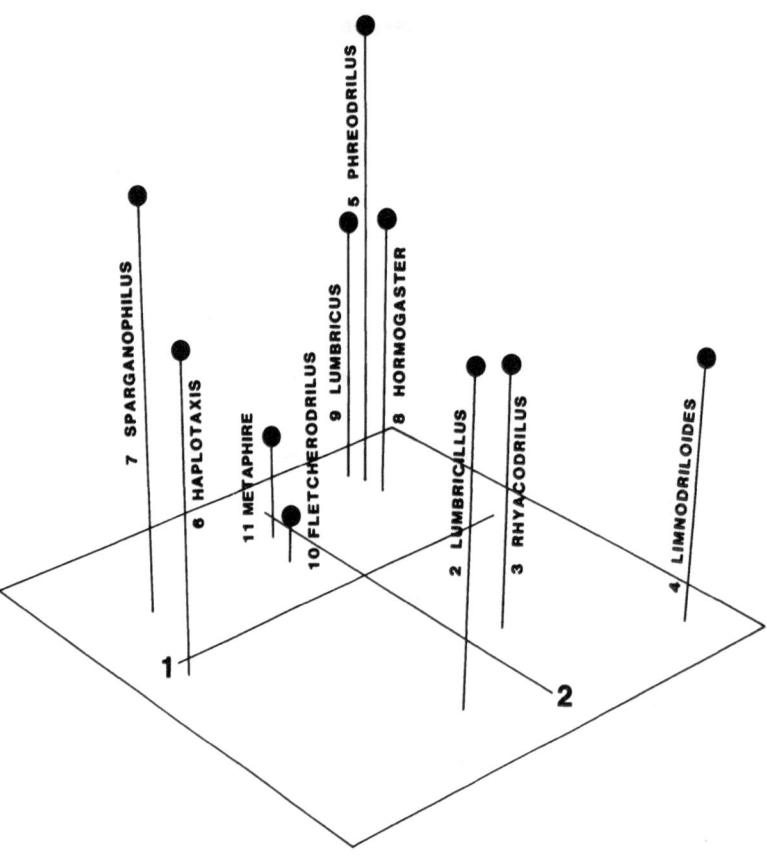

Fig. 10. Principal coordinates diagram of spermatozoal similarities, from same matrix as Fig. 4, showing vectors 1 to 3.

ultrastructure with taxonomic groupings. As acceptable phylogenies both phylograms, like the phenograms, are distorted by association of *Phreodrilus* not with the Tubificidae, with which it is conventionally placed in the Tubificida, but with one or both lumbricoids (Figs. 6 and 7). From general somatic and genital anatomy, there seems no question that *Phreodrilus* is a member of the Tubificida with sperm morphology convergent with that of megadriles rather than the opposite.

The only important divergence of the nearest neighbour phylogram from the phenograms lies in the position of *Haplotaxis,* a key genus, not least as the type-genus of the order Haplotaxida. In the phylogram (Fig. 7) it intervenes between the Tubificidae and the Enchytraeidae. This divergence disappears in the phylogram shown in Fig. 8: when the polarities of N17 and N19 are reversed and a short, uncoiled midpiece is regarded as plesiomorph and if, in addition, N15 is reversed so that a short secondary tube is considered plesiomorph, phylograms

result in which groupings are identical with the phenogram shown in Fig. 1 with the exception that *Lumbricus* (9) links first with *Fletcherodrilus* + *Metaphire* (10 and 11) (Fig. 8). In these phylograms *Haplotaxis* once more lies at the base of the opisthopores and is not associated with the Tubificida.

For the divisive analysis (Fig. 9) the polarity recognized for each attribute in the basic Gower analysis is retained but each attribute, including numerics, is represented by a shorter series of numbered states (Table 2), split into a series of binary states, in place of the raw data. The method differs notably from a strict hennigian analysis in utilizing plesiomorph states (coded as zero) but is hennigian in weighting synapomorphies. These are scored cumulatively for each attribute shared between two taxa. The method produces a dendrogram of monothetically defined groups.

The divisive analysis (Fig. 9) corresponds in many respects with the hennigian analyses using the Gower metric (Figs. 6 and 7) and detailed comparison is unnecessary. *Sparganophilus* and *Phreodrilus* remain linked with the higher opisthopores, as in the phenograms, but *Sparganophilus* is linked with the megascolecids, not an acceptable placing phylogenetically (see Jamieson, 1978a). *Haplotaxis* shows a still closer association with the Tubificida as it constitutes the sister-group of the enchytraeid.

It is thus seen that various cladistic strategies

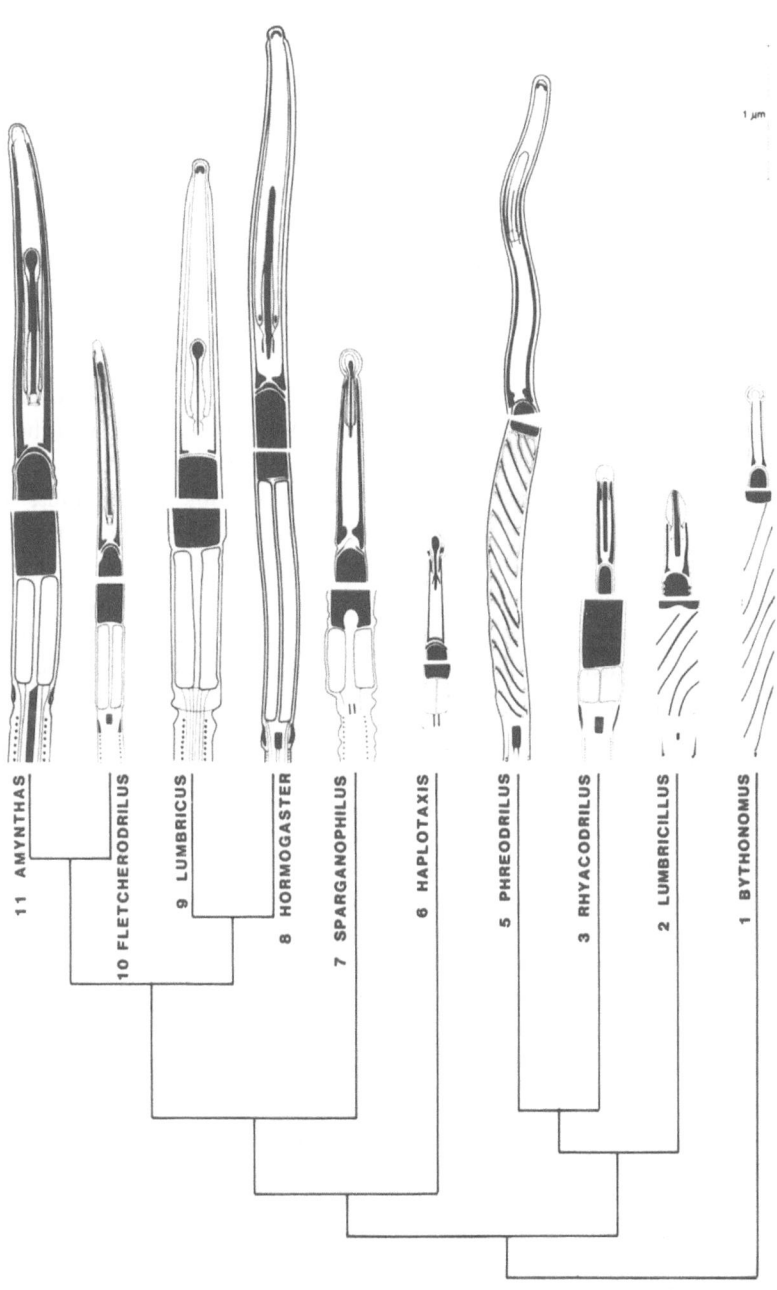

Fig. 11. Holomorphological phylogram (based on general, non-spermatozoal anatomy) for the families investigated in the present study, with sperm drawings (all to same scale) superimposed. *Pheretima s. lat.* is here represented by *Amynthas gracilis,* the sperm of which, though having a longer acrosome, closely resembles that of *Metaphire californica.* The portion from *Haplotaxis* to *Amynthas* is drawn from a Hennigian analysis of Jamieson (1978a), the remainder, for microdriles, is based on intuitive considerations (Jamieson, 1978a, 1981a). (From Jamieson, 1983).

produce phylograms with only limited mutual consistency although considerable similarity to conventional taxonomic groupings and to previously derived phylogenies based on holomorphological data is still demonstrated.

We may, alternatively to deriving phylograms from spermatozoal ultrastructure alone, examine congruence of this with the earlier non-spermatological holomorphological phylogeny and attempt to deduce the pathways which evolution of sperm structure has taken from the hypothetical plesiomorph sperm which was proposed above. This examination was made by Jamieson (1983) and is summarized in Fig. 11. The holomorphological phylogeny, originally for eighteen families, is represented in Fig. 11 for the five of these families for which sperm data are available. Added to the phylogram are tubificids, phreodrilids and enchytraeids in accordance with phylogenetic proposals by Jamieson (1978a, 1981a). It has many points of similarity with the phenograms. For a detailed discussion of origin of the spermatozoa of the various families from the plesiomorph precursor, see Jamieson (1983). The chief trends there deduced are summarized in the discussion below. These trends may be recognized independently of whether the holomorphological phylogram be accepted or not.

Discussion

The relationships in spermatozoal ultrastructure which have been demonstrated from phenetic and phylogenetic numerical analyses show close congruence in many respects with classification of the component oligochaete genera, representing their respective families, obtained by intuitive means or in holomorphological hennigian phylogenetic systematics (Jamieson, 1978a).

Congruence of the phenetic systems mutually and with the hennigian analysis of general morphology includes delimitation of the opisthoporous oligochaetes (*Sparganophilus, Hormogaster, Lumbricus, Fletcherodrilus* and *Metaphire*) as a distinct grouping; cohesion within these of the lumbricoids (*Hormogaster* and *Lumbricus*) as a group distinct from but the nearest affinity or sister-group of the well-defined megascolecid grouping (*Metaphire* and *Fletcherodrilus*); linkage of *Sparganophilus* basally with the lumbricoid-megascolecoid assemblage; and

cohesion of the Tubificidae and linkage with these of the Enchytraeidae, represented by *Lumbricillus,* as the nearest-affinity (sister) group. In the phenetic analyses *Haplotaxis* forms a pair-group with *Sparganophilus* at the base of the opisthopores.

A notable exception to congruence with previous systems is seen in linkage of *Phreodrilus* with the lumbricoids, reflecting a remarkable divergence of its sperm structure from that of the Tubificida, to which it belongs, and convergence towards that of 'megadriles' or, more specifically, the lumbricoids. Phreodrilids evidently constitute an exception to the general rule that sperm structure and related fertilization biology correspond with taxonomic groupings erected on non-spermatological criteria. Phreodrilid fertilization biology is under investigation. From inadequate spermatozoal data, the prosoporous *Bythonomus* (Lumbriculidae) groups with the enchytraeid or with the Tubificidae.

The cladistic procedures (Gower and divisive) produce variable phylograms, though these have major areas of agreement with the phenograms particularly with regard to placement of lumbricoids, megascolecids and *Phreodrilus*. The Gower metric is noteworthy in giving a very plesiomorph status to the Enchytraeidae; it and the divisive procedure link *Haplotaxis* more or less closely to the Tubificida. However, when a short, uncoiled midpiece and short secondary tube are regarded as plesiomorph, *Haplotaxis* retains its association with the opisthopores; enchytraeids, while again the most plesiomorph group, do not appear as distinctly basal as in the first Gower phylogeny, *Sparganophilus* has very equivocal affinities in the cladistic analyses, linking with the opisthopores or with *Haplotaxis* and hence the Tubificidae or (divisively) with the megascolecids. It must be stressed that the fact that *Haplotaxis* always has *Sparganophilus* as its nearest neighbour in phenetic and cladistic Gower spermatological analyses supports the conclusion that *Haplotaxis* is closer to the alluroidid through megascolecid assemblage in the Haplotaxida *sensu* Jamieson (1978a) than to the Tubificida or Lumbriculida. To elevate the Haplotaxidae to ordinal rank, and create a separate order Lumbricida for the related assemblage, as Brinkhurst (1982b, 1984) has done would therefore necessitate elevation of the Tubificida to a rank above the level of order.

A hypothetical plesiomorph spermatozoon has been proposed by Jamieson (1983). The chief trends from its plesiomorphies have been elongation of the acrosome and its tube; withdrawal of the PAV and the axial rod into the acrosome tube and development of a capitulum; development of connectives from the secondary tube to the axial rod (though there is some possibility that the reverse, absence of connectives, is apomorph); detorting and shortening of the midpiece (again, conceivably to be reversed) with an increase in numbers of mitochondria from the plesiomorph four to eight; modification of the base of the tube to form a limen of variable form; and, in one line (lumbricids) flattening of the tip of the nucleus and correspondingly of the limen.

The exceedingly plesiomorph condition of the enchytraeid sperm, with regard not only to the Lumbricina but also to their closest relatives, the Tubificina strengthens support for regarding the enchytraeids as the most plesiomorph oligochaetes with the possible exception of the Lumbriculidae the sperm of which is imperfectly known. Peculiar features of enchytraeids which are presumably autapomorphies are the double secondary tube and spirally flanged nucleus in the sperm (though requiring confirmation from further species) and, in the adult, the unique glandular postseptal region of the male funnels, and possibly the multiple batonsetae. Whether the frequent confluence of the spermathecae with the gut (also seen in some Tubificidae) is apomorph is debatable. The glandular funnels are here considered to support the view that absence of glands adding secretion to the seminal fluid is plesiomorph in the Oligochaeta and that development of such glands has proceeded in two distinct ways, specialization of the sperm funnels in enchytraeids or development of prostate glands, usually associated with atrial chambers of various forms, in other Tubificida and in some members of all groups of the Haplotaxida (with the notable exception of the Haplotaxidae).

It has been suggested previously (Michaelsen, 1928; Brinkhurst & Jamieson, 1971; Jamieson, 1980, 1981a) that the Lumbriculidae are the most plesiomorph living oligochaetes, the grounds for this being chiefly their prosoporous male genitalia. Unfortunately, sperm data have not been sufficient to aid a decision as to the validity of this view. On the few data available they show affinity with the Tubificida rather than basally with oligochaetes as a whole. Brinkhurst (1984), in a profound and significant analysis, has abandoned our earlier view of the basal position of the Lumbriculida. He suggests that the prosoporous condition is derived from the plesioporous condition seen in Haplotaxidae. As foreshadowed in Brinkhurst & Jamieson (1971), he regards the frequently multigonadial condition of lumbriculids as an aberration due to parthenogenesis and not as evidence for a multigonadial origin of the Oligochaeta. Hrabě (1984), like several earlier workers, also argues strongly for regarding the prosoporous condition as a secondary derivation from a primitive, plesioporous condition.

The double penetration by the vasa deferentia of the septum bearing the prosoporous male funnels in lumbriculids is certainly enigmatic but there remain reasons for caution in considering this to be strong evidence for former plesiopory, as Brinkhurst and Hrabě do. (If this were accepted double penetration of a septum anterior to the male pores in moniligastrids might have to be taken as evidence for earlier prosopory of the Moniligastridae). The presence in the lumbriculid *Rhynchelmis* of vestigial atria, without male ducts, in a segment preceding that with functional atria which receive two pairs of vasa, one pair from the same segment and one pair from the preceding segment with its vestigial atria, has been used for (Brinkhurst, in Brinkhurst & Jamieson, 1971) and against (Hrabě, 1984) acceptance of a former prosopory. Possibly the prosoporous condition in acanthobdellid leeches, in which the single pair of testes (there in segment 10) is associated with a pair of male ducts discharging posteriorly in the same segment, represents the primitive condition for clitellates, though not in numbers of such genital segments. Branchiobdellids (with the *Rhynchelmis*-type plesio- and prosopory in a single segment), lumbriculids and acanthobdellids were placed in the Lumbriculida by Michaelsen (1928). Brinkhurst (1984) supports regarding the plesiopore condition as primitive on the grounds that this is also the condition for the female ducts. However, the female pores are 'prosoporous' in questid polychaetes and aeolosomatids, a further support for the present author's view that gonoducts, as modified coelomoducts, primitively discharged in the segment of their gonads. The female ducts are virtually prosoporous in lumbriculids and the Tubificida while in the Haplotaxida, including *Haplotaxis* and 'megadriles' they are plesioporous. Supposed transformation of plesioporous male pores to the prosoporous condition in lumbriculids (Brinkhurst, 1983) is contrary to a well-defined and unequivocal trend for posterior migration of the pores in oligochaetes (with male pores in segments 12 and 13 in haplotaxids, 13 in alluroidids and some lumbricids, 15 in most lumbricids, 17 or 18 in many megascolecids and further posteriorly in other groups, including sparganophilids and lutodrilids, to cite some examples). There is no more reason to suppose that female pores have made the anteriorwards migration.

The view that plesiopory is the primitive condition of both male and female ducts probably stems from the common misconception that oligochaete gonoducts are modified nephridia. Thus Giere & Riser (1981) state, in their valuable paper on oligochaetoid polychaetes, that among the important synapomorphies characterizing the Clitellata is 'the gonoduct being derived from segmental organs', the latter, interestingly, an archaic term for nephridia. However, utilization of nephridia as gonoducts, though suspected in the Aeolosomatidae and seen in some polychaetes is not a basic annelid, or clitellate, condition. Some capitellid polychaetes have separate coelomoducts (as gonoducts) and nephridia in the same segment, many other polychaetes have combinations of the two (references in Schroeder & Hermans, 1975) and, more importantly, male and female funnels and their ducts coexist intrasegmentally in oligochaetes (e.g. Jamieson, 1970a, b; Reynolds, 1980). The fact that in most of these forms the coelomoducts are also 'plesioporous' does not militate against regarding the plesiomorph coelomoduct to have opened to the exterior in the segment of its funnel as argued above. Certainly, the fiction that clitellate gonoducts are nephridia must be finally discarded.

With regard to number of gonads in ancestral oligochaetes, Brinkhurst's contention that opisthopores and the Tubificida are derivable from octogonadial forms is persuasive. However, it also seems reasonable to envisage that early annelids, including the first oligochaetes, had longer series of gonads, as in lumbriculids (see Jamieson, 1981a). In support of this is the multigonadial condition of many polychaetes (with no suggestion that polychaetes were ancestral to oligochaetes), of aeolosomatids (Aphanoneura) and leeches.

To conclude, the ultrastructure of spermatozoa unifies the oligochaetes and, while setting them apart from the Branchiobdellida and the Hirudinea, indicates their closer relationship to these groups than to other annelids. The spermatozoal ultrastructure of many oligochaete families, of the Acanthobdellidae and the aeolosomatids awaits investigation, however.

Acknowledgements

Dr. G. Bonomi, other members of the organizing committee of this symposium, and the Italian National Research Council, are most sincerely thanked for the invitation and support to give this paper. Dr. R. O. Brinkhurst is thanked for allowing me to see the draft of his accompanying paper. Permission to use our unpublished micrographs was given by Dr. M. Ferraguti (*Hormogaster*) and Drs. K. S. Richards, T. Fleming and C. Erséus (*Lumbricus*). Special thanks are due to Dr. M. Dale of CSIRO for stimulating discussions and for arranging computations on CSIRONET, with the kind assistance of Mr. D. Ross. Continuing support of Mrs. L. Daddow in preparation of sections for electron microscopy is also gratefully acknowledged as is the indispensable support of the Australian Research Grants Scheme.

References

Baker, H. R. & R. O. Brinkhurst, 1981. A revision of the genus Monopylephorus and redefinition of the subfamilies Rhyacodrilinae and Branchiurinae (Tubificidae, Oligochaeta). Can. J. Zool. 59: 939–969.

Brinkhurst, R. O., 1982a. In: S. P. Parker (ed.), Synopsis and Classification of Living Organisms. McGraw-Hill, New York.

Brinkhurst, R. O., 1982b. Evolution in the Annelida. Can. J. Zool. 60: 1043–1059.

Brinkhurst, R. O., 1984. The position of the Haplotaxidae in the evolution of oligochaete annelids. In: G. Bonomi & C. Erséus (eds.), Aquatic Oligochaeta. Proceedings of the Second International Symposium on Aquatic Oligochaete Biology. Developments in Hydrobiology (this volume).

Brinkurst, R. O. & W. Fulton, 1981. On Haplotaxis ornamentus sp. nov. (Oligochaeta, Haplotaxidae) from Tasmania. Rec. Queen Vict. Mus. 72: 1–8.

Brinkhurst, R. O. & B. G. M. Jamieson, 1971. The Aquatic Oligochaeta of the World. Oliver & Boyd, Edinburgh, Toronto. 860 pp.

Chekanovskaya, O. V., 1962. The aquatic Oligochaeta of the USSR. Opred. Faune SSSR 78: 1–411.

Erséus, C., 1982. Taxonomic revision of the marine genus Limnodriloides (Oligochaeta: Tubificidae). Verh. naturwiss. Ver. Hamburg (N.F.), 25: 207–277.

Ferraguti, M. & B. G. M. Jamieson, 1984. Spermiogenesis and spermatozoal ultrastructure in Hormogaster (Hormogastridae, Oligochaeta, Annelida). J. submicrosc. Cytol. 16: 307–316.

Ferraguti, M. & G. Lanzavecchia, 1977. Comparative electron microscopic studies of muscle and sperm cells in Branchiobdella pentodonta Whitman and Bythonomus lemani Grube (Annelida, Clitellata). Zoomorph. 88: 19–36.

Giere, O. W. & N. W. Riser, 1981. Questidae – polychaetes with oligochaetoid morphology and development. Zool. Scr. 10: 95–103.

Hennig, W., 1966. Phylogenetic Systematics. University of Illinois Press, Urbana. 263 pp.

Hrabě, S., 1984. Two atavistic characters of some Lumbriculidae and their importance for the classification of Oligochaeta. In G. Bonomi & C. Erséus (eds.), Aquatic Oligochaeta. Proceedings of the Second International Symposium on Aquatic Oligochaete Biology. Developments in Hydrobiology (this volume).

Jamieson, B. G. M., 1970a. A revision of the earthworm genus Woodwardiella with descriptions of two new genera (Megascolecidae: Oligochaeta). J. Zool. 162: 99–144.

Jamieson, B. G. M., 1970b. Two new sympatric species of the earthworm genus Digaster (Megascolecidae: Oligochaeta) from Queensland. Proc. r. Soc. Qld 82: 35–46.

Jamieson, B. G. M., 1977. On the phylogeny of the Moniligastridae with description of a new species of Moniligaster (Oligochaeta: Annelida). Evol. Theory 2: 95–114.

Jamieson, B. G. M., 1978a. Phylogenetic and phenetic systematics of the opisthoporous Oligochaeta (Annelida: Clitellata). Evol. Theory 3: 195–233.

Jamieson, B. G. M., 1978b. A comparison of spermiogenesis and spermatozoal ultrastructure in megascolecid and lumbricid earthworms (Oligochaeta: Annelida). Aust. J. Zool. 26: 225–240.

Jamieson, B. G. M., 1980. Preliminary discussion of an Hennigian analysis of the phylogeny and systematics of opisthoporous oligochaetes. Revue Ecol. & Biol. Sol, 17: 261–275.

Jamieson, B. G. M., 1981a. Historical biogeography of Australian Oligochaeta. In: A. Keast (ed.), Ecological biogeography in Australia. Dr W. Junk, The Hague: 885–921.

Jamieson, B. G. M., 1981b. Ultrastructure of spermatogenesis in Phreodrilus (Phreodrilidae, Oligochaeta, Annelida). J. Zool. 194: 393–408.

Jamieson, B. G. M., 1981c. The Ultrastructure of the Oligochaeta. Academic Press, London, New York. 462 pp.

Jamieson, B. G. M., 1982. The ultrastructure of the spermatozoon of Haplotaxis ornamentus (Annelida, Oligochaeta, Haplotaxidae) and its phylogenetic significance. Zoomorph. 100: 177–188.

Jamieson, B. G. M., 1983. Spermatozoal ultrastructure: evolution and congruence with a holomorphological phylogeny of the Oligochaeta (Annelida). Zool. Scr. 12: 107–114.

Jamieson, B. G. M. & L. Daddow, 1979. An ultrastructural study of microtubules and the acrosome in spermiogenesis of Tubificidae (Oligochaeta). J. Ultrastruct. Res. 67: 209–224.

Jamieson, B. G. M., L. Daddow & J. Bennett, 1978. Ultrastructure of the tubificid acrosome (Annelida, Oligochaeta). Zool. Scr. 7: 115–118.

Jamieson, B. G. M., K. S. Richards, T. P. Fleming & C. Erséus, 1983. Comparative morphometrics of oligochaete spermatozoa and egg-acrosome correlation. Gamete Res. 8: 149–169.

Jamieson, B. G. M., K. S. Richards & T. P. Fleming, 1982. An ultrastructural study of spermatogenesis in Sparganophilus tamesis (Sparganophilidae, Oligochaeta, Annelida). J. Zool. 196: 63–79.

Michaelsen, W., 1928. Oligochaeta. Handb. Zool. 2: 1–118.

Reynolds, J. W., 1980. The earthworm family Sparganophilidae (Annelida, Oligochaeta) in North America. Megadrilogica 3: 189–204.

Richards, K. S., T. P. Fleming & B. G. M. Jamieson, 1981. Aberrant spermatozoa and spermatids, and observations on phagocytosis in Sparganophilus tamesis (Sparganophilidae, Oligochaeta): an ultrastructural study. Int. J. Invert. Reprod. 4: 181–191.

Schroeder, P. C. & C. O. Hermans, 1975. Annelida: Polychaeta. In A. C. Giese & J. S. Pearse (eds.), Reproduction of Marine Invertebrates. Annelids and echiurans. Academic Press, New York, pp. 1–213.

Timm, T., 1981. On the origin and evolution of the aquatic Oligochaeta. Eesti NSV Tead. Akad. Toim. Biol. 30: 179–181.

Webster, P. & K. S. Richards, 1977. Spermiogenesis in the enchytraeid Lumbricillus rivalis (Oligochaeta: Annelida). J. Ultrastruct. Res. 61: 62–77.

Two atavistic characters of some Lumbriculidae and their importance for the classification of Oligochaeta

Sergěj Hrabě
Stojanova 9, CS-602 00 Brno, Czechoslovakia

Keywords: aquatic Oligochaeta, Lumbriculidae, phylogeny

Abstract

The rudimentary atria, and the posterior sperm funnels and sperm ducts, which occur in some species of the Lumbriculidae, are discussed. It is shown that the posterior location of funnels and sperm ducts is the result of a forward shift of the atria, which refutes Stephenson's supposition that the Lumbriculidae is the most archaic family of the present-day Oligochaeta.

Two atavistic anatomical characters occurring in some lumbriculids are significant for the establishment of phylogenetic succession of genera in this family. These are (1) the rudimentary atria in some species of the genus *Rhynchelmis,* and (2) the sperm ducts entering, in numerous lumbriculids, through the dissepiment into the postatrial segment and returning back into the atrial segment before opening into the atria.

Michaelsen (1901) showed that *Lamprodrilus,* which has male ducts that are provided with only posterior male funnels (Fig. 1A), is phylogenetically older than lumbriculid genera, in which the male ducts have both anterior and posterior funnels (Fig. 1B, C). [*Dorydrilus,* which has been assigned to a separate family by Cook (1968), but which in my opinion is a member of the Lumbriculidae, is furnished with anterior funnels only (Fig. 2A).] According to Michaelsen, the male ducts of *Rhynchelmis* and the other genera with two funnels and two sperm ducts arose from an ancestor with two pairs of male ducts provided with posterior funnels by the opening of the first (anterior) pair of sperm ducts into the second pair of atria (Fig. 1B). The atria of the first pair remain as one pair of rudimentary atria without sperm ducts in *Rhynchelmis brachycephala,* as an unpaired rudimentary atrium in

the majority of *Rhynchelmis* species (Fig. 1B), and disappear entirely in *R. vejdovskyi, R. orientalis* and *R. granuensis* (Fig. 1C). Beyond a doubt, the male duct with two funnels and sperms ducts evolved in this way secondarily from the male duct with only the posterior funnel and sperm duct.

The second atavistic character (cf. above) illustrates how the male duct with only the posterior funnel and sperm duct arose from an ancestor with male ducts provided with an anterior funnel characteristic of Plesiopora (Archioligochaeta). Vejdovský (1884; table XI, figs. 19, 23) illustrated, for *Trichodrilus pragensis* and *Bythonomus lemani* (= *Claparedilla meridionalis*), and Hrabě & Černosvitov (1927) described and depicted, for *Stylodrilus* (= *Anastylus*) *parvus,* another significant atavistic feature, which, according to my later observations, occurs in numerous lumbriculids. In these species, the sperm duct of each male duct enters through the posterior dissepiment of the atrial segment, forms some loops in the following segment before it turns back into the foregoing segment to end in the atrium (Fig. 2B–D). This situation is confirmed by the studies by Světlov (1936), Cook (1967, 1971, 1975), and Holmquist (1976). The described course of the sperm ducts in numerous species excludes the possibility of an accidental origin or a diversion in

Hydrobiologia 115, 15–17 (1984).
© Dr W. Junk Publishers, Dordrecht.

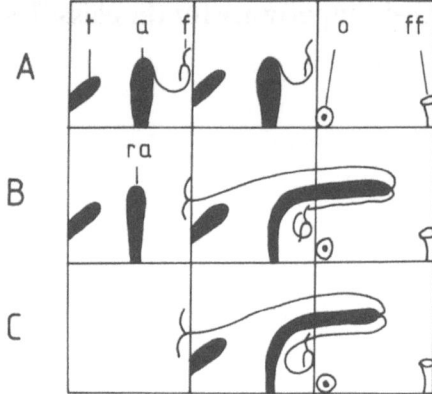

Fig. 1. Derivation of the *Rhynchelmis* male ducts. A. Ancestor with two pairs of male ducts with posterior funnels. B. The majority of *Rhynchelmis* species. C. *Rhynchelmis vejdovskyi, R. orientalis* and *R. granuensis.* (a = atrium, f = male funnel, ff = female funnel, o = ovarium, ra = rudimentary atrium, t = testis)

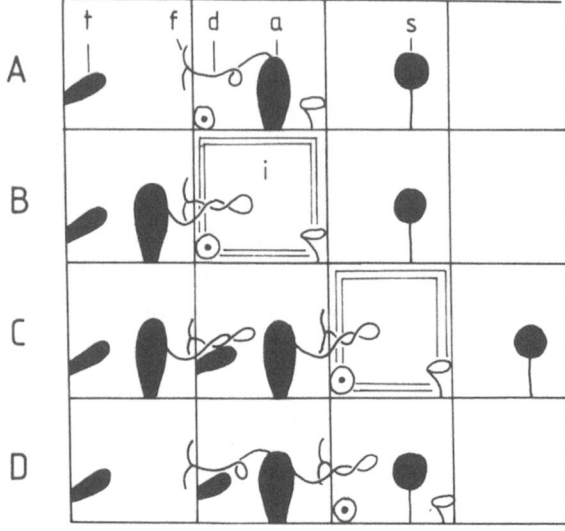

Fig. 2. Genital organs of some lumbriculids. A. *Dorydrilus* (Prosopora); male duct with anterior funnel. B. *Teleuscolex;* male duct with posterior funnel. C. *Lamprodrilus;* male ducts with posterior funnels. D. *Trichodrilus;* male duct with anterior and posterior funnels. (a = atrium, d = sperm duct, f = male funnel, i = intervening segment, s = spermatheca, t = testis)

response to a physiological requirement. It rather demonstrates that the posterior sperm duct of the Lumbriculidae has its origin in a male duct with an anterior funnel, in which the atrium has shifted forward into the testicular segment (Fig. 2A–B).

The portion of the sperm duct retained in the postatrial segment reflects the earlier course of the whole sperm duct.

The species and genera of the Lumbriculidae with sperm ducts fully contained within the atrial segment (as in Fig. 1A) are phylogenetically younger than those with sperm ducts entering through the dissepiment into the postatrial segment (as in Fig. 2B–C). The genera with posterior funnels (*Lamprodrilus, Teleuscolex, Agriodrilus*) are again younger than the genera furnished with only anterior funnels occurring in the suborder Plesiopora (Archioligochaeta). According to the observations mentioned above it is easy to see how the Lumbriculidae arose from an ancestor with male ducts with only anterior funnels and sperm ducts, which is the situation found in the Plesiopora.

In the genera in which each pair of male ducts only has posterior funnels and sperm ducts (*Lamprodrilus, Teleuscolex, Agriodrilus*), the spermathecae are separated from the anteriorly situated atrial segment by one intervening segment (with ovaria and female funnels only) (as in Fig. 2B–C). This is contrary to the situation in the genera with both anterior and posterior funnels and sperm ducts (Fig. 2D). The intervening segment indicates a forward shift of the atria and supports my interpretation of evolutionary events. It refutes Stephenson's (1930; 705) supposition that the Lumbriculidae is the most archaic family of the present-day Oligochaeta and, at the same time, the classification of the Oligochaeta in the papers by Yamaguchi (1953) and Brinkhurst (1971) based on it.

References

Brinkhurst, R. O., 1971. Phylogeny and classification, 1. In R. O. Brinkhurst & B. G. M. Jamieson (eds.), Aquatic Oligochaeta of the world. Oliver & Boyd, Edinb. 165–177.

Cook, D. G., 1967. Studies on the Lumbriculidae in Britain. J. Zool., Lond. 153: 353–368.

Cook, D. G., 1968. The genera of the family Lumbriculidae and the genus Dorydrilus (Annelida, Oligochaeta). J. Zool., Lond. 156: 273–289.

Cook, D. G., 1971. Trichodrilus allegheniensis n.sp. (Oligochaeta, Lumbriculidae) from a cave in Southern Tennessee. Trans. am. microsc. Soc. 90: 381–383.

Cook, D. G., 1975. Cave-dwelling aquatic Oligochaeta (Annelida) from the Eastern United States. Trans. am. microsc. Soc. 94: 24–37.

Holmquist, C., 1976. Lumbriculids (Oligochaeta) of Northern Alaska and Northwestern Canada. Zool. Jb. Syst. Ökol. Geogr. Tiere 103: 377-431.

Hrabě, S. & L. Černosvitov, 1927. Über eine neue Lumbriculiden-Gattung Anastylus parvus n.g. n.sp. aus Karpathorussland. Zool. Anz. 71: 203-207.

Michaelsen, W., 1901. Oligochaeten der zoologischen Museen zu St. Petersburg and Kiew. Bull. Acad. Sci. Pétersbourg 16: 137-215.

Světlov, P., 1936. Lamprodrilus isoporus Mich. aus dem Ladoga- und dem Onegasee. Zool. Anz. 113: 87-93.

Stephenson, J., 1930. The Oligochaeta. The Clarendon Press, Oxford, 978 pp.

Yamaguchi, H., 1953. Studies on the aquatic Oligochaeta of Japan, 6. J. Fac. Sci. Hokkaido Univ. 11: 277-342

Vejdovský, F., 1884. System und Morphologie der Oligochaeten. Prag., 166 pp.

Phylogenetic and taxonomic problems in freshwater Oligochaeta with special emphasis on chitinous structures in Tubificinae

Michael Dzwillo

Zoologisches Institut und Museum der Universität Hamburg, Martin-Luther-King-Platz 3, D-2000 Hamburg 13, Federal Republic of Germany

Keywords: aquatic Oligochaeta, Tubificinae, phylogeny, taxonomy, chitinous penis sheaths

Abstract

Setae and chitinous penis sheaths are the main characters used to distinguish genera and species of the subfamily Tubificinae. Genital setae and penis sheaths are of functional importance to facilitate copulation. Similar structures in different genera may be homologous or products of parallel evolution (homoiologous). Form and dimensions of the penis sheaths of many tubificine species are very variable. Transspecific overlap of quantitative species characters can make the determination of specimens in the genus *Limnodrilus* difficult. The configuration of the distal ends of the penis sheaths is an important character to distinguish *Limnodrilus* species. The definition of the intraspecific variability of this morphological character is problematic.

Freshwater Oligochaeta of the subfamily Tubificinae are important in ecological research. They represent a considerable part of the bottom fauna of rivers, lakes and other bodies of freshwater.

'The supraspecific classification of the Tubificidae is by convention based upon the principal organisation of the male duct and its accessories' (Erséus, 1980). It is open to discussion whether or not the phylogenetic base of this convention is well founded. It will be necessary to see if similar configurations are homologous or products of convergent evolution. Different authors (e.g. Timm, 1981; Brinkhurst, 1982) have discussed this problem recently with regard to higher taxonomic levels.

In distinguishing Tubificinae at species level, chitinous structures such as setae and penis tubes are important characters. The determining work of many biologists is often confined to these structures. Setae and chitinous structures of the male copulation organs can be identified without great effort as far as preparation is concerned. However, wide intraspecific and transspecific variabilities often make the determination work difficult. A further question is: are the characteristic structures in

their respective configurations homologous? Often homology is simulated, while it is really only a homoiology. The investigation of male genital organs of mature specimens show that the similarity of somatic setal characters of *Potamothrix hammoniensis* and some forms of *Tubifex tubifex* is a product of convergence. Holmquist (1978) showed that *Peloscolex* is an artificial genus, the characteristic papillate cutaneous cover having been formed several times independently.

The tendency to develop specific chitinous structures in the genital regions is based on the functional importance of such structures for copulation in these hermaphroditic animals. Up to now, however, no one has seen the actual act of copulation of a tubificid. It is true that this was claimed by Ditlevsen (1904), but from his description no real conclusion can be drawn. We only have indirect structural evidence for the mode of copulation in this family. Genital setae (penial and spermathecal setae) fix the copulants to each other in the correct position. Chitinous penis sheaths or tubes facilitate insemination, i.e. the introduction of the penes into the spermathecae. The distal structures of these tubes,

Hydrobiologia 115, 19–23 (1984).

20

mostly dilatations, fix the penes to the spermathecal pores.

The homology of the specific shape of genital setae in most tubificine genera is beyond doubt, but there is no homology in the modification of ventral setae to genital setae in different genera. On the other hand the absence of genital setae in *Limnodrilus* may not be a primitive character. The function of the genital setae may be transferred to the distal structures of the penis sheaths. Genital setae may be altered back to normal somatic setae.

Chitinous penis tubes are found in different genera and species of the subfamily Tubificinae. These structures are of importance for the recognition of species. We do not know if the existence of a chitinous penis sheath is a synapomorphic character of the Tubificinae, or if it evolved independently in different genera. In the first case we have to assume convergent reduction of this structure in different genera. The convergent reduction of organs has been demonstrated in various animal groups; Giere *et al.* (1984), for instance, have shown that the gut

has been reduced independently in different marine Tubificidae. However, major differences in the shape of the penis sheaths, for instance in the genera *Limnodrilus* and *Psammoryctides*, indicate that these structures evolved in parallel or convergently, although, as a rule, there seems to be no doubt about the homology of the penis sheaths within one genus.

In every key to the Tubificidae we read that the species of the genus *Limnodrilus* are characterized by the length/breadth relation of the penis tubes. Kennedy (1969), Barbour *et al.* (1980) and others showed that there is a considerable variation in this ratio within the species. Figure 1 shows the normal distribution of the length/breadth ratios of the penis sheaths of common *Limnodrilus* species. A transspecific overlap can be seen. Many specimens are not identifiable by this feature. New investigations show that the variability of this length/breadth relation in *Limnodrilus* species is to some degree dependent on the absolute length of the penis tube, an allometric relationship being suspected.

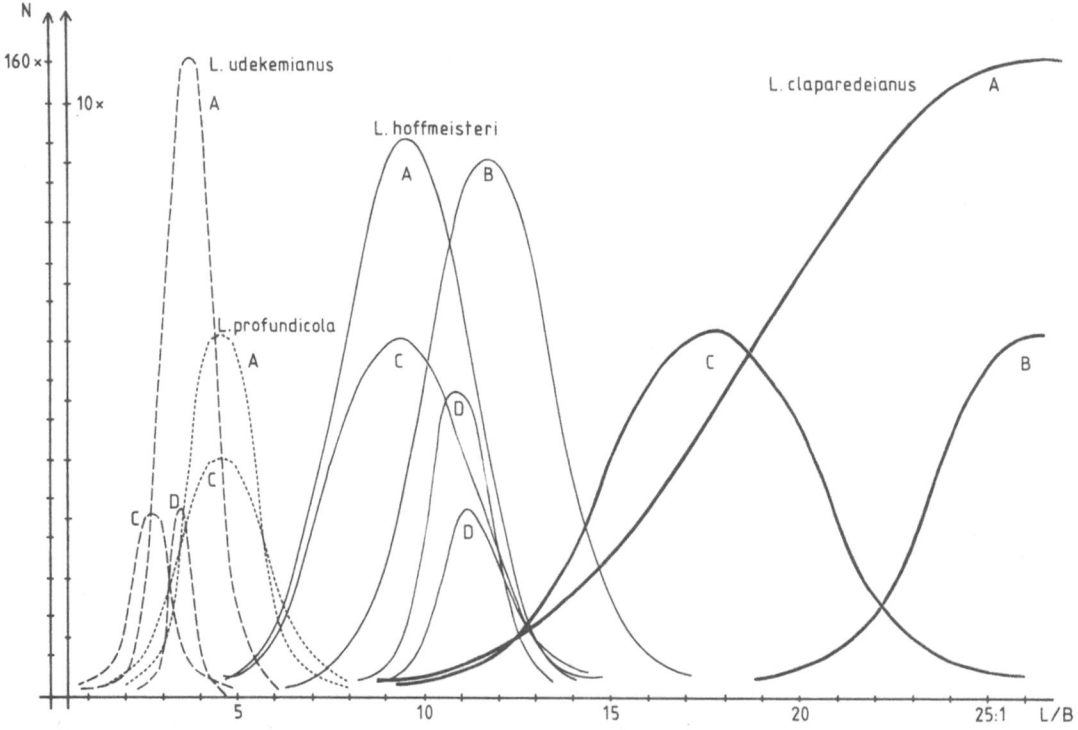

Fig. 1. Distribution curves for length/breadth relationships of penis sheaths of common *Limnodrilus* species. Sources: (A) Kennedy (1969); (B) Barbour *et al.* (1980); (C) measurements taken by B. Mentel; (D) Gavrilov & Paz (1949, 1950).

An important character for distinguishing *Lim-nodrilus* species in the shape of the distal end of the penis sheaths. In many animal groups, for instance different arthropods and viviparous fish, the male genital organs have manifold specific structures. It was asserted, first by Dufour (1844), that these structures in copulation act like lock and key. It was believed that this lock and key mechanism is an isolating one. However, experiments by Şengün (1944, 1949) on *Bombyx* and poeciliid fish showed that, at least in many cases, these structures do not function as mechanical isolation factors as indicated above. Destroyed and atypically regenerated distal regions of penes were able to perform normal copulation and fertilization. Nevertheless, the complex structures of male copulation organs in many animals are constant characteristics enabling us to recognize species. Looking at figures of ectal ends of penis tubes of *Limnodrilus* and other tubificine species in systematic monographs and identification keys, one may suppose that there is a similar specific consistency. It is true that Kennedy (1969), Pfannkuche (1977) and others showed that there is

a considerable intraspecific variability in the shape of the distal part of the penis sheaths, but the variability demonstrated by these authors seems to be limited, so that separation of species using this character may still be possible. One does realize that extreme difficulties exist in demonstrating the variations within a species, and that only an 'ideal' form can be shown without difficulty in identification work.

Studies of many specimens have shown that the variability in different species of tubificine oligochaetes is greater than is often supposed. In Figs. 2–4, the two penis sheaths of several specimens of *Tubifex tubifex*, *Limnodrilus udekemianus* and *Limnodrilus hoffmeisteri* are depicted. The figures show that there is not only a great intraspecific variability but also a considerable intraindividual difference of the penis tubes. Because mature specimens of the same population often show a wide variation in the shape of the penis sheaths, factors other than ecological ones must be responsible for the variability. Nevertheless, the influence of ecological factors on taxonomically relevant

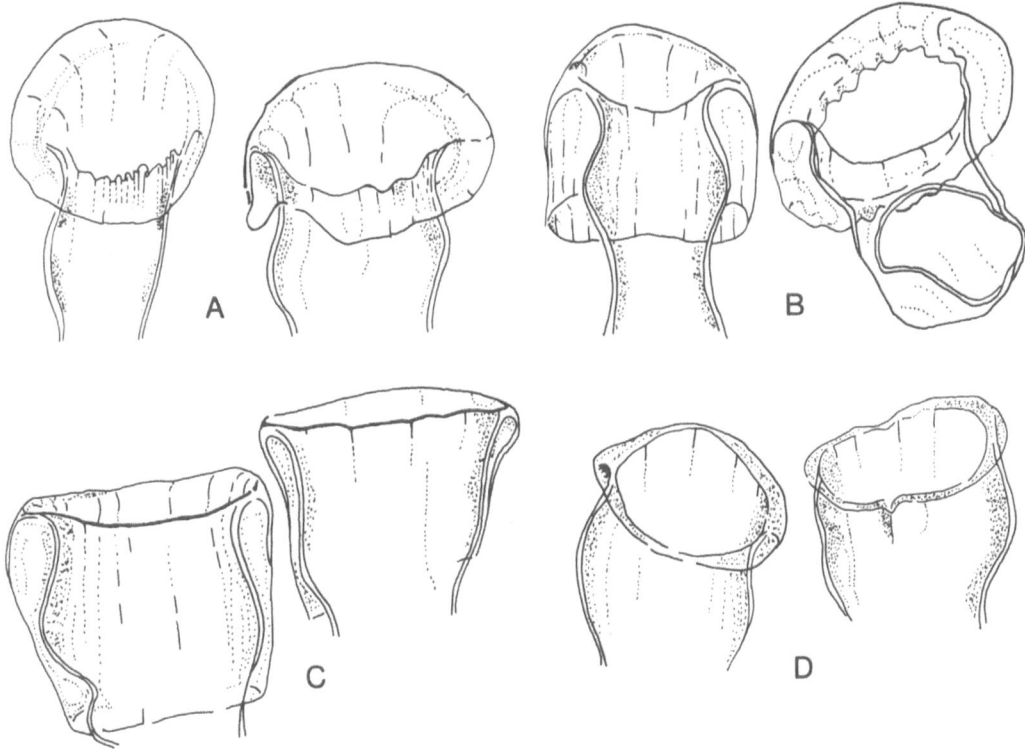

Fig. 2. Penis sheaths of four specimens (A–D) of *Tubifex tubifex*.

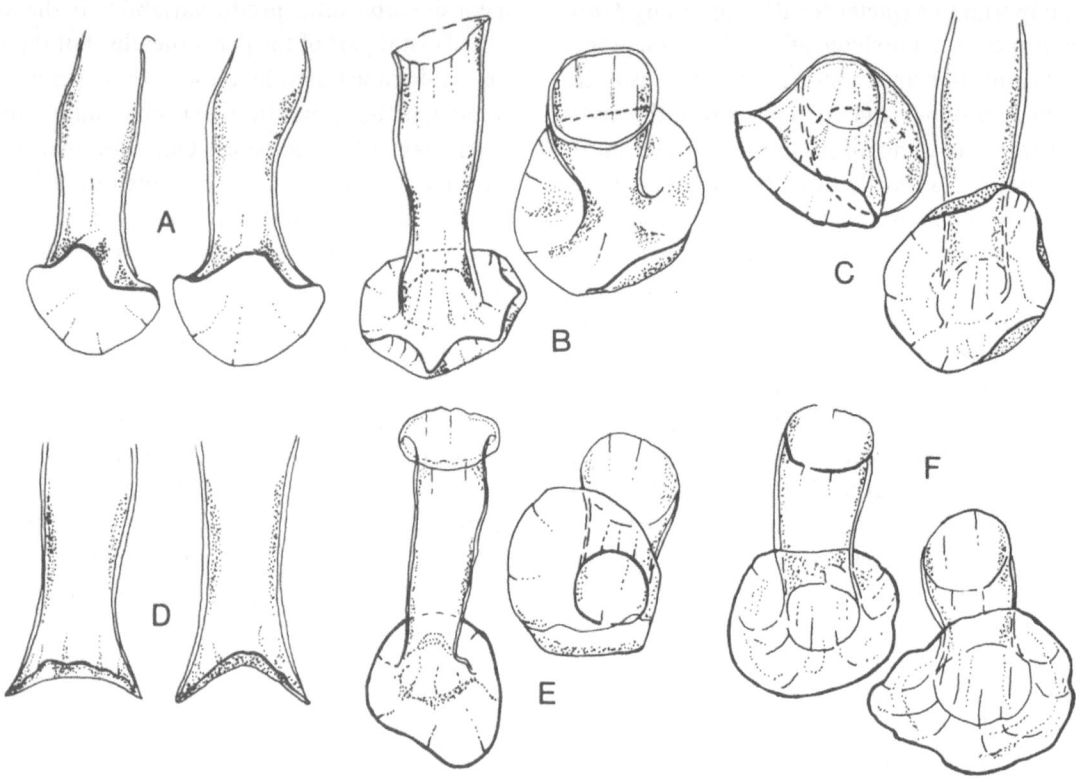

Fig. 3. Penis sheaths of six specimens (A–F) of *Limnodrilus udekemianus.*

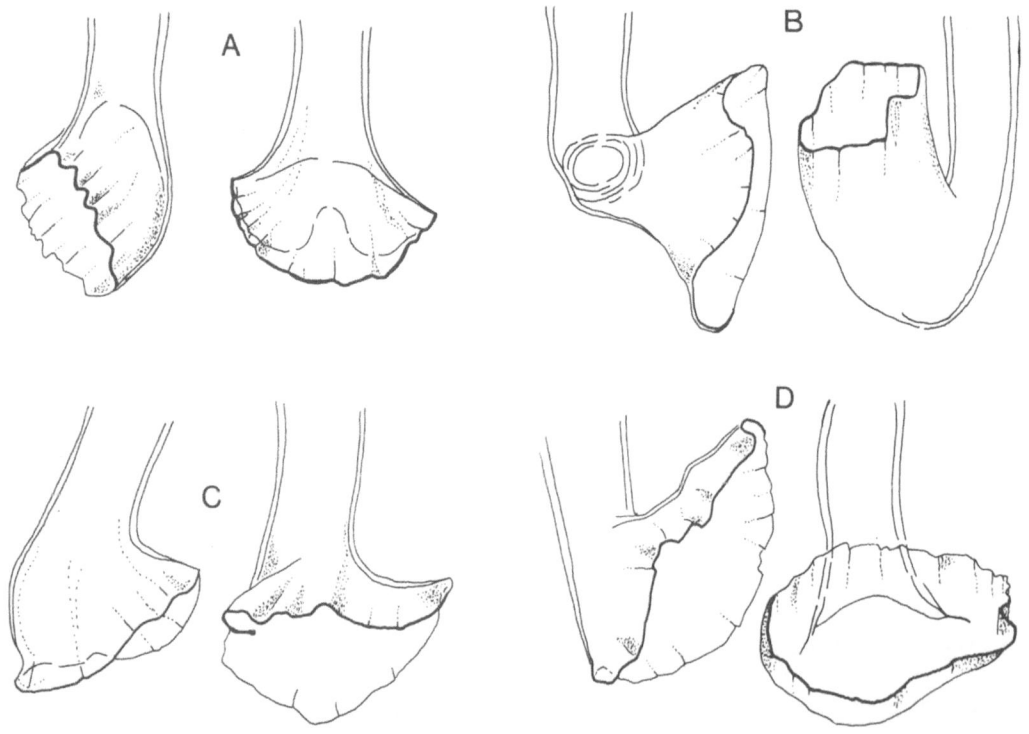

Fig. 4. Ectal ends of penis sheaths of four specimens (A–D) of *Limnodrilus hoffmeisteri.*

characters have to be investigated. Loden & Harman (1980) in their experiments on naidids demonstrated the extreme modificability of setal characteristics induced by several ecological factors.

The aim of this presentation is not to discourage biologists who intend to determine common Tubificidae, but to show that a great variability of taxonomically relevant structures does not prevent an exact determination of single specimens.

Acknowledgements

I would like to thank Mr. B. Mentel and Mr. P. Stiewe for producing the illustrations for this paper.

References

Barbour, M. T., D. G. Cook & R. S. Pomerantz, 1980. On the question of hybridization and variation in the oligochaete genus Limnodrilus. In R. O. Brinkhurst & D. G. Cook (eds.), Aquatic Oligochaete Biology. Plenum Press, New York: 41–53.

Brinkhurst, R. O., 1982. Evolution in the Annelida. Can. J. Zool. 60: 1043–1059.

Ditlevsen, A., 1904. Studien an Oligochäten. Z. wiss. Zool. 77: 398–480.

Dufour, L., 1844. Anatomie générale des Diptères. Annls Sci. nat. 1: 244–264.

Erséus, C., 1980. Specific and generic criteria in marine Oligochaeta, with special emphasis on Tubificidae. In R. O. Brinkhurst & D. G. Cook (eds.), Aquatic Oligochaete Biology. Plenum Press, New York: 9–24.

Gavrilov, K. & N. G. Paz, 1949. Limnodrilus inversus n.sp. y su reproducción uniparental. Acta zool. lilloana 8: 537–565.

Gavrilov, K. & N. G. Paz, 1950. Nota adicional sobre la reproducción de Limnodrilus. Acta zool. lilloana 9: 533–568.

Giere, O., H. Felbeck, R. Dawson & G. Liebezeit, 1984. The gutless oligochaete Phallodrilus leukodermatus Giere (Tubificidae), a tubificid of structural, ecological and physiological relevance. In G. Bonomi & C. Erséus (eds.), Aquatic Oligochaeta. Proceedings of the Second International Symposium on Aquatic Oligochaete Biology. Developments in Hydrobiology (this volume).

Holmquist, C., 1978. Revision of the genus Peloscolex (Oligochaeta, Tubificidae). 1. Morphological and anatomical scrutiny; with discussion on the generic level. Zool. Scr. 7: 187–208.

Kennedy, C. R., 1969. The variability of some characters used for species recognition in the genus Limnodrilus Claparède (Oligochaeta: Tubificidae). J. nat. Hist. 3: 53–60.

Loden, M. S. & W. J. Harman, 1980. Ecophenotypic variation in setae of Naididae (Oligochaeta). In R. O. Brinkhurst & D. G. Cook (eds.), Aquatic Oligochaete Biology. Plenum Press, New York: 33–39.

Pfannkuche, O., 1977. Ökologische und systematische Untersuchungen an naidomorphen Oligochaeten brackiger und limnischer Biotope. Dissertation Universität Hamburg, 138 pp.

Şengün, A., 1944. Experimente zur sexuell-mechanischen Isolation. Istanb. Üniv. Fen Fak. Mecm. (Seri B) 9: 239–253.

Şengün, A., 1949. Experimente zur sexuell-mechanischen Isolation, 2. Istanb. Üniv. Fen Fak. Mecm. (Seri B) 14: 114–128.

Timm, T., 1981. On the origin and evolution of aquatic Oligochaeta. Eesti NSV Tead. Akad. Toim. (Biol.) 30: 174–181.

The position of the Haplotaxidae in the evolution of oligochaete annelids

R. O. Brinkhurst

Institute of Ocean Sciences, P.O. Box 6000, Sidney, British Columbia, Canada V8L 4B2

Keywords: aquatic Oligochaeta, annelids, evolution

Abstract

The Haplotaxidae have all the characteristics to support the hypothesis that they are the living descendents of the stem forms from which all of the Oligochaeta Clitellata (Orders Lumbriculida, Haplotaxida, Lumbricida, Tubificida) can be derived. The Aphanoneura are distinct from the Clitellata and are raised to a separate Class. There is no evidence to support the view that the elaborate setae of many Tubificida are derived from a polychaete ancestry; both are held to be independent modifications to aquatic life derived from a simple burrowing protoannelid with lumbricine setae.

Introduction

After more than thirty years experience with systematics of the Oligochaeta, Michaelsen (1930) concluded that two alternative phylogenetic systems were equally credible, and that it was impossible to choose between them. According to one scheme, the family Haplotaxidae could be visualised as ancestral to both the terrestrial and aquatic lines of oligochaete evolution; the alternate hypothesis being that those aquatic forms which may have complex dorsal setae (the modern Tubificida; see Brinkhurst, 1982) were derived from polychaete ancestors other than those that gave rise to the haplotaxids and thence the Lumbricida (the earthworms of familiar terminology).

Prior to that date, a linear evolutionary sequence was visualised beginning with the archiannelids which supposedly provided the ancestors of first the Aeolosomatidae and from them the Naididae, Tubificidae, thence to the Lumbriculidae (with their bifid but paired instead of abundant setae) and from them to the haplotaxids and hence to the earthworms. This sequence has been thoroughly discredited at every step of the supposed sequence (Brinkhurst, 1982 reviews the literature). Indeed, the later accounts attempted to reverse the sequence to read Lumbriculidae-Tubificidae-Naididae-Aeolosomatidae, which idea contains the most important but unrecognised premise that the complex dorsal setae of most Tubificida were derived from forms with lumbricine paired setae. Stephenson (1930) was in fact inconsistent in this respect in first claiming the polychaete ancestry of tubificid hair setae, but electing the lumbriculids as the ancestral group a few pages later. Yamaguchi (1953) also saw the lumbriculids as ancestral (becuse of the variability of the male reproductive system) but Clark (1978) was more aware of the problems inherent in this selection. These authors missed the alternative proposition that the Haplotaxidae might provide the stem forms for all of the Oligochaeta, later restated by Brinkhurst & Jamieson (1971). This was largely because Michaelsen chose to classify the haplotaxids with the opisthoporous megadriles despite the fact that they are plesioporous microdriles (see Brinkhurst, 1982 for definitions). Their position as ancestors of the terrestrial forms was recently confirmed by Jamieson (1977, 1978, 1981) who also concluded that the Tubificida could not be

Hydrobiologia 115, 25–36 (1984).

26

interposed between the haplotaxids and their terrestrial descendents but could be independently derived from the haplotaxids but not the Lumbriculida. I concur with these findings with the exception that I now regard the prosopore condition of the lumbriculids to be derived from the plesiopore condition of the haplotaxids (see Brinkhurst, 1982).

The most recent phylogenetic account (Timm, 1981) seeks to return to the concept that hair setae in oligochaetes are derived from an ancestral polychaete and that the lumbricine setal state is derived from forms with more complex setae by a progressive simplification akin to the progressive specialisation of segmental organ systems seen throughout phyla such as the annelids, arthropods and chordates. I am indebted to Dr. Timm for the many exchanges of correspondence in regard to this and other issues, and trust I have quoted his views accurately in the following account.

Origin from proto-haplotaxids

Brinkhurst & Jamieson (1971) focused attention on the sequential arrangement of gonads in existing oligochaetes and demonstrated that all of the multigonadal forms were parthenogenetic and that all oligochaetes can be derived from a worm with four pairs of gonads in successive segments with the testes in front of the ovaries (G.I–G.IV Fig. 1). The segmental position of the gonads varies but this is a secondary condition, not necessarily derived from a multigonadal ancestor. Eleven of the 18 living haplotaxid species have all eight gonads whereas this condition is only found in two relatively primitive species of the Lumbricida. The male and female gonoducts of the earliest oligochaete may be supposed to have been very similar, just as they are in the haplotaxids. In all other families the male ducts become larger and more elaborate than the female

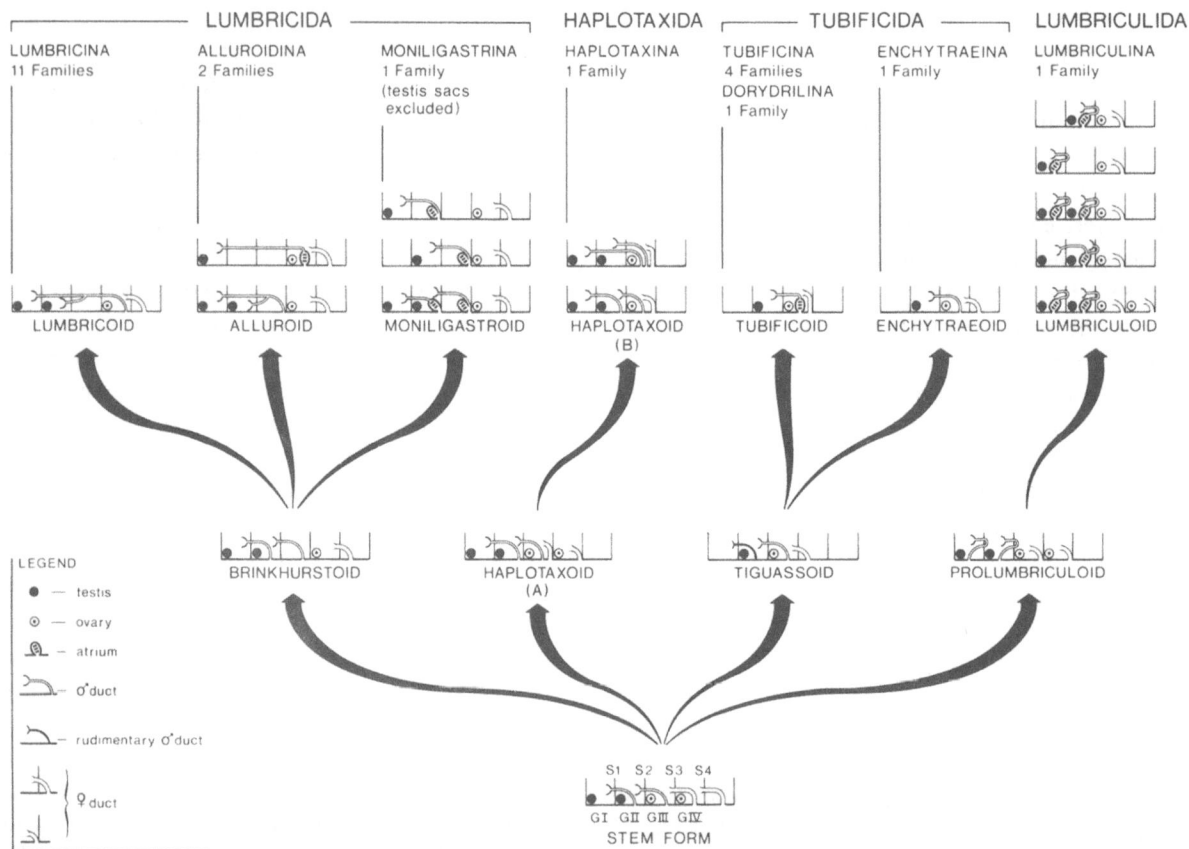

Fig. 1. Phylogeny of the Oligochaeta based on gonad sequence, form of gonoducts and presence of atria (from Brinkhurst, 1982). Dorydrilidae should be shown as equivalent to Tubificoid and Enchytraeoid stems (but see Fig. 2 and Brinkhurst, 1984).

ducts. The latter tend towards the prosopore condition but interestingly retain the plesiopore condition in many supposedly ancient families, including the Enchytraeidae. The male ducts develop some form of sperm storage and feeding organs, which are termed atria (with or without prostates) in microdriles and prostates in the megadriles. The haplotaxids are the only group that totally lacks such structures, the male ducts opening to the exterior via very short ectodermal invaginations that are the precursors of the atria/prostates.

The opisthoporous state of the megadrile male ducts is quite clearly secondary and I now believe that this is true of the prosopore state of the Lumbriculidae (Brinkhurst, 1982). Modern representatives of earlier, less specialized, intermediate forms between the octogonadal haplotaxids and two of the three other Orders (which show various patterns of gonad reduction) are to be seen in *Haplotaxis brinkhursti* and *Tiguassu*. It is not surprising to find that such living forms show advanced characteristics as well as traces of their ancestry. The latter for example has a proboscis and a simple gizzard but still has the eversible pharyngeal roof of all aquatic oligochaetes. It reveals its ancestry by retaining non-functional male funnels in G.I. with ducts in G.II, but there are no testes in G.I. Similar rudimentary organs are found in many Lumbriculidae in which the presence of non-functional atria in G.I. assures us that the two pairs of male ducts associated with atria in G.II must have been derived from the prosopore condition. It is for this reason that the Dorydrilidae can never have been derived from the lumbriculids, as their atria lie in G.III in the plesiopore condition only found in the Tubificida (Fig. 2). All other superficial similarities to the Lumbriculidae are of necessity convergent (Brinkhurst, 1984).

The rearward extension of the anterior pair of male ducts in *Haplotaxis violaceus*, resulting in both pair of male pores opening in G.III, is evidence that at least one living haplotaxid shows some signs of the development of the opisthopore condition. The direct ancestor of the opisthopores would have had a different gonadal reduction sequence though as the ovary of *H. violaceus* is in G.III not G.IV as it is in the Lumbricida. Traces of a prosopore state can be seen in haplotaxids like *H. smithii*, *H. hologynous*, and *H. ornamentus* in which the second pair of male ducts open close to the anterior border of G.II if not in G.I.

While many college texts refer to the similarity of oligochaete and polychaete setae, there has been very little careful comparison of them. Brinkhurst (1982) reviewed setal form and function and revealed, among other things, that most of the Enchytraeidae, as well as all of the Lumbricida and Lumbriculida, have the lumbricine setal condition (Fig. 3). These are usually simple-pointed, but they may be bifid with simple upper teeth in many lumbriculids. Other aquatic families usually adopt fully bifid rather than simple-pointed setae and more complex setae are found in the dorsal bundles of many species. This suggests a substrate crawling antecedent with long dorsal setae for protection from predators. The multiple simple-pointed setal condition is found in few Tubificina (the ancient *Telmatodrilus* shows such a tendency) and in the recent perichaetine earthworms. There is no sign of it in forms I consider to be ancestral by virtue of other anatomical and zoogeographic features, whereas Timm (pers. corresp.) sees the ancestral annelid as essentially perichaetine.

The wide taxonomic distribution of the lumbricine state indicates it is the basic pattern which conforms with the theories of Clark concerning fundamental annelid form. The most useful disposition of a few setae in an ancestral active burrower might well be at the 'corners'; a further modification to allow for slight dorso-ventral specialization would permit surface crawling through loose particles. So widespread a phenomenon as this setal positioning cannot be seriously attributed solely to the last pale shadow of the former glory of parapodia without reference to selective advantage to maintain it. Polychaete setae are not in fact similar to oligochaete setae in form, number or disposition (Brinkhurst, 1982).

Setae in the Haplotaxidae demonstrate that the family has within it the capability of forming all of the setal types of oligochaetes. While most species have lumbricine setae, there are paired, simple-pointed and bifid setae in *H. glandularis* (foreshadowing the situation of ventral setae in the Phreodrilidae) and bifid, if not pectinate, setae in *H. denticulatus*. The setae of the familiar *H. gordioides* and similar *H. heterogyne* are highly modified for life in coarse substrates (see also *Grania*, Enchytraeidae). A recently discovered South American haplotaxid has somatic setae like those of *H. gordioides*, but also has hair-like genital setae

28

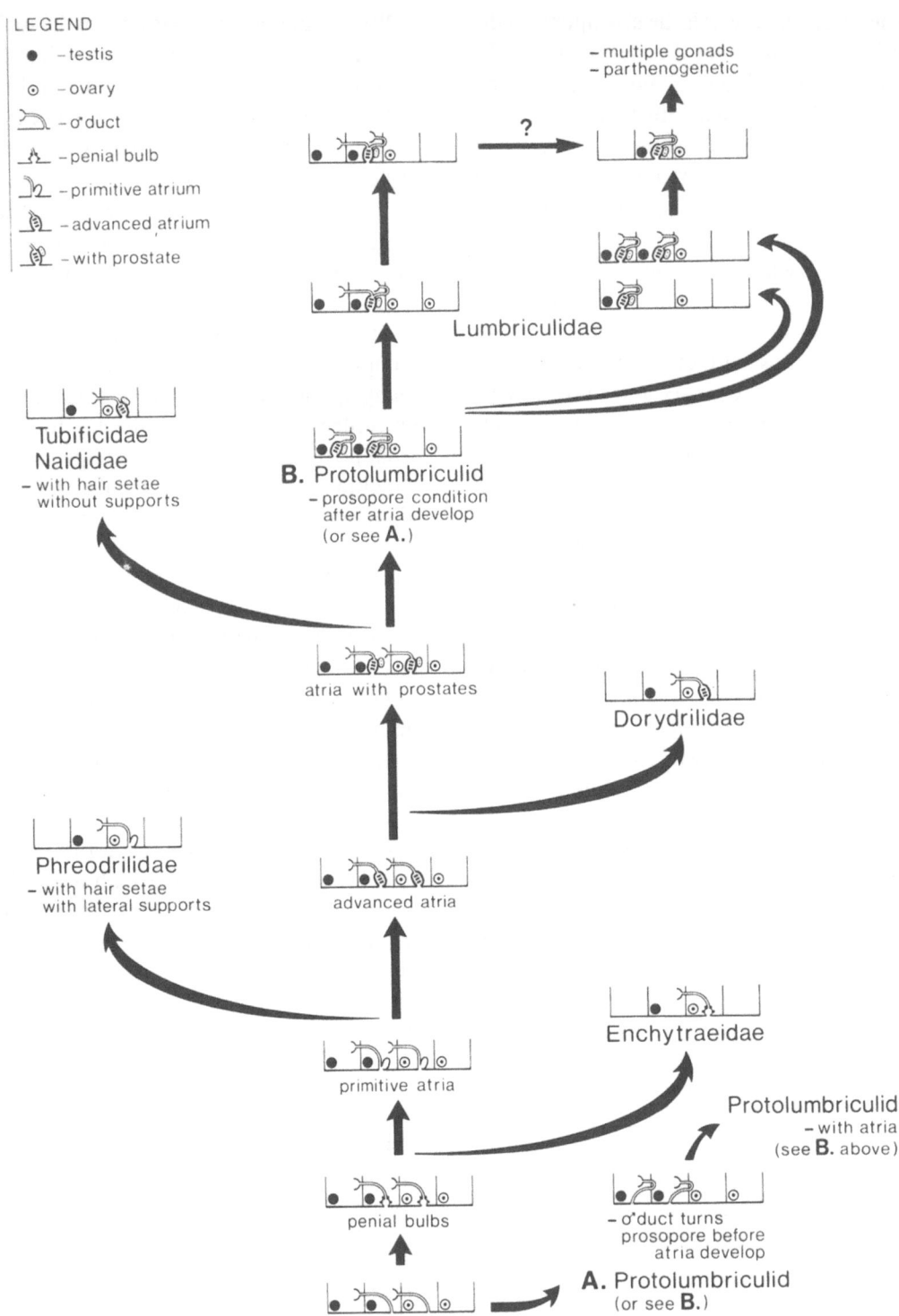

Fig. 2. Alternative origins of Lumbriculidae. If alternative A is adopted, all forms above the haplotaxids and lumbriculids should have gonads reduced to GII · GIII. If alternative B is adopted atria and prostates are evolved once, the main stem retains paired setae and, therefore, the hair setae of phreodrilids are derived independently from those of the naidids/tubificids, the lumbriculids can be derived from a theoretical proto-dorydrilid. (See also Brinkhurst, 1984 for modified version.)

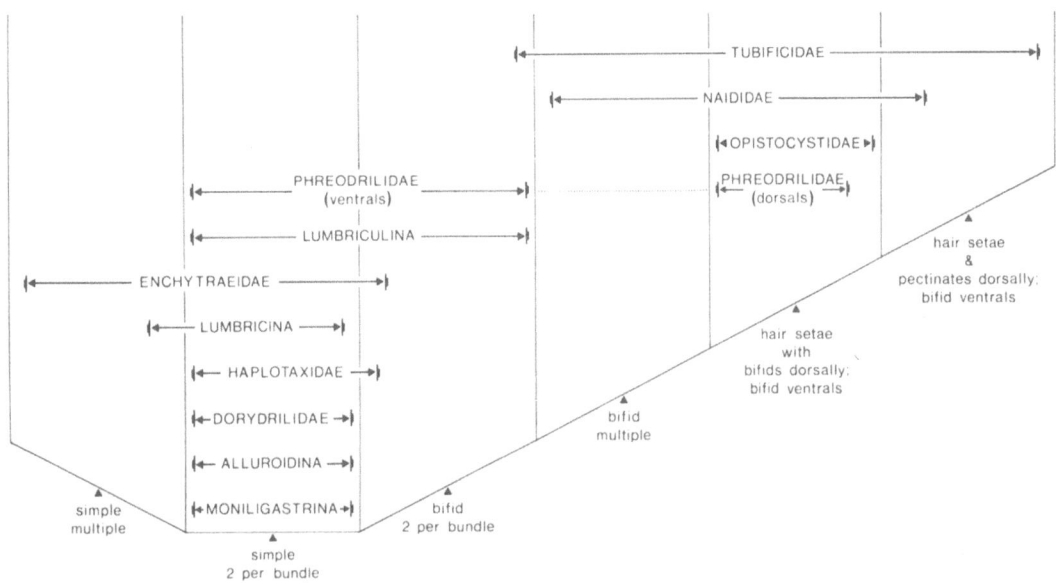

Fig. 3. Distribution of setal form and number in oligochaetes (from Brinkhurst, 1982).

which, while not ancestral to those of the Tubificida by any stretch of the imagination, demonstrate that the single cells that secrete setae can form a hair-like structure even within the haplotaxids (Brinkhurst, unpubl.). Hair setae with minute bifid tips in tubificids and naidids suggest there is no fundamental difference in form in any case.

The extreme anatomical simplicity of the other organ systems of the haplotaxids render them acceptable as ancestors of the other three orders. An independent study of nephridia for instance, led to the suggestion that the haplotaxids were the stem forms of oligochaetes (see Cekanovskaya, 1962). The haplotaxids have the discontinuous global distribution pattern of a very ancient group, and occupy refuge habitats in centers renowned for their relic faunas, such as Tasmania. Most other families clearly date from after the break up of continents and are northern or southern in origin. In order to emphasize the stem position of the haplotaxids my scheme differs from more recent classifications in erecting the Order Haplotaxida for the modern descendents of the stem group.

There is no impediment to the acceptance of the Haplotaxidae as modern descendents of the ancestors of the oligochaetes, not too far removed from the descendents of all annelids, which were themselves derived from unsegmented coelomate burrowing forms represented to-day by a few scattered

groups like the sipunculids. It should be emphasized that no modern living form is here seen as a direct antecedent to other modern living forms. In many early treatments there is an unfortunate tendency to expect the living forms to be derivable one from another with the only concessions to the possibility of extinction being to create theoretical intermediate types that could add or delete segments at will in order to pass from one form to another. No evidence for the existence of such intermediate forms was ever presented and the concept of the need for selective advantage of every new development was ignored. These problems also tainted the earlier views of simultaneous evolution of the coelom and segmentation, whereas such pitfalls were avoided by Clark (1946). His brilliant exposition of the independent origin of both leads us to the presumption of a simplified ancestor to both oligochaetes and polychaetes with earthworm-like anatomy apart from its reproductive system (Brinkhurst, 1982).

Evolution within the Tubificida

It is necessary to touch on one or two implications of this theory to the evolution of the various families within the Tubificida.

The Enchytraeidae are now seen to progress from

very simple forms with lumbricine setae and simple glandular precursors of the atria to the very advanced aquated *Propappus* with its bifid setae and well developed atria. *Propappus* has been claimed as derivative of an enchytraeid ancestor to fit the concept that the family must have been derived from the Tubificidae. While there is no intrinsic evidence yet available for reading the evolutionary sequence in one direction or the other, the common and widespread form of the enchytraeids should be regarded as basic rather than the rare and peculiar as expressed in *Propappus*. Recent evidence (K. Coates, pers. comm.) shows that *Propappus* can no longer be classified with the Enchytraeidae.

The Phreodrilidae/Opistocystidae seem to be (?earlier) southern hemisphere parallels to the originally northern Naididae/Tubificidae. The ventral setae of phreodrilids are primitive, as are their atria. These more nearly resemble megadrile prostates in that the male ducts do not open through them, and they lack externalized prostate cells. Elaborate penes have developed here as well as in the tubificids and the lumbriculids, evidence that such ectodermal invaginations as penes, atria and megadrile prostates have been evolved independently many times. The Opistocystidae are a very poorly known family.

The Dorydrilidae, formerly seen as the link between the Tubificidae and the lumbriculids en route to the haplotaxid-megadrile line, are clearly tubificine. Their atria lie in G.III which cannot be derived from nor lead to the lumbriculid condition, their gonad sequence is reduced. Other similarities are quite clearly convergent, but they were classified as a suborder along with the Enchytraeina and Tubificina (Tubificidae, Naididae, Phreodrilidae, Opistocystidae) by Brinkhurst (1982). The ancestral form interpolated between points of origin of the Phreodrilidae and Dorydrilidae (labelled 'advanced atria') must be classified as a dorydrilid and so the direct ancestors of that family might well lie along the main line of descent of lumbriculids from a proto-tubificine condition if we accept option B for the evolution of the lumbriculids (Fig. 4). This would then be in accord with the views of S. Hrabě as expressed elsewhere in this volume except that the various tubificine groups would have (independently) reduced the gonad series, as shown in Fig. 2. (Brinkhurst, 1984 elaborated this.)

The Naididae and Tubificidae are closely related,

with the former showing some advanced adaptations to aquatic life, such as locomotory methods, eyes and more forms of gill than in the other family, whereas all of the tubificids are obligate burrowers with more highly developed male ducts. If hair setae were a defensive adaptation in forms that gave rise to the four families in the Tubificida that still possess them, it is difficult to see the adaptive significance of the dorsal hair setae in those forms such as the tubificids and the naidid genus *Dero* that burrow. Some successful genera such as *Limnodrilus* totally lack hair setae, as do some species within genera that otherwise possess them (e.g. *Potamothrix*). We cannot assume that the tubificids are caught in the act of disposing of their hair setae as they have existed for far too long for this still to be an ongoing process. As species with and without hair setae successfully co-exist in large numbers, it is hard to see any adaptive significance to their possession. Those species which possess them may be represented by forms that lack them in certain biotopes, particularly those in which the water has a high conductivity (see *Tubifex* and *Ilyodrilus*). It is Timm's contention (1981) that burrowers lacking hair setae represent an important evolutionary step between the species with complex dorsal setae and the lumbricine types that are derived by further degeneration of the setal equipment. It is even suggested that this takes place by neoteny, a process commonly evoked in early phylogenetic speculations when no particular selective advantage could be identified. There is no evidence that immature stages of Tubificida possess earthworm-like setae and so Timm's suggestion is untenable. Secondary simplification clearly does take place, as evidence by the setae in *Clitellio* and the Phallodrilinae.

The Aeolosomatidae and their allies (Class Aphanoneura) must be similarly derived from the ancestral annelid pool or perhaps not far from the archiannelids, a specialised polychaete group with which they have much in common. This may again be regarded as convergence as they are anatomically totally distinct from the Oligochaeta, even to the point of lacking a clitellum, and they are no longer included within that Class. The ventral reproductive gland of the aelosomatids is not a clitellum in any sense of the word, although it should not be referred to as a copulatory gland either (see Brinkhurst, 1982).

Comparison with the alternative hypothesis of Timm

The views of Timm (1981 and pers. commun.) are summarized in Figs. 4–6. Space does not permit a detailed consideration of this scheme beyond pointing out the salient criticisms of it in light of the haplotaxid origin theory. The diagrams will enable a detailed comparison to be made. Referring first to Fig. 4 I would suggest that there is little or no evidence of successful freshwater polychaetes that could provide an ancestor to the Tubificida. Apart from the reversed polarity of the evolution in the Enchytraeidae and the doubtful positioning of the Lycodrilidae* in terms of our ignorance of that family, the rest of the developments within the Tubificida do not require further comment. Both

* The Lycodrilidae are now seen to be attributable to other extant families (Brinkhurst, in press; Can. J. Zool., 1984).

schemes accept the Haplotaxidae as ancestors of the Lumbricida. The ancestor of the Lumbricomorpha postulated by Timm must be rather similar to my protohaplotaxid ancestor, which I also see as giving rise to the Lumbriculidae. The supposedly 'extraordinarily variable, shattered morphology of the genital system' of lumbriculids and moniligastrids referred to by Timm (1981) and earlier authors (e.g. Yamaguchi, 1953), depends upon undue focus on a few parthenogenetic species in the former and misunderstanding of the latter (see Brinkhurst, 1982; Jamieson, 1977). The Dorydrilidae, however, cannot be derived from the Lumbriculidae by any stretch of the imagination and the leeches are normally derived from the Lumbriculidae (see, e.g., Clark, 1969), in contrast to their position in Timm's scheme. They have recently been classified as Hirudinoidae (Richardson, 1970).

In Timm's theory the terrestrial species are all said to have lost the atria of the Tubificida, which

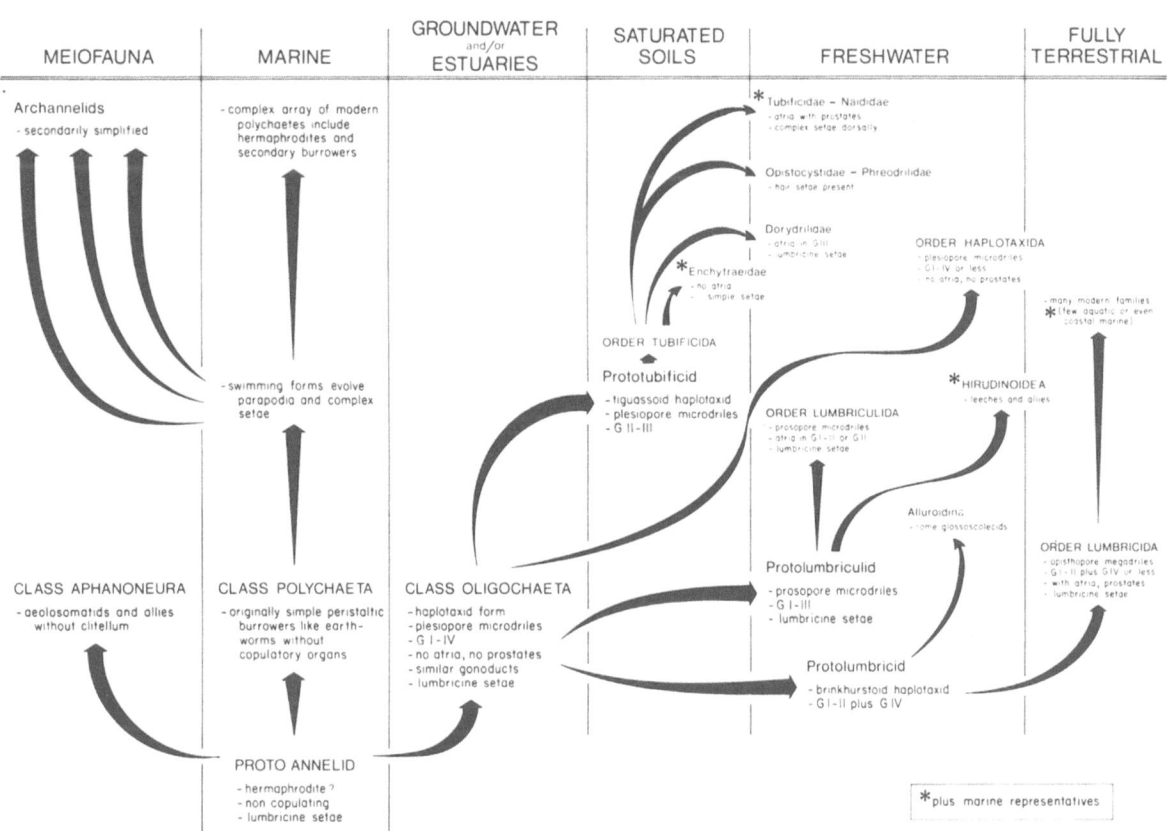

Fig. 4A. Oligochaete phylogeny: Haplotaxid origin theory (some details such as marine enchytraeids and tubificids, freshwater polychaetes omitted where clearly secondary).

32

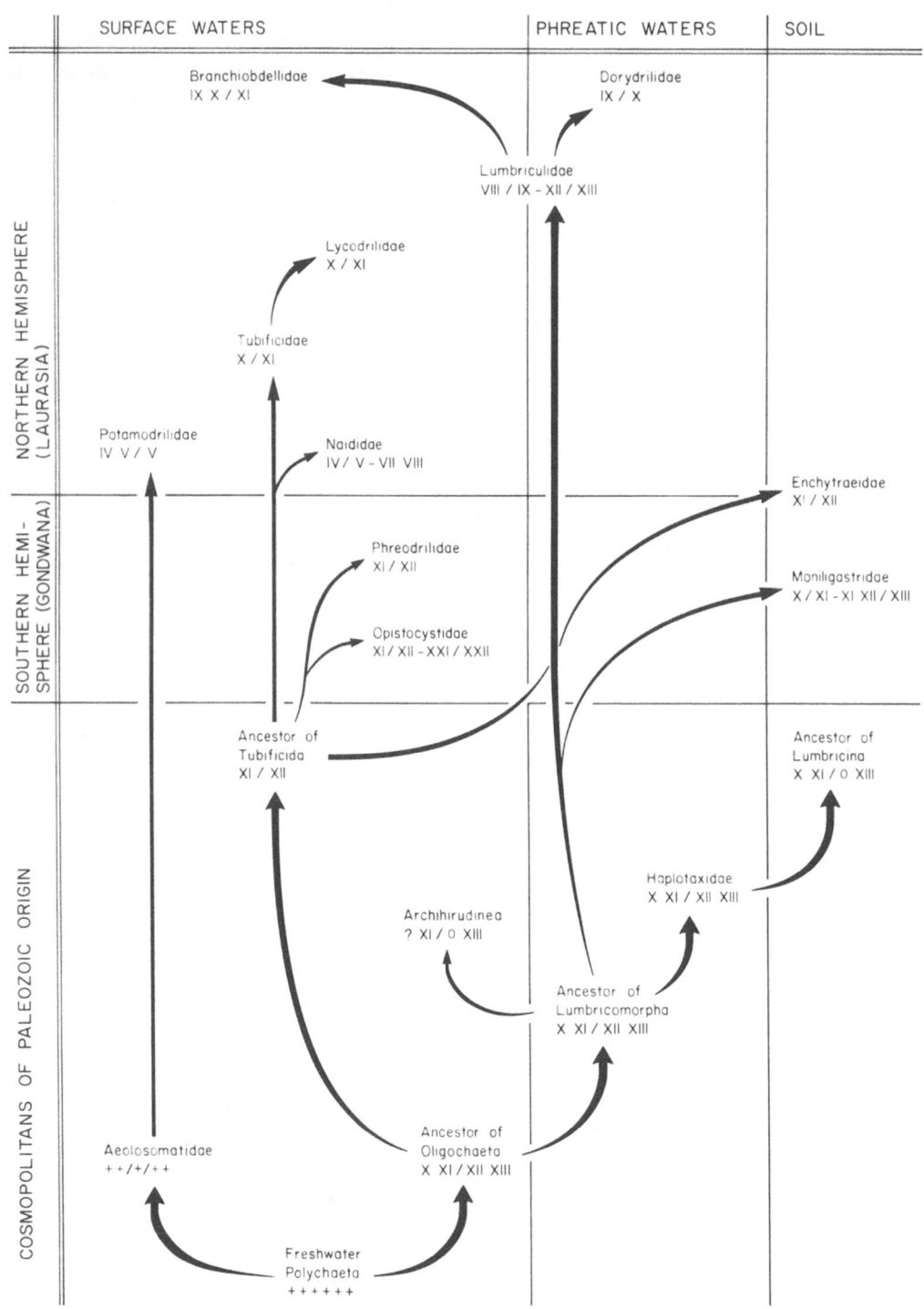

Fig. 4B. Oligochaete phylogeny: Theory from Timm (1981).

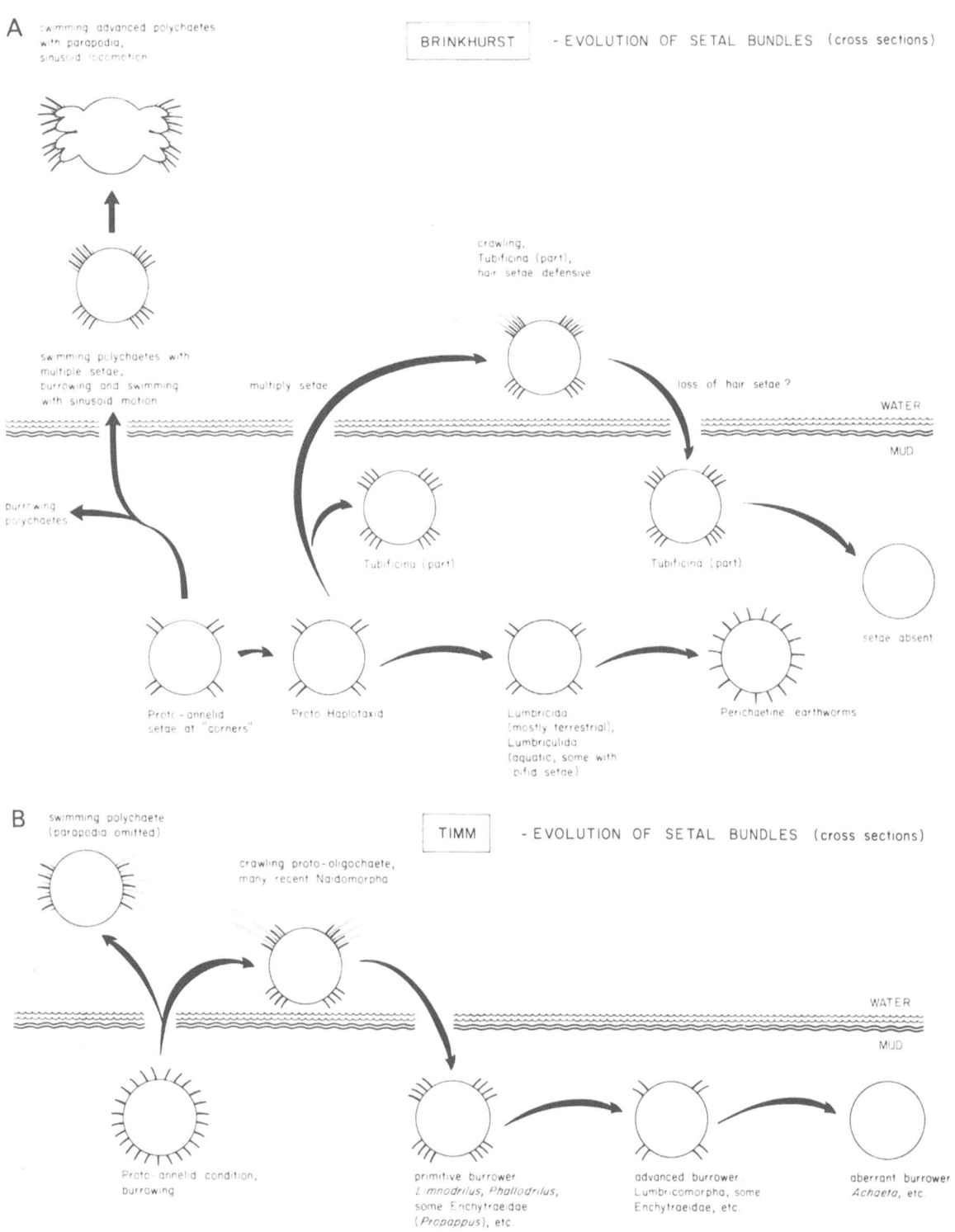

Fig. 5A. Comparison of theories of evolution of setal bundles: Haplotaxid origin.

Fig. 5B. Comparison of theories of evolution of setal bundles: Theory of Timm (pers. commun.).

34

process is seen 'in progress' in the enchytraeids. This ignores the evolution of the many forms of prostates in megadriles which are functional analogues of the atria in microdriles. The atria/prostates are even present in a simple form in the Lumbricidae and are only absent in haplotaxids. In fact, Timm makes relatively little use of the evidence based on reproductive organs as preference is given to the arguments based on the origin of oligochaete setae from those of polychaetes.

The two theories in regard to setal evolution are documented in Figs. 5 and 6. I would claim in support of my views that they are more parsimonious than those of Timm, they accept the wide taxonomic distribution and hence antiquity of the lumbricine condition, they agree with the general notion that the perichaetine state is a rare experiment in extant megadriles and that tubificids do not have larvae with lumbricine setae as required by the concept of neoteny. The small upper tooth of the bifid setae of many species in the (unrelated) Tubificidae and Lumbriculidae may have much to do with the nature of the substrate the species inhabits but cannot seriously be viewed as representing a stage in the conversion of bifid to simple-pointed setae as Timm contends. Furthermore, there is little

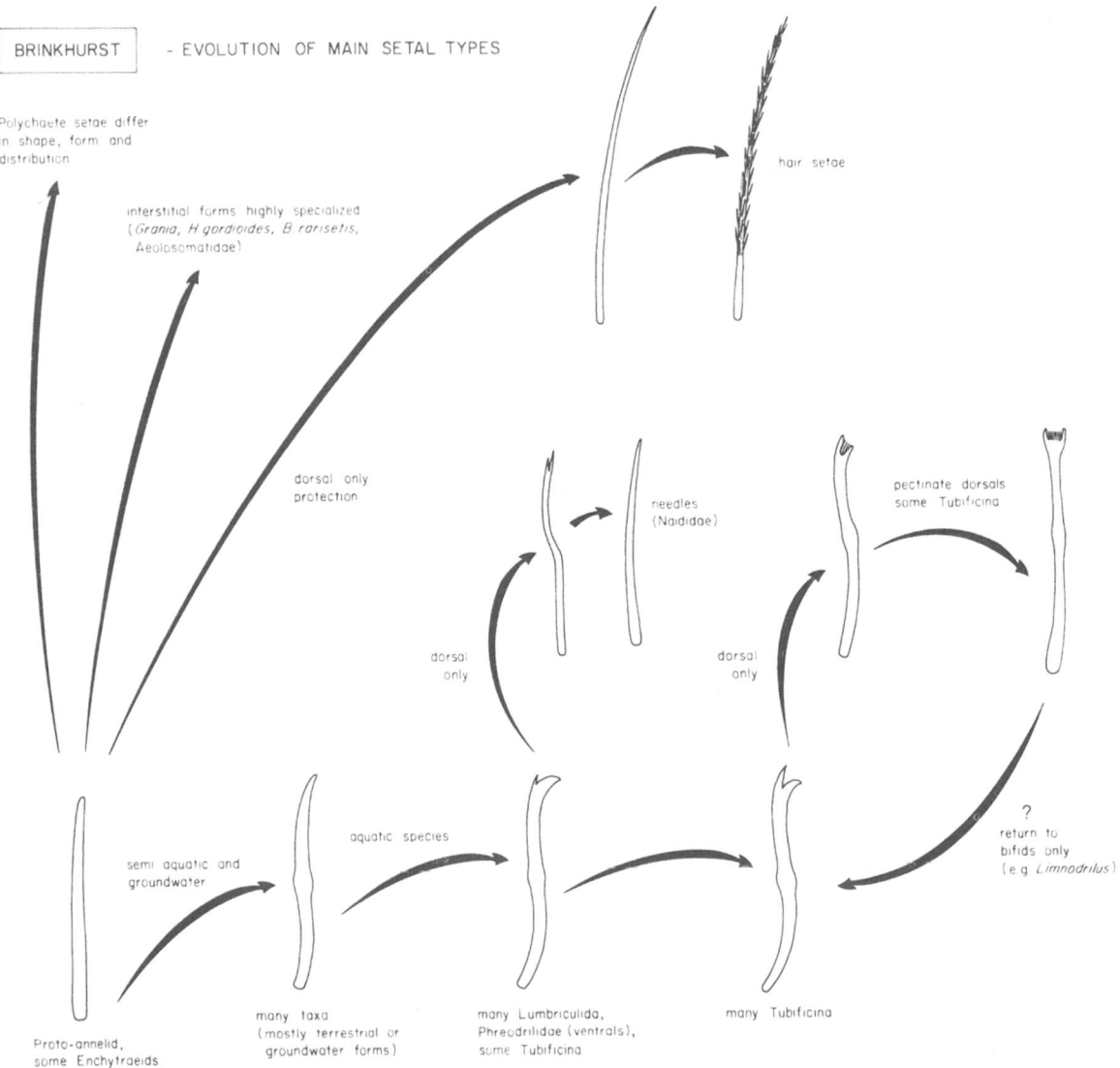

Fig. 6A. Comparison of theories of evolution of setal form: Haplotaxid origin.

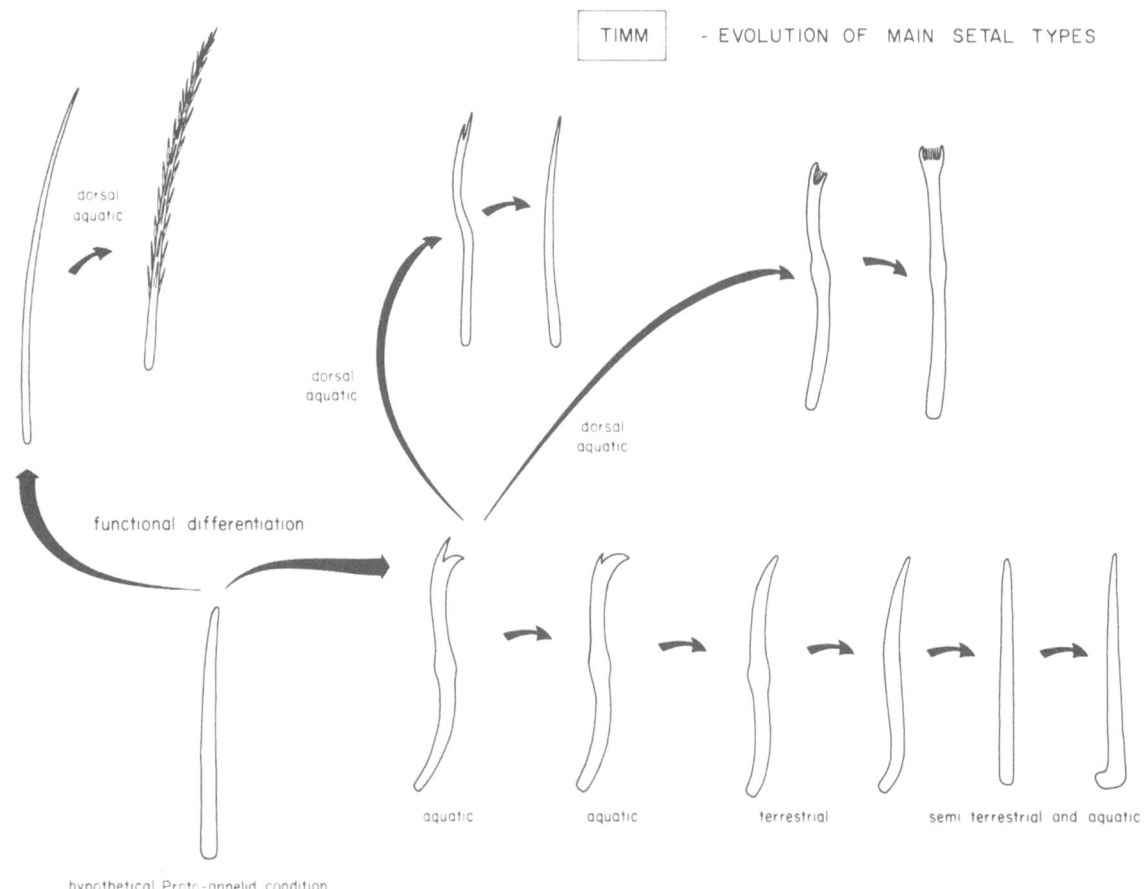

Fig. 6B. Comparison of theories of evolution of setal form: Theory of Timm (pers. commun.).

evidence of the octogonadal form with hair setae postulated by Timm, whereas there is evidence of the possible existence of the intermediate forms claimed by me. In fact, Timm's whole hypothesis is compromised from the outset by the assumption of a polychaete ancestry for hair setae, and that the paired lumbricine arrangement is an echo of the presence of parapodia. Clark has shown that parapodia interrupt the cylindrical body wall musculature and that they must be an adaptation for swimming in a form that is no longer dependent upon peristaltic burrowing. The oligochaetes retain the original intact cylindrical body wall, and I see the lumbricine setal placement as advantageous rather than accidental in an early annelid that became segmented in response to the burrowing habit that preceded all the later annelid life styles.

I would claim that the haplotaxid origin theory is not compromised by any unsupportable assumptions, that it fits all of the known facts, including those fragments of evidence derived from traces of former reproductive structures among the lumbriculids and others and that it is based on the range of gonadal and setal plans found in the modern descendents of the original pool of haplotaxid ancestral forms. It is a more parsimonious theory than the alternative and, while it relies less on evidence of geological history than that proposed by Timm, it does not seem to be in conflict with that evidence.

Acknowledgements

I am indebted to Dr. T. Timm for long and frank exchanges of opinion on this subject, and to Ms. K. Coates for pointing out my assumptions regarding enchytraeids. Ms. Coates and Mr. H. R. Baker critically reviewed the manuscript. The illustrations were prepared by Techni-Graphics Ltd. The manuscript was prepared by M. Stone.

36

References

Brinkhurst, R. O., 1982. Evolution in the Annelida. Can. J. Zool. 60: 1043–1059.

Brinkhurst, R. O., 1984. Comments on the evolution of the Annelida. Hydrobiologia 109: 189–191.

Brinkhurst, R. O. & B. G. M. Jamieson, 1971. Aquatic Oligochaeta of the world. Oliver & Boyd, Edinb., 860 pp.

Cekanovskaja, O. V., 1982. The aquatic Oligochaeta of the USSR. Opred. Faune SSSR 78: 1–411 (transl. U.S. Dep. Interior Nat. Sci. Fndtn Amerind, New Delhi, 1981).

Clark, R. B., 1964. Dynamics in metazoan evolution. Clarendon Press, Oxford, 313 pp.

Clark, R. B., 1969. Systematics and phylogeny: Annelida, Echiura, Sipincula. Chem. Zool. 4: 1–68.

Clark, R. B., 1978. Composition and relationships. In P. J. Mill (ed.), Physiology of annelids. Academic Press, Lond.; N.Y. S Francisco: 1–32.

Jamieson, B. G. M., 1977. On the phylogeny of the Moniligastridae with description of a new species of Moniligaster (Oligochaeta, Annelida). Evol. Theory 2: 95–114.

Jamieson, B. G. M., 1978. Phylogenetic and phenetic systematics of the opisthopore Oligochaeta (Annelida, Clitellata) Evol. Theory 3: 195–233.

Jamieson, B. G. M., 1981. Historical biogeography of Australian Oligochaeta. In A. Keast (ed.), Ecological biogeography of Australia. Dr. W. Junk b.v. Publishers, The Hague; Boston; Lond.: 889–921.

Michaelsen, W., 1930. Nachtrag 2. Oligochaeta. Handb. Zool. 2: 116–118.

Richardson, L. R., 1970. Towards the new hirudinology. J. Parasit. 56: 237.

Stephenson, J., 1930. The Oligochaeta. Clarendon Press, Oxford, 978 pp.

Timm, T., 1981. On the origin and evolution of aquatic Oligochaeta. Eesti NSV Tead. Akad. Toim. Biol. 30: 174–181.

Yamaguchi, H., 1953. Studies on the aquatic Oligochaeta of Japan. IV. A systematic report with some remarks on the classification and phylogeny of the Oligochaeta. J. Fac. Sci. Hokkaido Univ., Ser. 6 Zool. 11: 277–342.

Aspects of the phylogeny of the marine Tubificidae

Christer Erséus
Swedish Museum of Natural History, Stockholm, and (postal address:) Department of Zoology, University of Gothenburg, Box 25059, S-400 31 Göteborg, Sweden

Keywords: aquatic Oligochaeta, Tubificidae, phylogeny

Abstract

A tentative phylogeny of the oligochaete family Tubificidae, with emphasis on the marine representatives, is presented. The scheme is based on the morphology and arrangements of prostate glands and the setal patterns. The rhyacodriline, more or less diffuse prostates are regarded as a primitive stage in prostate evolution, preceded only by the aprostate condition assumed for the ancestor of the family. An early split of the subfamily Rhyacodrilinae supposedly led to (1) a marine branch, from which evolved the highly diverse, exclusively marine subfamilies Phallodrilinae and Limnodriloidinae, and (2) a freshwater branch, which later divided into the Telmatodrilinae, Tubificinae and Aulodrilinae. The marine subfamilies invariably lack hair setae, whereas about half of the species within the other, freshwater subfamilies possess such setae in their dorsal bundles. Some marine genera, such as *Monopylephorus* (Rhyacodrilinae), *Tubificoides* and *Clitellio* (both Tubificinae) are regarded as recent off-shoots from the main freshwater stock.

The families Naididae and Opistocystidae are considered likely to have evolved from rhyacodriline Tubificidae, whereas Phreodrilidae, the fourth family within the suborder Tubificina, is regarded as a sister group to the Tubificidae.

Introduction

The Tubificidae have traditionally been considered as a predominantly limnic oligochaete family (cf. Brinkhurst & Jamieson, 1971). This view is rapidly changing, as new marine species are discovered at an accelerating rate. It can be expected that the number of known marine species will soon exceed that of known freshwater species.

When few marine forms were known, it was easy to regard them as occasional, apparently recent off-shoots from a freshwater stock (cf. e.g. Timm, 1980). We should now consider the possibility of a reverse situation, or at least, we have reasons to analyze in more detail the possible phylogenetic relationships between the limnic and marine groups of Tubificidae. Such a study may also add to our conception of the evolution of oligochaete families other than the Tubificidae.

Different views of the evolution of aquatic Oligochaeta have recently been accounted for by Brinkhurst (1982, 1984), Jamieson (1978a, 1981) and Timm (1980, 1981), but none of these authors has discussed, at any length, evolution within the Tubificidae. Brinkhurst (1982) recognizes three suborders within the order Tubificida. One of these, the Tubificina, is the only oligochaete group that exhibits hair setae in the somatic bundles. It contains the families Tubificidae, Naididae, Opistocystidae and Phreodrilidae. The species of the first three of these families generally have prostate glands on their atria. The present account will deal primarily with the Tubificidae. The relations between this family and the other members of the Tubificina will be merely touched upon.

Hydrobiologia 115, 37–44 (1984).

The subfamilies within the Tubificidae

Different attempts at dividing the Tubificidae into subfamilies have been made (see Brinkhurst & Jamieson, 1971 for a review of earlier literature). I will here briefly characterize the subfamilies as recognized by recent authors (Baker & Brinkhurst, 1981; Erséus, 1982).

RHYACODRILINAE. A cosmopolitan, diverse group of both freshwater and marine genera. Prostates diffuse (the free ends of elongated prostatic cells arise from the inner atrial epithelium, penetrate the atrial wall individually, and form a diffuse layer covering a great part of the atrial ampulla), or irregularly scattered in two or more clusters of cells broadly attached to the atrial surface, or absent. Dorsal setal bundles complex (with hairs and pectinates) in about 50% of the freshwater species; only bifid setae in the marine forms (except *Jolydrilus* and a few *Monopylephorus* species) and the remaining freshwater species. Some genera (*Branchiura, Bothrioneurum, Macquaridrilus*) with elaborate, highly modified male ducts. Coelomocytes generally abundant and conspicuous. Spermatozeugmata not found in spermathecae. Body wall always smooth.

PHALLODRILINAE. An extremely diverse, cosmopolitan marine group, with a few freshwater representatives. Prostates generally solid and stalked, two per atrium. In some genera, one or both prostates absent, and in one genus (*Discordiprostatus;* see Baker, 1982) anterior prostate diffuse. Somatic setae bifid, secondarily single-pointed in a few species. Genital setae sometimes highly modified. Coelomocytes not abundant. Sperm generally loose in spermathecae, but forming roundish spermatozeugmata(?) in a few species (*Bathydrilus*). Body wall smooth in all species except one (*Duridrilus tardus;* see Erséus, 1983a).

TELMATODRILINAE. A small group of freshwater species, with discontinuous distribution (Holarctic and Tasmania). Prostates solid and stalked, more than two per atrium (possibly only two in the poorly known *Telmatodrilus bifidus*). Hair and pectinate setae common in dorsal bundles. Coelomocytes not abundant. Spermatozeugmata(?) found in spermathecae of some species; sperm loose in other species. Body wall often encrusted with foreign matter, and bearing scattered sensory papillae.

AULODRILINAE. A small group of freshwater species inhabiting tubes, and generally using posterior (more or less unsegmented) ends as gills. Prostates solid and stalked (may be broadly attached in some species, but this has been little studied), one per atrium. Hair setae, and other modified setae (pectinates, oar-shaped bifids, etc.), often present in dorsal bundles. Setae always numerous. Many species reproduce asexually by fragmentation, some species have genital organs shifted forwards. Coelomocytes not abundant. Sperm in loose, random masses in spermathecae. Body wall always smooth.

LIMNODRILOIDINAE. A large, cosmopolitan group of marine genera, most of which occur in muddy sands of warmer seas. Prostates solid, one per atrium, broadly attached to ental part of atrium (atrial ampulla). Somatic setae always bifid. Oesophagus modified in segment IX of most species; either bearing a pair of diverticula, or enlarged and glandular. Coelomocytes not abundant. Sperm bundled, or as spermatozeugmata(?) in spermathecae. Body wall papillate (encrusted with foreign matter) in some species (*Tectidrilus*).

TUBIFICINAE. Most of the freshwater genera, and a few marine ones. Appears cosmopolitan, at least in freshwater, but this is probably a reflection of the distribution of some highly peregrine species (*Tubifex tubifex, Limnodrilus hoffmeisteri*). Prostates solid, stalked, one per atrium, when present; absent in a few species. Hair and pectinate setae in dorsal bundles of many species, more so among limnic than among marine forms. Penes generally well developed, often with cuticular sheaths. Coelomocytes not abundant. Spermatozeugmata found in spermathecae. Body wall papillate in some species.

The morphology and occurrence of setal types in the Tubificidae

The predominant setae in the Tubificidae are the bifid crotchets. In a few species, one of the two teeth of these setae are very inconspicuous or even absent, but this can, in all instances, be regarded as a derived character.

Hair setae occur in the dorsal bundles of about

half of the freshwater species of Tubificidae, but they are extremely rare among the marine forms. When present, the hairs are often accompanied by small pectinate setae, i.e. bifid setae with one or more, generally very inconspicuous, intermediate teeth. The adaptive significance of the dorsal hairs and pectinates is still not understood (see Brinkhurst, 1984, for a more detailed discussion). It can only be concluded from their scattered occurrence throughout the Tubificidae that they have existed for a long time in the family, and it is clear that they have been secondarily lost on several occasions. The plesiomorphic state of the hair setae in the Tubificidae is also supported by the fact that hairs are present also in the closely related families Naididae, Opistocystidae and Phreodrilidae. The presence or absence of hair setae thus should not, per se, be used as a fundamental key for inferring phylogenetic relationships. However, the consistent absence of hairs in the Phallodrilinae, Limnodriloidinae and virtually all marine Rhyacodrilinae can be regarded as supportive evidence of these groups being long separated from the Tubificinae, Aulodrilinae, Telmatodrilinae and freshwater Rhyacodrilinae (cf. below). In the latter groups, hairs occur in about one half of the species.

Modified genital setae, called penial setae if located near the male pores, spermathecal setae if located near the spermathecal openings, are found throughout the family Tubificidae. Penial setae are the rule rather than the exception in the Rhyacodrilinae, Phallodrilinae and Telmatodrilinae, but are rare in the other subfamilies. The absence of penial setae is generally correlated with the terminalia of the male efferent ducts being modified into copulatory organs (penes or elaborate pseudopenes). This indicates that penial setae in some way aid in copulation. In most instances, they are certainly used only as hooks, for the mutual holding of concopulants, but in particular genera (*Adelodrilus, Inanidrilus, Discordiprostatus;* and probably some species of *Nootkadrilus*), individual setae within the penial bundles are modified into structures, which appear to be transmittant organs for sperm. With very few exceptions, spermathecal setae are modified bifid setae with greatly elongated teeth. They do exist in species with penes developed, but they are more frequent in genera in which the penes are 'soft' and not equipped with thick cuticular sheaths. The spermathecal setae thus also appear to aid in the transfer of sperm.

The great functional significance of the genital setae and their dependence on structural modifications of the male terminalia lay them open to selective pressures. Therefore, they should not be used as indicators of phylogenetic relationships between higher taxa.

A tentative phylogeny of the Tubificidae (Fig. 1)

The tubificid atrium is composed of an inner (invaginated) ectodermal epithelium and an outer mesodermal layer of muscles. The prostatic tissues are specialized secretory cells of the atrial epithelium. In a few species, these cells are wholly retained within the epithelium, but in most tubificids, the nucleate, secretion-producing portions of the prostatic cells penetrate the muscle layer of the atrium to form distinct glands; these outgrowths are the actual prostate glands. Fleming (1981) has described the ultrastructure of the prostate glands of *Tubifex tubifex*.

The atria of *Clitellio arenarius* and some species of *Potamothrix* (all Tubificinae) do not bear prostates, but this most certainly represents an apomorphic state versus the situation in the remaining species of the subfamily; the latter all have discrete prostate glands on their atria.

Coralliodrilus (Phallodrilinae) also lacks prostate glands. However, it is still open to question as to whether this is an ancestral or derived character. The same applies to four other aprostate genera, which are all assigned to the Rhyacodrilinae (largely based on their abundant coelomocytes), the brackish water-marine genera *Vadicola* and *Jolydrilus,* and the freshwater genera *Taupodrilus* (included in *Rhyacodrilus* by Brinkhurst & Jamieson, 1971) and *Epirodrilus*. It is possible that at least some of these genera are modern representatives of a lineage (or lineages) in which prostates have never developed. An aprostate ancestor of the whole family Tubificidae can be assumed, not only because some recent genera lack prostates, but simply because the ontogeny of these glands suggests it. It should be noted here, however, that *Jolydrilus* and *Epirodrilus* are regarded as highly derived genera by Baker & Brinkhurst (1981; cf. their fig. 13).

The most primitive prostate glands are probably the diffuse ones found in many Rhyacodrilinae. These glands cover most of the atrial surface, and there is no differentiation into distinct clusters of

40

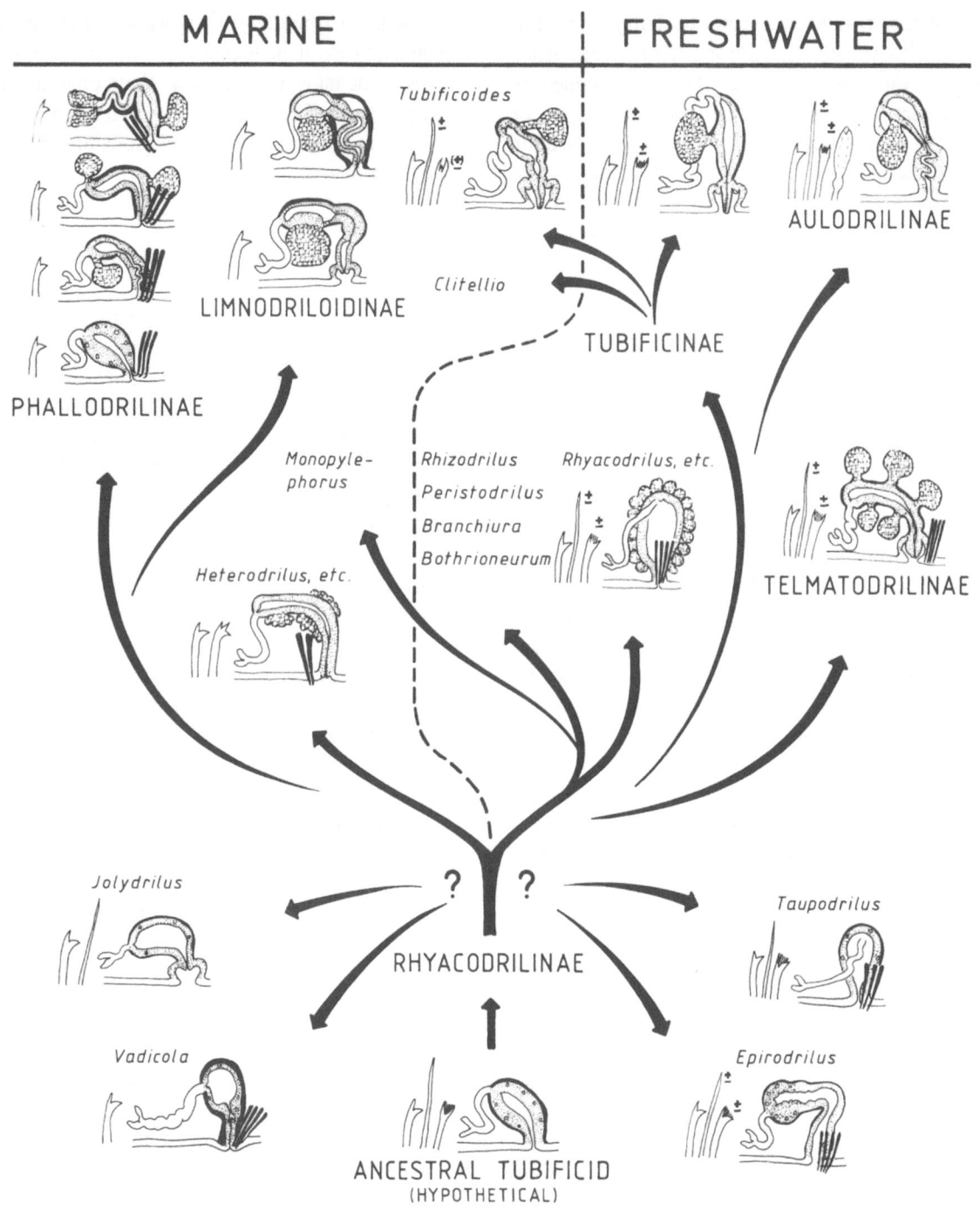

Fig. 1. A tentative phylogeny of the Tubificidae based on setal patterns and arrangements of prostate glands.

cells. However, a lumping of prostate cells is definitely present in some supposedly derived marine Rhyacodrilinae, such as *Heterodrilus* and *Giereidrilus* (see Erséus, 1981 b), or the limnic *Macquari-*

drilus, which is assumed to be closely related to marine forms (see Erséus & Jamieson, 1981). Lumping has become strongly pronounced in the subfamily Telmatodrilinae. It appears highly probable that the Rhyacodrilinae split early into two main branches. One of these is marine, and it has led to the *Heterodrilus* and *Macquaridrilus* complexes (see Erséus, 1981b). The other one is limnic and has led, on the one hand, to the highly derived rhyacodriline genera *Peristodrilus, Rhizodrilus, Branchiura, Bothrioneurum* and *Monopylephorus* (see Baker & Brinkhurst 1981, fig. 13), the last-mentioned of which has become marine secondarily, and on the other hand, to the subfamily Telmatodrilinae. *Rhyacodrilus* and *Protuberodrilus* (and probably also *Rhyacodriloides*) are little derived genera within the freshwater Rhyacodrilinae.

The setal patterns found in all these groups suggest that hair setae dissappeared early from the main marine lineage, but were retained in many species among the freshwater Tubificidae.

In the subfamilies Phallodrilinae, Limnodriloidinae, Tubificinae and Aulodrilinae, the prostates are with very few exceptions solid and discrete bodies, and they are never more than one or two per atrium. The selective advantage of prostate lumping is still not fully understood, but it appears feasible that the emersion of a prostate gland from a restricted area of the atrial surface will leave the atrial musculature more intact and thus render it an efficient pump, for the ejection of sperm or for the discharge of secretion produced by the atrium. Differentiation into two types of prostates, spaced on the atrium (as in, e.g., *Nootkadrilus*), will facilitate the separate discharges of different secretory products.

The question then arises: did the solid prostates evolve more than once? A definite answer can not be given at present. The morphology of the male ducts of the Phallodrilinae, Limnodriloidinae and Tubificinae varies greatly within the respective subfamilies, but it is difficult to interpret the homologies, and thus the phylogenetic clues, when comparing the ducts of the different subfamilies. The male genitalia of the Aulodrilinae, however, show some clear affinities to those of the Tubificinae.

This makes it difficult to infer the phylogeny of higher Tubificidae based only on male duct morphology, and I therefore here return to the setal patterns. The Phallodrilinae and the Limnodriloidinae invariably lack hair setae, and this would ally them with the marine rhyacodrilines. The Tubificinae and Aulodrilinae, however, both contain many haired species, and thus they appear to be linked with the freshwater group of the Rhyacodrilinae. This would imply that the rhyacodriline (more or less diffuse) prostates have become concentrated into solid bodies on more than one occasion during the evolution of higher Tubificidae.

The basic prostate pattern within the Phallodrilinae appears to be two more or less pedunculate prostate glands per atrium; at any rate, this is the situation found in most genera. Some marine rhyacodrilines (*Giereidrilus* and some *Heterodrilus*) exhibit tendencies towards a lumping of the prostate cells into two more discrete bodies, although these are still 'diffusely' attached to the atrium (see Erséus, 1981b). This actually foreshadows the situation in most phallodrilines. It is true that it may be a phenomenon of convergence, but for the time being it is a fact that suggests a bridge between the marine Rhyacodrilinae and the Phallodrilinae. Taking this standpoint, however, one has to regard the aprostate phallodriline genus *Coralliodrilus* as a group that has lost its prostates secondarily.

In the Limnodriloidinae, each atrium bears only one prostate gland, which is broadly attached to the ental part of the atrial ampulla. Ectal to the ampulla is the atrial duct, a part of which is granulated in most of the species. The histology of the epithelium of the atrial ampulla, which stores the secretion produced by the prostate gland, is always different from the granulated part of the atrial duct. It seems feasible to regard this differentiation as a specialization of an arrangement of two prostates per atrium, in which the posterior gland has become wholly retained within the inner epithelium of the atrial duct. This would support the view that the Limnodriloidinae are closely related to the Phallodrilinae, i.e. the view that is illustrated in Fig. 1. It should be mentioned, however, that one could argue for the Limnodriloidinae all being hairless descendants of a tubificine or aulodriline ancestor. This is to say that the limnodriloidine prostate gland may actually be homologous to that of the Tubificinae and Aulodrilinae.

It is difficult to judge whether the single, stalked prostate gland on the atrium of the Tubificinae-Aulodrilinae is the result of a further lumping or partial reduction of a telmatodriline type of prostates, or whether it evolved from rhyacodriline prostates independently of the Telmatodrilinae. The first alternative is somewhat supported by the fact that some telmatodriline species (*Telmatodrilus papilla-*

tus and *T. bifidus*) have a reduced number of prostate glands (2–3 per atrium), which are located close together near the ental end of the atrium. However, the two species mentioned are still little studied, and the exact morphology of their prostates is not established (Brinkhurst & Fulton, 1979).

Most tubificine genera are confined to freshwater; only *Tubificoides* (sensu Brinkhurst & Baker 1979), the monotypic *Clitellio* and some individual species of other genera are found in truly marine habitats.

The problem of spermatozeugmata

Spermatozeugmata are particular sperm aggregations in the spermathecae, and they have long been crucial for the definition of the subfamily Tubificinae. The spermatozeugmata of *Tubifex tubifex* has been described in detail by Braidotti, Ferraguti & Fleming (1980) and by Braidotti & Ferraguti (1982). It is composed of an axial cylinder of normal, fertilizing spermatozoa, and a peripheral layer of helically wound modified spermatozoa with degenerated nuclei. In many species of the Limnodriloidinae, sperm arrangements very similar to the tubificine spermatozeugmata are found (Erséus, 1982). Slender spermatozeugmata are also reported for some species of Telmatodrilinae (Brinkhurst & Jamieson, 1971; Brinkhurst & Fulton, 1979; Holmquist, 1974). Roundish sperm packages with an outer hyaline layer are found in the spermathecae of some species of *Bathydrilus* within the Phallodrillinae (Erséus, 1979, 1981a). In *Rhizodrilus* (Rhyacodrilinae), the sperm are organized in distinctly oriented 'bundles' (Baker & Brinkhurst, 1981).

Jamieson (1978b) has questioned the use of spermatozeugmata for the definition of subfamilies and other taxa. His doubts are well founded, as the tubificid sperm appear to have a general potential for formation of spermatozeugmata throughout the subfamilies. This is probably associated with the presence of two sperm lines, which may be a feature shared by all Oligochaeta (cf. Braidotti & Ferraguti, 1982).

The spindle-shaped, generally very slender tubificine spermatozeugmata can still be regarded as quite distinct, and they can definitely be used as one of the defining characters of the Tubificinae. However, for the time being, the presence of similar sperm structures in particular species of the remaining Tubificidae should not be used as evidence for their close phylogenetic relationships with that subfamily.

Discussion and conclusions

The tentative phylogeny of the Tubificidae outlined here (Fig. 1) can always be disputed, as it involves many details in which the direction of change of the character states cannot be ascertained. Undoubtedly, there are many instances of parallel and convergent evolution in the phylogenetic history of the family (loss of hair setae, modification of genital setae, development of penes and other copulatory organs, formation of spermatozeugmata, etc.). These will, of course, obscure the interpretation of homologies. I have not been able to relate my 'branching tree' to the time axis; I have only referred to time in a relative sense.

However, a fundamental conclusion is that most of the marine Tubificidae (perhaps with a reservation for the Limnodriloidinae) appear to have evolved quite independently of the freshwater groups of the family. The great diversity of the marine forms indicates that the Tubificidae have existed in the sea for a considerable length of time. For instance, among the marine tubificids, there are species that lack a normal alimentary system (see Giere, 1981; Giere et al., 1984; Erséus, 1981a). Furthermore, the Phallodrilinae appear to be well established in the deep sea (e.g. see Erséus, 1983b).

More recent invasions of the salt-water habitats are probably those reflected by the modern genera *Monopylephorus, Tubificoides* and *Citellio*. Today, freshwater genera are still in the process of 'evolving' into brackish-water habitats, e.g. *Potamothrix* and *Psammoryctides* in the Caspian Sea (Timm, 1980). Examples of the reverse situation are known, although they are few; a few species closely related to marine forms have succeeded in inhabiting fresh water (see Cook & Hiltunen, 1975; Giani & Martinez-Ansemil, 1981; Erséus & Jamieson, 1981).

The simultaneous possession of hair and pectinate setae, diffuse prostates and coelocytes is shared by many Rhyacodrilinae, and the freshwater families Naididae and Opistocystidae. The two latter were probably derived from rhyacodriline tubificids. It is true that naidids and opistocystids show drastic shifts in the location of the genital organs, but this can be understood in the light of the reproductive strategies used by most of these small worms. They generally reproduce by budding or fragmentation, only occasionally by means of sexual interbreeding. Once a shift in the location of genitalia has taken place due to some mutation, cloning may rapidly increase the proportion of such a genotype in a population (cf. Brinkhurst & Jamieson, 1971, p. 174).

The phreodrilids lack prostate glands, just like the hypothetical ancestor of the Tubificidae (cf. Baker & Brinkhurst, 1981, fig. 14). The shared possession of bifid setae supports the view of common ancestry of the two families, but as to other setal comparisons problems remain. There are hair setae, and even finely pectinated (serrate) setae, in the phreodrilids too (Brinkhurst & Fulton, 1979), but they are generally different from those found in the Tubificidae. It is therefore possible that they have evolved in parallel with the corresponding tubificid types of setae.

The aim of this paper has been to put forward some hypotheses towards the understanding of tubificid phylogeny, and that of marine forms in particular. These theories, which indeed are preliminary, now have to be tested by subsequent studies, taxonomic as well as more detailed morphological ones. For instance, it would be most important to scrutinize, preferably using electron microscopy, the various kinds of spermatozeugmata reported from throughout the family. The histology of male efferent ducts in different genera should also be more clearly described, as most of the existing taxonomic accounts are very superficial with regard to these organs.

It is my contention that the still very fragmentarily known marine Tubificidae will reveal a number of important facts for the inference of tubificid evolution as a whole. At present, it is not possible to settle definitely the question of whether the first tubificids were marine or limnic, but the enormous diversity and apparent antiquity of many of the marine forms do not, as yet, contradict the hypothesis that the Tubificidae actually originated in the sea.

Acknowledgements

I am deeply indebted to Dr. Ralph O. Brinkhurst, Mr. H. Randy Baker, and Ms. Kathryn Coates (all at the Institute of Ocean Sciences, Sidney, B. C., Canada), for most stimulating discussions on tubificid phylogeny over the years; I am particularly grateful for their valuable criticism on a preliminary draft on the present paper.

References

Baker, H. R., 1982. Two new Phallodriline genera of marine Oligochaeta (Annelida; Tubificidae) from the Pacific northeast. Can. J. Zool. 60: 2487–2500.

Baker, H. R. & R. O. Brinkhurst, 1981. A revision of the genus Monopylephorus and redefinition of the subfamilies Rhyacodrilinae and Branchiurinae (Tubificidae: Oligochaeta). Can. J. Zool. 59: 936–965.

Braidotti, P., M. Ferraguti & T. P. Fleming, 1980. Cell junctions between spermatozoa flagella within the spermatozeugmata of Tubifex tubifex (Annelida: Oligochaeta). J. Ultrastruct. Res. 73: 299–309.

Braidotti, P. & M. Ferraguti, 1982. Two sperm types in the spermatozeugmata of Tubifex tubifex (Annelida, Oligochaeta). J. Morph. 171: 123–136.

Brinkhurst, R. O., 1982. Evolution in the Annelida. Can. J. Zool. 60: 1043–1059.

Brinkhurst, R. O., 1984. The position of the Haplotaxidae in the evolution of oligochaete annelids. In: G. Bonomi & C. Erséus (eds.), Aquatic Oligochaeta. Proceedings of the Second International Symposium on Aquatic Oligochaete Biology. Developments in Hydrobiology (this volume).

Brinkhurst, R. O. & H. R. Baker, 1979. A review of the marine Tubificidae (Oligochaeta) of North America. Can. J. Zool. 57: 1553–1569.

Brinkhurst, R. O. & W. Fulton, 1979. Some aquatic Oligochaeta from Tasmania. Rec. Queen Vict. Mus. 64: 1–8.

Brinkhurst, R. O. & B. G. M. Jamieson, 1971. Aquatic Oligochaeta of the world. Oliver & Boyd, Edinburgh. 860 pp.

Cook, D. G. & J. K. Hiltunen, 1975. Phallodrilus hallae, a new tubificid oligochaete from the St. Lawrence Great Lakes. Can. J. Zool. 53: 934–941.

Erséus, C., 1979. Taxonomic revision of the marine genus Bathydrilus Cook and Macroseta Erséus (Oligochaeta, Tubificidae), with descriptions of six new species and subspecies. Zool. Scr. 8: 139–151.

Erséus, C., 1981a. Taxonomic studies of Phallodrilinae (Oligochaeta. Tubificidae) from the Great Barrier Reef and the Comoro Islands, with descriptions of ten new speices and one new genus. Zool. Scr. 10: 15–31.

Erséus, C., 1981b. Taxonomic revision of the marine genus Heterodrilus Pierantoni (Oligochaeta, Tubificidae). Zool. Scr. 10: 111–132.

Erséus, C., 1982. Taxonomic revision of the marine genus Limnodriloides (Oligochaeta, Tubificidae). Verh. naturwiss. Ver. Hamburg (NF) 25: 207–277.

44

Erséus, C., 1983a. Duridrilus tardus gen. et sp.n., a marine tubificid (Oligochaeta) from Bermuda and Barbados. Sarsia 68: 29–32.

Erséus, C., 1983b. Deep-sea Phallodrilus and Bathydrilus (Oligochaeta, Tubificidae) from the Atlantic Ocean, with descriptions of ten new species. Cah. Biol. mar. 24: 125–146.

Erséus, C. & B. G. M. Jamieson, 1981. Two new genera of marine Tubificidae (Oligochaeta) from Australia's Great Barrier Reef. Zool. Scr. 10: 105–110.

Fleming, T. P., 1981. The histochemistry and ultrastructure of the glandular vas deferens of Tubifex tubifex (Annelida: Oligochaeta). J. Zool. 195: 311–330.

Giani, N. & E. Martinez-Ansemil, 1981. Observaciones acerca de algunos Tubificidae (Oligochaeta) de la Peninsula Ibérica, con la descripcion de Phallodrilus riparius n. sp. Annls Limnol. 17: 201–209.

Giere, O., 1981. The gutless marine oligochaete Phallodrilus leukodermatus. Structural studies on an aberrant tubificid associated with bacteria. Mar. Ecol. Prog. Ser. 5: 353–357.

Giere, O., H. Felbeck, R. Dawson & G. Liebezeit, 1984. The gutless oligochaete Phallodrilus leukodermatus Giere, a tubificid of structural, ecological and physiological relevance.

In G. Bonomi & C. Erséus (eds.), Aquatic Oligochaeta. Proceedings of the Second International Symposium on Aquatic Oligochaete Biology. Developments in Hydrobiology (this volume).

Holmquist, C., 1974. On Alexandrovia onegensis Hrabě from Alaska, with a revision of the Telmatodrilinae (Oligochaeta, Tubificidae). Zool. Jb. (Syst.) 101: 249–268.

Jamieson, B. G. M., 1978a. Phylogenetic and phenetic systematics of the opisthoporous Oligochaeta (Annelida: Clitellata). Evol. Theory 3: 195–233.

Jamieson, B. G. M., 1978b. Rhyacodrilus arthingtonae a new species of freshwater oligochaete (Tubificidae) from North Stradbroke Island, Queensland. Proc. R. Soc. Qd 89: 39–43.

Jamieson, B. G. M., 1981. Historical biogeography of Australian Oligochaeta. In A. Keast (ed.), Ecological Biogeography of Australia. Dr W. Junk bv Publishers, The Hague-Boston-London: 887–921.

Timm, T., 1980. Distribution of aquatic oligochaetes. In R. O. Brinkhurst & D. G. Cook, Aquatic Oligochaete Biology. Plenum Press, New York and London: 55–77.

Timm, T., 1981. On the origin and evolution of aquatic Oligochaeta. Eesti NSV Tead. Akad. Toim. (Biol.) 30: 174–181.

Specific criteria in *Grania* (Oligochaeta, Enchytraeidae)

Kathryn A. Coates
*Department of Biology, University of Victoria, P.O. Box. 1700, Victoria, British Columbia,
Canada V8W 2Y2*

Keywords: aquatic Oligochaeta, oligochaete taxonomy, Enchytraeidae, *Grania*

Abstract

The structure of the penial bulb and male efferent duct system of *Grania* species may be used in addition to setal pattern and spermathecal shape to distinguish species. Six penial bulb types are distinguished: (1) a simple, small, glandular bulb surrounding the male pore; (2) a small, glandular bulb, with a large, associated, dorso-medial gland mass; (3) a small glandular bulb, medial to the male pore, with an elongate male bursa (the aglandular sac), the vas deferens exiting directly into the invaginated male pore; (4) a glandular bulb with an aglandular sac and a small, cuticular stylet embedded in the bulb, extending from the ectal end of the vas deferens; (5) a glandular bulb and an aglandular sac with a long stylet extending from the vas deferens, through the bulb into the sac; and (6) glandular bulb reduced or absent, with or without an aglandular sac; with a long stylet and other prominent modifications, usually muscular, of the vas deferens. The details of the male duct structure were consistent within specimens grouped on the basis of setal distribution and shape and detailed spermathecal structure. Diverse male duct patterns are found within the polytypic species *G. macrochaeta* and *G. postclitellochaeta*. The positions of the spermathecal and male pores in their respective segments are distinctive for some species.

Introduction

With increased study of marine Oligochaeta (Tubificidae, Naididae and Enchytraeidae) during the past twenty years, the number of species known has greatly increased and their significance in marine intertidal and subtidal communities has been recognized (Giere, 1975; Coates & Ellis, 1980). The importance of the male ducts as generic and specific criteria in most groups of aquatic microdriles has long been established (Brinkhurst & Jamieson, 1971). For the Enchytraeidae the location of the male pores is used as a familial characteristic and attempts have been made (Eisen, 1904; Welch, 1914; Černosvitov, 1937) to establish general structural types of penial bulbs as primary subfamilial characteristics. Due to the instability of the taxonomy of the Enchytraeidae, the subfamilial classifications formulated remain open to debate. The details of the structure of the male duct and glandular bulb, however, are not commonly used in the Enchytraeidae as either generic or specific criteria.

Grania Southern, 1913 is an exclusively marine genus, primarily defined on the basis of the unique setae. The number of recognized taxa in *Grania* has increased from two in 1913 to six in 1967, and to seventeen in 1980 (Fig. 1) (Kennedy, 1966; Lasserre, 1966, 1967, 1971; Erséus, 1974, 1977, 1980; Erséus & Lasserre, 1976, 1977; Lasserre & Erséus, 1976; Jamieson, 1977; Shurova, 1979; Coates & Erséus, 1980). Approximately twenty-five additional, undescribed species are in my new collections.

When only a few *Grania* species were recognized, setal numbers and distributions sufficed for identification, but details of other systems must now be used to separate the known species. Following is an

Hydrobiologia 115, 45–50 (1984).

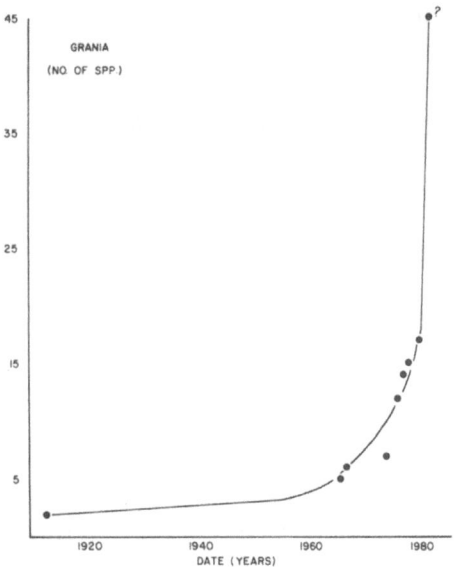

Fig. 1. Cumulative number of *Grania* species known versus year (date); plotted to show the tremendous increase in information from the mid-1960's to the present.

initial account of the possibilities of using the male ducts of *Grania* to assist in species discrimination.

Materials and methods

A collection of more than 500 specimens of *Grania* was examined. The material was from localities in western North America, eastern Australia, Atlantic deep sea, eastern North America and western and northern Europe (in rank order). Type specimens have been borrowed, wherever possible, in order to confirm the identities of the North American and Australian specimens identified by me. Dr. C. Erséus sent identified European specimens, which he has confirmed from types, and loaned me numerous specimens of unidentified deep sea species. K. Kossmagk-Stephan sent material of *Grania postclitellochaeta postclitellochaeta* from the type locality.

Most specimens were stained in a carmine stain and mounted whole in Canada balsam. Some well-preserved specimens of Australian and western North American species were sectioned (penial bulb type 3).

Results

Six types of penial bulb and male duct were recognized.

Type 1. Penial bulb small, glandular, surrounding a simple invaginated male pore. Vas deferens communicating directly with pore (Fig. 2).

Representative taxa: three undescribed species.

Distribution of material: deep sea off France; Argentine Basin; and Swedish shallow marine.

Type 2. Penial bulb small, glandular, surrounding simple, invaginated male pore; a large, almost separate, gland surrounding the penial bulb dorsomedially. Vas deferens exiting directly into the male pore (Fig. 3).

Representative taxa. *G. postclitellochaeta postclitellochaeta* (Knöllner, 1935) (Fig. 3A) and three undescribed species.

Distribution of material: France; northern Norway; West Germany; Heron Island, and Masthead and Erskine Reefs, Australia; and Bermuda.

Type 3. Penial bulb small, glandular, attached medially to male pore, with an elongate lateral outpocketing of the male bursa (aglandular sac). Aglandular sac with strong muscular attachments to the body wall (Figs. 4A, 4C and 5A); walls of aglandular sac formed by cuboidal, non-granular cells.

Representative taxa: *G. paucispina* (Eisen, 1904), *G. incerta* Coates & Erséus, 1980 (Fig. 4), *G. macrochaeta trichaeta* Jamieson, 1977 (Fig. 5A), *G. pacifica* Shurova, 1979 (Fig. 5B) and eight undescribed species.

Distribution of material: northest Pacific (California to British Columbia); Atlantic deep sea of France, Bermuda, Ireland, South Africa, Barbados and Brazil; Italy; Sweden; and Victoria Harbor (Victoria) and Heron Island (Queensland), Australia.

Type 4. Penial bulb relatively large, glandular, with a short, cuticular stylet (?solid or hollow) and an aglandular sac (Fig. 6).

Representative taxa: *G. variochaeta* Erséus and Lasserre, 1976 (Fig. 6) and two undescribed species.

Distribution of material: Sweden; Norway; and Heron Island, Australia.

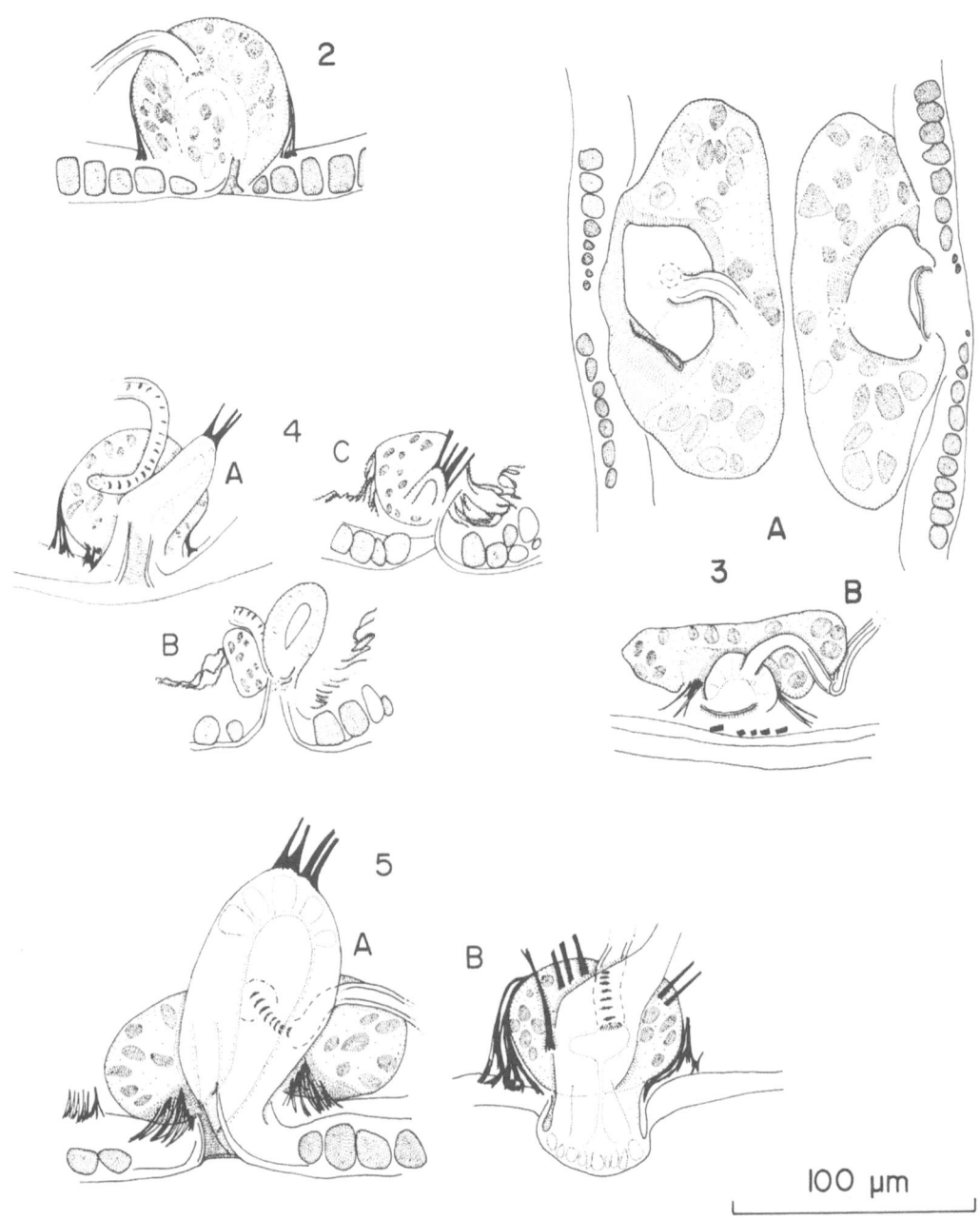

Fig. 2–5. 2. Penial bulb Type 1; an undescribed species from Sweden, lateral view. 3. Penial bulb Type 2; (A) *Grania postclitellochaeta postclitellochaeta*, ventral view of XII; and (B) an undescribed species from Heron Island, Australia, lateral view. 4–5. Penial bulb Type 3; (4A) *Grania incerta*, lateral view of entire penial bulb; (4B–C) *Grania incerta*, longitudinally sectioned bulb (B showing the entrance of the vas deferens into the male invagination, C a more medial section, showing the muscular attachments (solid black), to the body wall, of the aglandular sac.); (5A) *Grania macrochaeta trichaeta*, lateral view; and (5B) *Grania pacifica*, lateral view, showing the male pore protruded.

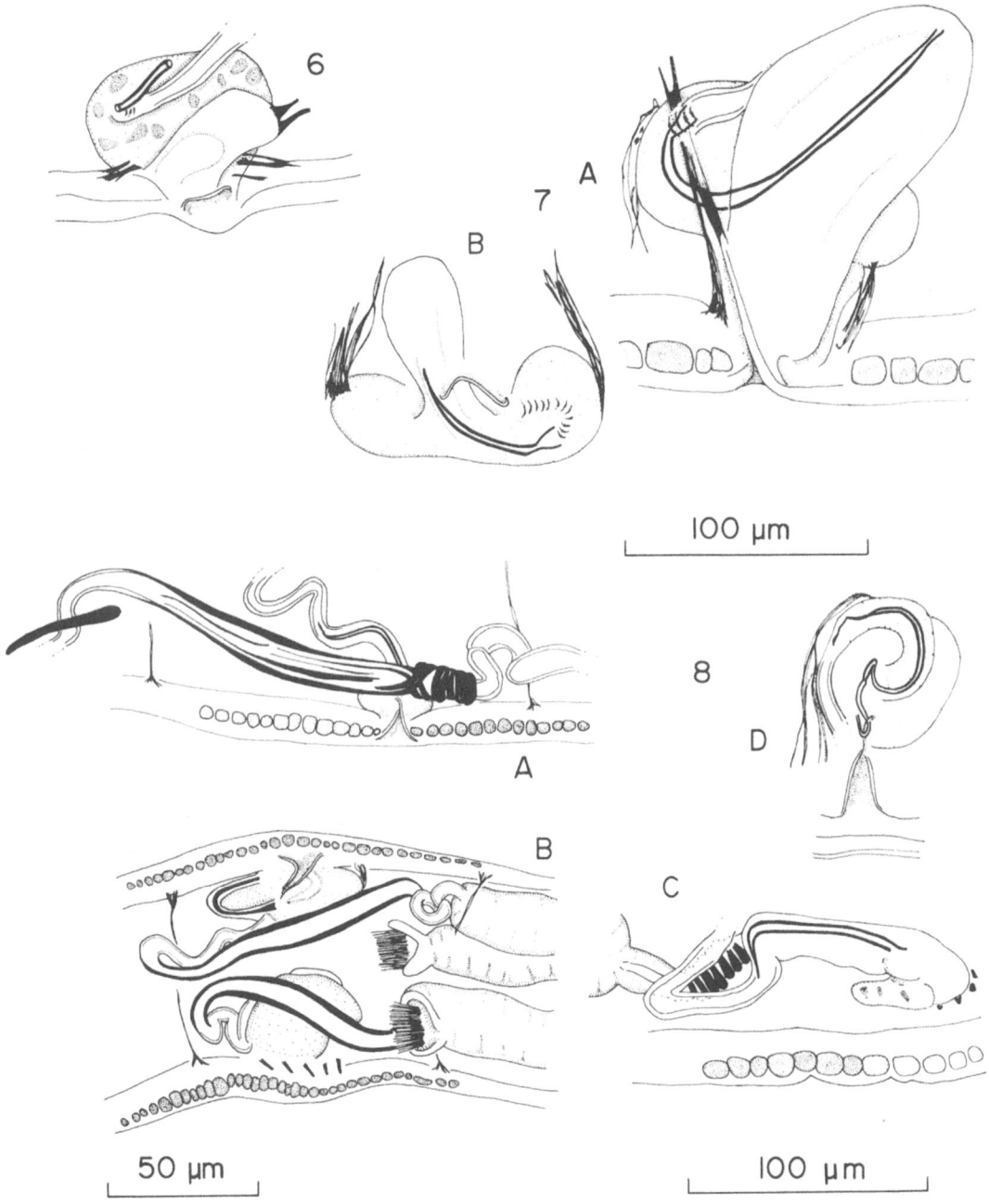

Figs. 6–8. 6. Penial bulb Type 4; *Grania variochaeta*, lateral view, showing the short stylet originating at the ectal end of the vas deferens. The stylet is surrounded by penial bulb tissue. 7. Penial bulb Type 5; (7A) *Grania maricola*, lateral view; and (7B) an undescribed species from a deep sea habitat off South West Africa, ventral view (the lateral side is uppermost). 8. Penial bulb Type 6; (8A) *Grania macrochaeta bermudensis*, lateral view of XII; (8B) *Grania monospermatheca*, dorsal view of XII; (8C) an undescribed species from South Carolina, lateral view of XII; and (8D) an undescribed species from Heron Island, Australia, lateral view of the ectal end of the vas deferens and male pore. Muscular tissues are solid blackened. The stylets are drawn with thick lines.

Type 5. Penial bulb large, glandular, surrounding medial ectal part of the aglandular sac; with a long, narrow stylet, extending into the aglandular sac (Fig. 7).

Representative taxa: *G. macrochaeta pusilla* Erséus, 1974, *G. roscoffensis* Lasserre, 1967, *G. postclitellochaeta longiducta* Erséus & Lasserre, 1976, *G. maricola* Southern, 1913 (Fig. 7A), and five undescribed species.

Distribution of material: Sweden; Norway; Ireland; Atlantic deep sea off Scotland, France and South West Africa; eastern North America; Victoria, Australia; and Italy.

Type 6. Glandular penial bulbs reduced or absent, aglandular sac often absent; with a long stylet plus considerable modifications of other parts of the vasa deferentia (Fig. 8).

Representative taxa: *G. macrochaeta bermudensis* Erséus & Lasserre, 1976 *sensu stricto* (Fig. 8A), *G. monospermatheca* Erséus & Lasserre, 1976 (Fig. 8B), and two undescribed species (Figs. 8C–D).

Distribution of material: eastern North America, including estuarine habitats in South Carolina; Bermuda; and Heron Island, Australia.

Discussion and conclusions

Male duct and penial bulb patterns within the polytypic species *G. macrochaeta* fall into Types 3, 5 and 6. Similarly, the patterns of the polytypic *G. postclitellochaeta* fall into Types 2 and 5. This confirms (see Erséus, 1977) that these species definitions are too broad and reflects the variability in detailed spermathecal structure formerly allowed between the subspecies of a *Grania* species (*sensu* Erséus & Lasserre, 1976). This variability was notably broader than traditionally allowed in other enchytraeid genera.

The penial bulb types recognized are not restricted geographically although some areas such as the northeast Pacific appear to harbor species with only one bulb type, e.g. Type 3. The worldwide distribution and diversity of the genus suggests a long period of evolution in the marine environment.

The large range of penial bulb and male structures observed and their consistency within groups previously determined on the basis of setal and spermathecal characteristics show that these can (and should) be used in future systematic studies of *Grania*. No attempt has been made to establish phylogenetic relationships between the Types recognized. Type 6 is, in fact, a very artificial group and includes more structural diversity than the other Types described.

Before decisions about the status of *Grania* and *Grania* species can be made, more attention must be given to additional characteristics such as the locations of the spermathecal and penial pores, presence of accessory glands on the nerve cord posterior to the clitellum, and the structure of the sperm funnel. A greater variety of spermathecal structure has been found within the genus than was previously described and several species have been found which have glands associated with the spermathecal ducts.

The diversity of male duct structure in *Grania* seems to be greater than that known in other enchytraeid genera (excepting perhaps *Mesenchytraeus* Eisen, 1878). However, not much detail is available for the other genera so that the superspecific, taxonomic significance of male duct structure is presently impossible to evaluate.

Acknowledgements

I am greatly indebted to C. Erséus for the extended loan of numerous specimens and for discussions about much of this information. I am also indebted to J. Phillips of the Victoria Museum of Australia for the loan of material from their collections and to K. Kossmagk-Stephan for sending specimens of *Grania postclitellochaeta postclitellochaeta*. R. O. Brinkhurst and H. R. Baker kindly provided valuable criticisms of the manuscript.

References

Brinkhurst, R. O. & B. G. M. Jamieson, 1971. Aquatic Oligochaeta of the World. Oliver & Boyd, Edinburgh. 860 pp.

Cernosvitov, L., 1937. System der Enchytraeiden. Zap. nauchno-issled. Ob"ed. russk. svob. Univ. Prage 5: 262–295.

Coates, K. A. & D. V. Ellis, 1980. Enchytraeid oligochaetes as marine pollution indicators. Mar. Pollut. Bull. 11: 171–174.

Coates, K. A. & C. Erséus, 1980. Two species of Grania (Oligochaeta, Enchytraeidae) from British Columbia. Can. J. Zool. 58: 1037–1041.

Eisen, G., 1904. Enchytraeidae of the west coast of North America. In Harriman Alaska expedition series. Vol. XII. Annelida. Smithsonian Institution, New York, pp. 1–166.

Erséus, C., 1974. Grania pusilla sp.n. (Oligochaeta, Enchytraeidae) from west coasts of Norway and Sweden. With some taxonomic notes on the genus Grania. Sarsia, 56: 87–94.

Erséus, C., 1977. Marine Oligochaeta from the Koster Area, west coast of Sweden, with description of two new enchytraeid species. Zool. Scr. 6: 293–298.

Erséus, C., 1980. A new species of Grania (Oligochaeta, Enchytraeidae) from Ascension Island, South Atlantic. Sarsia, 65: 27–28.

Erséus, C. & P. Lasserre, 1976. Taxonomic status and geographic variation of the marine enchytraeid genus Grania Southern (Oligochaeta). Zool. Scr. 5: 121–132.

Erséus, C. & P. Lasserre, 1977. Redescription of Grania monochaeta (Michaelsen) a marine enchytraeid (Oligochaeta) from South Georgia (SW Atlantic). Zool. Scr. 6: 299–300.

Giere, O., 1975. Population structure, food relations and ecological role of marine oligochaetes with special reference to meiobenthic species. Mar. Biol. 31: 139–156.

Jamieson, B. G. M., 1977. Marine meiobenthic Oligochaeta from Heron and Wistari Reefs (Great Barrier Reef) of the genera Clitellio, Limnodriloides, and Phallodrilus (Tubificidae) and Grania (Enchytraeidae). Zool. J. Linn. Soc. 61: 329–349.

Kennedy, C. R., 1966. A taxonomic revision of the genus Grania (Oligochaeta, Enchytraeidae). J. Zool. 148: 399–407.

Lasserre, P., 1966. Oligochètes marins des côtes de France. I: Bassin d'Arcachon: Systématique. Cah. Biol. mar. 7: 295–317.

Lasserre, P., 1967. Oligochètes marins des côtes de France. II: Roscoff, Penpoull, Étangs saumâtres de Concarneau: systématique, ecologie. Cah. Biol. mar. 8: 273–293.

Lasserre, P., 1971. The marine Enchytraeidae (Annelida, Oligochaeta) of the eastern coast of North America, with notes on geographical distribution and habitat. Biol. Bull. mar. biol. Lab., Woods Hole, 140: 440–460.

Lasserre, P. & C. Erséus, 1976. Oligochètes marins des Bermudes. Nouvelles espèces et remarques sur la distribution géographique de quelques Tubificidae et Enchytraeidae. Cah. Biol. mar. 17: 447–462.

Shurova, N. M., 1979. Enchytraeids (Oligochaeta) of the far eastern seas of the USSR. Laboratories of Chorology EVM, Vladivostock, pp. 75–89.

Southern, R., 1913. Clare Island Survey. Part 48. Oligochaeta. Proc. R. Ir. Acad. (B) 31: 1–14.

Welch, P. S. 1914. Studies on Enchytraeidae of North America. Bull. Ill. St. Lab. nat. Hist. 10: 123–211.

Capilloventer atlanticus gen. et sp.n., a member of a new family of marine Oligochaeta from Brazil

Walter J. Harman & Michael S. Loden[1]
Department of Zoology, Louisiana State University, Baton Rouge, LA 70803, U.S.A.
[1]*Environmental and Development Control Department, 3600 Jefferson Highway, Jefferson Parish, LA 70121, U.S.A.*

Keywords: aquatic Oligochaeta, *Capilloventer*, Capilloventridae, new family, new genus, Tubificida

Abstract

Specimens of a previously undescribed oligochaete from the Bay of Rio de Janeiro, Brazil were determined to belong to a new family of microdiles within the order Tubificida, suborder Enchytraeina. The specimens are characterized by dorsal and ventral hair setae, spermathecae in VII, testes in XI, ovaries in XII, and a clitellum. The number of primitive characters present in the new species indicate that it is a primitive member of the order that diverged early from the evolutionary line leading to the Enchytraeidae.

Introduction

Through the kind efforts of Dr. Christer Erséus, we were provided an opportunity to examine specimens of a previously undescribed oligochaete. The specimens were collected by Dr. Claude Jouin of the Université Paris under the direction of Dr. Jeanete Maron Ramos of the Universitá Santa Úrsula from the Bay of Rio Janeiro, Brazil, in August 1975. These specimens represent a new family of Oligochaeta with several phylogenetic implications.

Material and methods

Nine specimens were examined. All were received mounted on microscope slides and stained with paracarmine. One specimen was removed from the slide, serially sectioned, and stained with hematoxylin. Holotypic and paratypic material (2 specimens) was deposited in the Museum National d'Histoire Naturelle (MNHN), Paris; other paratypic material was placed in the U.S. National Museum of Natural History (USNM 80608-80610) and the Louisiana State University Museum of Zoology (LSU 3241).

Capilloventridae fam. n.

Dorsal and ventral setae two per bundle, one hair and one crotchet. Testes in XI. Ovaries in XII. Spermathecae paired in VII. Male ducts paired, unmodified, opening into copulatory chamber in XII. Female pores in XIII.

Capilloventer gen. n.

Setae beginning in III, one hair and one bifid seta per bundle. Ventral setae of XII modified to form penial setae. Gland cells associated with terminal portion of male duct.
Type species. Capilloventer atlanticus sp.n.
Etymology. capillus L., hair; *venter* L., belly; refers to the ventral placement of hair setae.

Capilloventer atlanticus sp.n. (Figs. 1-4)

Type Locality. Brazil: Bay of Rio de Janeiro, 23°03'S, 43° 17.3'W; August 6, 1975. C. Jouin.

Hydrobiologia 115, 51–54 (1984).
© Dr W. Junk Publishers, Dordrecht.

52

Etymology. Refers to the type-locality, the Bay of Rio de Janeiro, an inlet of the Atlantic Ocean.

Description. Length 2.5–3.8 mm (fixed), 19–32 segments. Diameter in whole-mounted, slightly compressed specimens: 0.09–0.14 mm. Prostomium rounded, slightly longer than width at peristomial junction. Clitellum thin, covering segments XI, XII, XIII. Somatic setae beginning in III, all bundles with 1 (occasionally 2) finely serrated hair, 133–185 μm long; and 1 (occasionally 2) bifid crochet seta, 37–49 μm long. Bifid crotchets all identical in structure, distal tooth much shorter and thinner than proximal. Ventral setae of XII modified to form penial bundles, each bundle with two types of setae: 3–4 thick setae, ca 150 μm long, hooked distally, elongate tip, coiled proximally in

spherical setal pouch, ca 50 μm diameter; 1–2 thin hair setae closely associated with each thick seta, ca 150 μm long. Distal end of penial setal bundle protruding into small copulatory chamber. Male pores paired in XII, in somatic setal planes.

Pharynx with dorsal pharyngeal plate composed of columnar cells. Paired salivary glands present. Esophagus covered with glandular tissue. Chlorogogue sparse. No pharyngeal, esophageal, or intestinal diverticula present. Esophagus and intestine finely ciliated.

Spermathecae paired in VII, ca 160 μm long, up to 50 μm diameter, thin-walled, extending posteriorly to VIII. Spermathecal pores opening laterally in intersegmental furrow 6/7, with associated gland cells. Sperm cells diffuse, not organized in bundles.

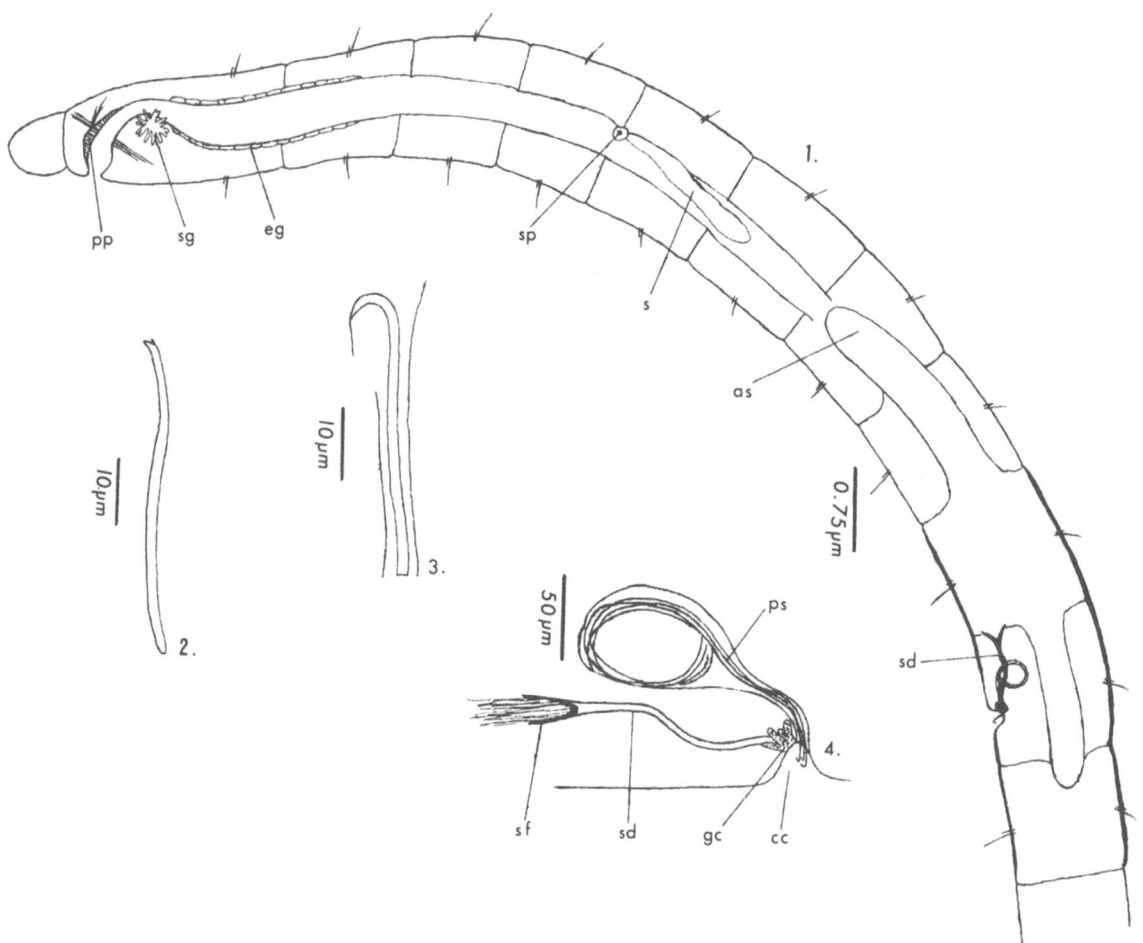

Figs. 1–4. Capilloventer atlanticus gen. et sp.n. 1. Reconstruction of anterior end of body (freehand drawing). 2. Bifid seta. 3. Distal ends of penial setae. 4. Male duct (freehand drawing). (as = anterior sperm sac; cc = copulatory chamber; eg = esophageal glandular tissue; gc = gland cells; pp = pharyngeal plate; ps = penial setae; s = spermatheca; sd = sperm duct; sf = sperm funnel; sg = salivary gland; sp = spermathecal pore)

Male reproductive system with all structures paired. Sperm funnels conical, 20 μm diameter, 40 μm long, on septum 11/12. Sperm duct ca 180 μm long, 10 μm diameter, not ciliated; without distensions, loops, or coils. Granular, elongate gland cells associated with termination of sperm duct at copulatory chamber. Testes in XI. Anterior sperm sac formed from septum 10/11, extending anteriorly to septum 8/9. Posterior sperm sac formed from septum 11/12, extending posteriorly to XIII.

Ovaries in XII. Female pores in XIII, posterior to intersegmental furrow 12/13.

Habitat. Marine, 22 m depth, coarse sand.

Discussion

One of the unique characters of *C. atlanticus* is hair setae in the ventral bundles. According to Brinkhurst (1982), the presence of ventral hair setae should exclude such an organism from the Oligochaeta. However, such features as a hermaphroditic condition, pharyngeal plate, restriction of the gonadal segments, a clitellum, spermathecae, and characteristics of the male reproductive apparatus clearly relegate *C. atlanticus* to the Oligochaeta, and in particular to the order Tubificida.

While the anterior location of the spermathecae appears to place the new family in the evolutionary line leading to the Enchytraeidae, divergence from this line apparently occurred quite early. We are, therefore, placing the Capilloventridae in the suborder Enchytraeina, following the scheme of Jamieson (1980). The number of plesiomorphic characters retained by *C. atlanticus* lead us to believe that it is the most primitive member of the Tubificida.

The simple, undifferentiated male ducts are not significantly different from those found in the Haplotaxidae, but the multiple gonadal condition typical of that family is lacking. The male funnel is simple, and more characteristic of the suborder Tubificina than the Enchytraeidae. The terminus of the male duct appears to provide some insight into the formation of the enchytraeid penial bulb, as well as the atrium of the Tubificina.

The structure of the distal portion of the male duct is such that several phylogenetic implications can be made. First, the terminal end is surrounded by elongated glandular cells (Fig. 4) that are similar in appearance and likely homologous to the pros-

tate tissue of the Tubificina. The copulatory chamber would need only a narrowing of the neck to be quite similar to the atrium of the Tubificina; we believe that a homology between these two structures is likely. There is, in addition, a possible homology of the copulatory chamber to the 'atrium' of the enchytraeid genus *Propappus* and the penial bulb of the remainder of the Enchytraeidae. We believe that the structure of the terminus of the male duct of *C. atlanticus* is a likely retention of an ancestral condition.

The digestive tract lacks any diverticula or pouches. A 'pharyngeal plate' (Brinkhurst & Jamieson, 1971) is present dorsally; the cells comprising the plate are columnar, and anterior and posterior muscle fibers provide for its apparent protrusion and retraction. Paired glandular structures ('salivary glands') are located at the junction of the pharynx and esophagus. The esophagus extends from the pharynx to segment IV; it is covered with a glandular tissue that is very similar in appearance to the tissue comprising the enchytraeid septal glands.

The presence of hair setae in the dorsal and ventral setal bundles in *C. atlanticus* suggests that hair setae may have been characteristic of the proto-Tubificida, as suggested by Timm (1981). Timm, however, suggested that the hypothetical ancestor of the Oligochaeta possessed hair setae only in the dorsal bundles. The new species tends to suggest that the ancestral condition was hair setae present in both dorsal and ventral setal bundles. The arrangement of somatic setae, one hair and one bifid seta per bundle (a second seta of one type or another being found occasionally, possibly as a replacement), is lumbricine in form, another likely ancestral character retained by *C. atlanticus*. Bifid setae are generally accepted as primitive.

The plesioporous condition of the female pores of *C. atlanticus* tends to corroborate Brinkhurst's (1982) view that the condition of female pores located in the intersegmental furrow is apomorphic to the Tubificina. The female duct of *C. atlanticus* is much reduced, a condition typical of other Tubificida.

With regard to two recently published phylogenetic schemes (Timm, 1981; Brinkhurst, 1982), we believe that the gonadal and reproductive inferences of the new species tend to support many of Brinkhurst's concepts. The setal arrangement appears to substantiate many of the concepts of

Timm. Regarding placement of *Propappus*, Brinkhurst felt that it is outside the main line of enchytraeid evolution. However, it appears intermediate between *C. atlanticus* and the remainder of the Enchytraeidae. We agree with Brinkhurst that the major difference between the evolutionary schemes of Brinkhurst and Timm is the placement of the Haplotaxida. The new species provides little insight into rectifying difference of opinion between the two authors other than suggesting that the lack of hair setae in the Haplotaxidae may be a derived condition.

References

Brinkhurst, R. O., 1982. Evolution in the Annelida. Can. J. Zool. 60: 1043–1059.

Brinkhurst, R. O. & B. G. M. Jamieson, 1971. Aquatic Oligochaeta of the world. Oliver & Boyd, Edingburgh, 860 pp.

Jamieson, B. G. M., 1980. Preliminary discussion of a Hennigian analysis of the phylogeny and systematics of opistoporous oligochaetes. Revue Ecol. Biol. Sol 17: 261–275.

Timm, T., 1981. On the origin and evolution of aquatic Oligochaeta. Eesti NSV Tead. Akad. Toim. (Biol.) 30: 174–181.

A method of identifying immature specimens of marine Enchytraeidae (Oligochaeta) by vital staining of epidermal glands

K.-J. Kossmagk-Stephan
II. Zoologisches Institut und Museum der Universität Göttingen, Berliner Str. 28, D-3400 Göttingen, Federal Republic of Germany

Keywords: aquatic Oligochaeta, marine Enchytraeidae, recognition of immature stages, epidermal glands, vital staining, Neutral Red

Abstract

Epidermal glands in Enchytraeidae are arranged in several transverse rows in each segment. These glands are more or less inconspicuous. It is possible to show up the gland cells by vital staining with Neutral Red (dilution 1:25 000 to 1:200 000 in tap- or sea-water, pH 6.0–6.8). Studies using this staining method have shown that there is a distinct pattern of epidermal glands for each species of marine Enchytraeidae investigated. The possibility of distinguishing immature specimens of different marine enchytraeid species, found in one particular locality, is demonstrated.

Introduction

Most authors involved in the taxonomy of Enchytraeidae mention the presence of epidermal glands, which are arranged segmentally in definite numbers of transverse rows (Nielsen & Christensen, 1959; Lasserre, 1967; and others); each species shows a more or less constant number of such rows per segment.

Richards (1977a, b) has investigated the ultrastructure and function of the epidermis in several species of the enchytraeid genus *Lumbricillus*. Three types of secretory cells were described: granular, acid mucous (uronic acid-containing) cells; small granular cells; and globular non-acid (neutral glycoprotein) cells. The acid mucous cells have not been recorded in littoral or terrestrial species of enchytraeid genera other than *Lumbricillus*.

As these epidermal glands are usually more or less inconspicuous in living and fixed worms, up to now it has been impossible to compare different patterns of glands at the species level.

The present study shows that, in sympatric populations of marine enchytraeids in a tidal beach of the Isle of Sylt (North Sea), each species investigated has a distinct pattern of glands. In this way we gain an additional character which can be used to distinguish between immature specimens of similar species. The different species can be identified after vital staining of the epidermal glands with Neutral Red.

Material and methods

Enchytraeid specimens were collected from a tidal beach at List/Sylt (North Sea). The beach has already been described by Schmidt (1968). Living worms were stained with Neutral Red (Romeis, 1948; Lasserre, 1967) in sea- or tap-water. The dilution of Neutral Red was 1:25 000 to 1:200 000 (orange form), pH 6.0–6.8. The worms were stained for 15–30 minutes, during which time the staining process was controlled under a dissecting scope to prevent staining of the whole animal.

Results

Patterns of epidermal glands in three species of the genus *Lumbricillus* and four *Marionina* species are shown in Figs. 1–16.

Hydrobiologia 115, 55–58 (1984).

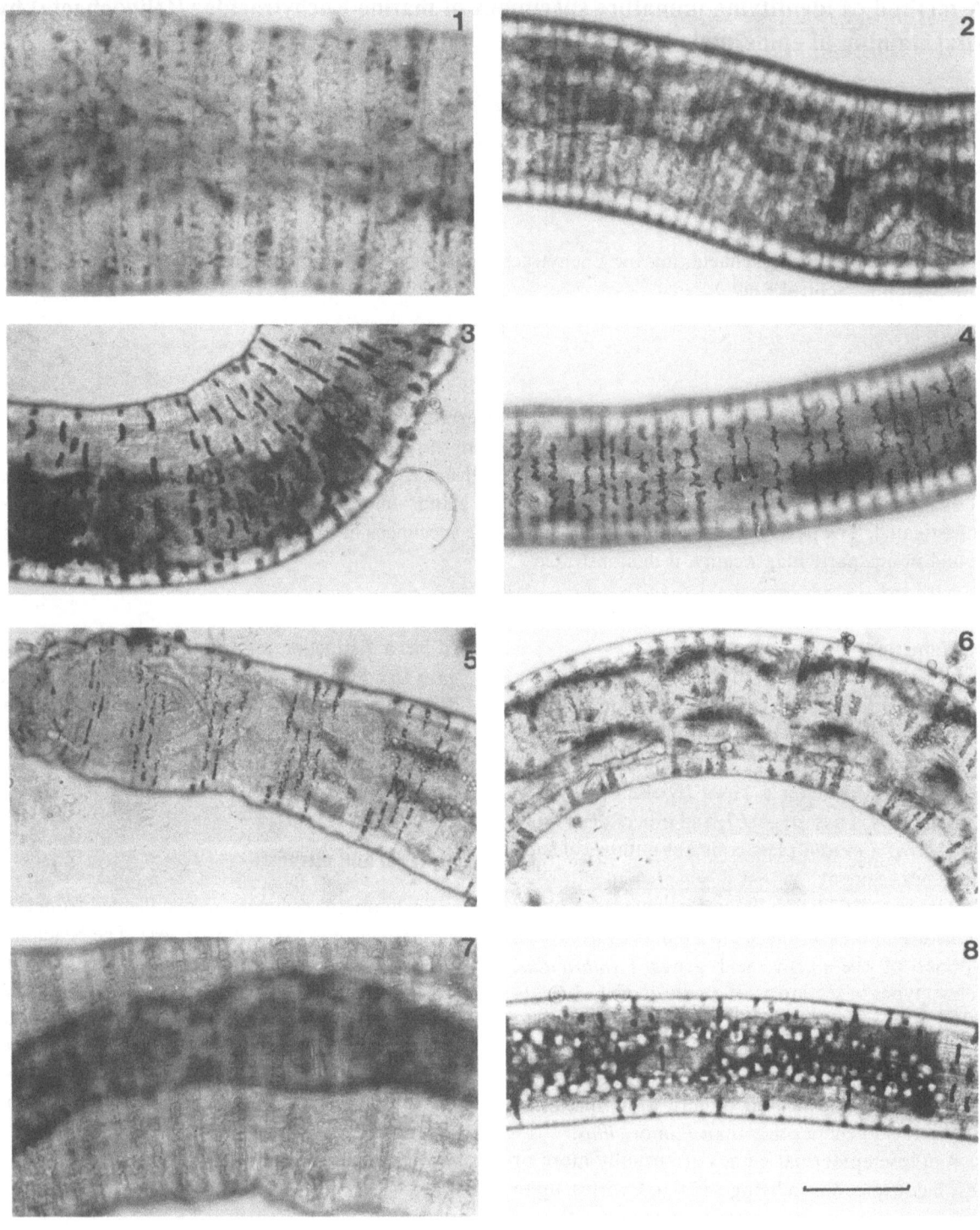

Figs. 1–8. Microphotos of epidermal gland pattern. 1. *Lumbricillus rubidus.* 2. *L. cervisiae.* 3. *L. westheidei.* 4. *Marionina subterranea* (ventral view). 5. *M. graefei* (dorsal view) 6. *M. graefei* (ventral view). 7. *M. spicula.* 8. *M. preclitellochaeta.* Scale line representing 100 μm for all figures.

Lumbricillus rubidus Finogenova & Streltzov, 1978 (Figs. 1 & 9) has 8–9 transverse rows of glands in each segment; within the dorsal rows there are 9–14 gland cells (length 15–45 μm, width 10–15 μm) of an irregular form and outline. In *L. cervisiae* Kossmagk-Stephan, 1983 (Figs. 2 & 10), the glands are arranged in 7–8 transverse rows per segment; within the dorsal rows there are 8–10 glands with a length of 10–35 μm and a width of 5–8 μm. This species has very prominent epidermal glands with an irregular, diffuse form, and they show a very characteristic pattern. *L. westheidei* Kossmagk-Stephan, 1983 (Figs. 3 & 11), a supralittoral species, also has a species-specific cell pattern: the glands are elongated, arranged in 4–5 rows per segment, each dorsal row containing 4–7 gland cells, which are 15–30 μm long, 3–5 μm wide and of a regular outline.

Marionina subterranea (Knöllner, 1935) (Figs. 4 & 12) has 7–8 transverse rows with 3–6 gland cells (length 20–30 μm, width 4–7 μm) in each segment; the cells have an irregular outline, and they are similar to those of *L. cervisiae*. *M. graefei* Kossmagk-Stephan, 1983 (Figs. 5, 6, 13 & 14) possesses 4–5 rows of epidermal glands per segment; each row contains 3–6 glands (length dorsally 20–30 μm, ventrally 15–25 μm; width dorsally 2–4 μm, ventrally 4–8 μm); the rows are clustered in the middle of each segment. *M. spicula* (Leuckhart, 1847) (Figs. 7 & 15) has 7–8 rows of glands per segment, each row containing 15–20 cells (length 7–15 μm, width 5–10 μm); the rows are evenly distributed over the segments. *M. preclitellochaeta* Nielsen & Christensen, 1963 (Figs. 8 & 16) is a species with very prominent gland cells of almost rectangular shape; the cells are 3–6 per row (length 15–20 μm, width 5–7 μm), and they are arranged in 4–5 transverse rows per segment.

Investigations of some of the species listed above (mainly *M. preclitellochaeta* and *M. spicula*) from other beaches of Sylt, showed that the gland pattern is very constant and does not depend on the habitat of different populations.

The pattern of epidermal glands remains constant throughout the life cycle of an animal.

Using the Neutral-Red-method it is also possible to stain clitellar glands and glandular parts at the orifice of the spermatheca. Clitellar gland cells appear to show different patterns in different species.

Differences in gland pattern exist between the dorsal and ventral side of an animal. Figs. 5, 13, 6 & 14 show the dorsal and ventral side of *M. graefei*. As a rule, the ventral glands are larger than the dorsal ones. This is also evident in other species and is especially prominent in *L. cervisiae*.

For one species investigated, *M. oligosetosa* Kossmagk-Stephan, 1983, occurring in the supralittoral region, the staining method failed.

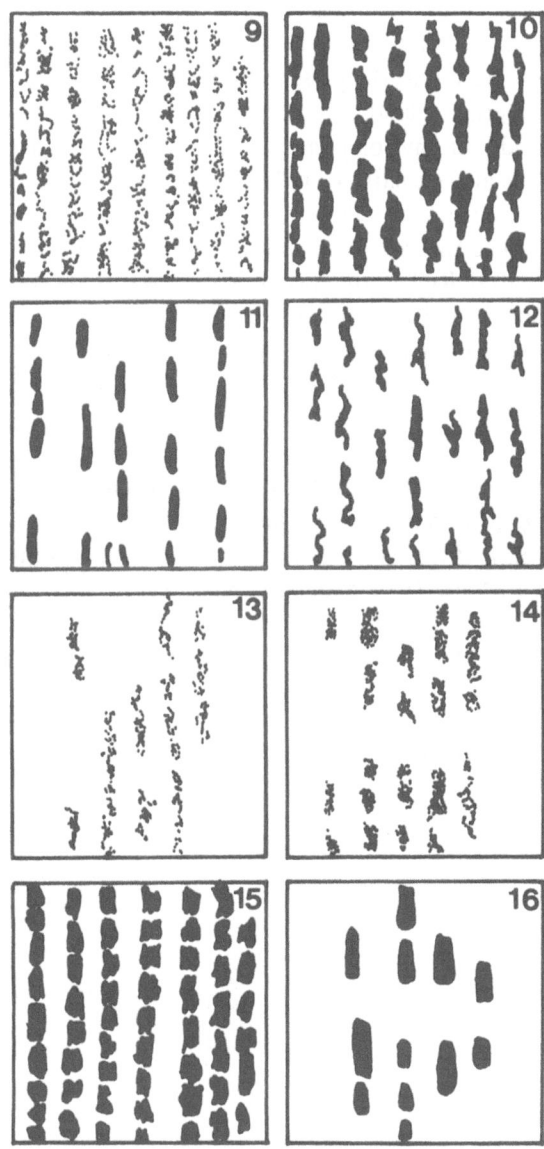

Fig. 9–16. Schematic gland pattern for one segment. 9. *L. rubidus*. 10. *L. cervisiae*. 11. *L. westheidei*. 12. *M. subterranea*. 13. *M. graefei* (dorsal view). 14. *M. graefei* (ventral view). 15. *M. spicula*. 16. *M. preclitellochaeta*.

58

Discussion

At present, it is not possible to say which gland type (Richards, 1977a) is stained by Neutral Red.

Figures 1–8 show that the glands of different species react with varying intensity to vital staining. This seems to depend on the different number of granules in the epidermal glands of the different species.

Terrestrial enchytraeids of the genus *Fridericia* show no positive reactions to Neutral Red (Kossmagk-Stephan, unpubl. obs.). Richards (1977a) suggested that the presence of the various types of epidermal glands in the genus *Lumbricillus* depends on environmental factors. This could perhaps also explain the failure of the staining in the supralittoral *M. oligosetosa*.

Nevertheless, staining of epidermal glands with Neutral Red is significant in two ways. (1) The morphological similarity between species of the genera *Lumbricillus* and *Marionina* is occasionally very high (Nielsen & Christensen, 1959; Coates & Ellis, 1981; Kossmagk-Stephan, 1983). It is therefore necessary to obtain additional taxonomic criteria to distinguish between these species. Two of the species mentioned here, *L. cervisiae* and *L. westheidei*, are very similar in their main taxonomic characters (Kossmagk-Stephan, 1983). Figures 2–3 show the large difference in the pattern of their epidermal glands, which is a good criterion for discriminating between living specimens of the two species, even when dealing with immature stages. (2) For ecologists investigating local populations of marine enchytraeids it is generally difficult to distinguish between immature specimens of similar species. Staining of the epidermal glands with Neutral Red will facilitate the recognition of the differences.

It would thus be most useful if taxonomists, when describing enchytraeids, would pay attention to the number and arrangement of the transverse rows of epidermal glands as well as to the shape of these glands.

References

Coates, K. & D. V. Ellis, 1981. Taxonomy and distribution of marine Enchytraeidae (Oligochaeta) in British Columbia. Can. J. Zool. 59: 2129–2150.

Kossmagk-Stephan, K.-J., 1983. Marine Oligochaeta from a sandy beach of the island Sylt (North Sea) with description of four new enchytraeid species. Mikrofauna Meeresboden 89: 1–28.

Lasserre, P., 1967. Contribution à l'étude des Oligochètes marins du Bassin d'Arcachon. Thésis Univ. Bordeaux, 126 pp.

Nielsen, C. O. & B. Christensen, 1959. The Enchytraeidae, critical revision and taxonomy of European species. Natura jutl. 8–9: 1–160.

Richards, K. S., 1977a. Structure and function in the oligochaete epidermis (Annelida). Symp. zool. Soc. Lond. 39: 171–193.

Richards, K. S., 1977b. The histochemistry and ultrastructure of the clitellum of the enchytraeid Lumbricillus rivalis (Oligochaeta: Annelida). J. zool., Lond. 183: 161–176.

Romeis, B., 1948. Mikroskopische Technik, 15th ed. Oldenbourg Press, München, 695 pp.

Schmidt, P., 1968. Die quantitative Verteilung und Populationsdynamik des Mesopsammons am Gezeiten-Sandstrand der Nordseeinsel Sylt. 1. Faktorengefüge und biologische Gliederung des Lebensraumes. Int. Revue ges. Hydrobiol. 53: 723–779.

Cytological aspects of oligochaete spermiogenesis

Marco Ferraguti

Dipartimento di Biología, Sezione di Zoologia e Citologia, Università di Milano, Via Celoria 26, I-20133 Milano, Italy

Keywords: aquatic Oligochaeta, spermatozoa, spermatogenesis, cell differentiation

Abstract

Spermiogenesis in Oligochaeta occurs with peculiar modalities, common to all the species studied in this group. Gonial cells, produced in the testes, drop into coelomatic cavities (usually seminal vesicles) where they undergo a series of mitotic divisions, without cytodieresis, and finally meiosis. In the seminal vesicles, a series of 'morulae' is present, composed of 2, 4, 8, 18, 32, ... cells. The number of spermatids produced is variable, but is constant for each species. The process of spermiohistogenesis involves many steps, including: nuclear shaping, chromatin condensation, production of an acrosome, reduction of the number of mitochondria. The present knowledge of the mechanisms of cell differentiation, and the problem of the presence of two different sperm lines in some species is discussed.

Introduction

Spermiogenesis in Oligochaeta has received much attention in the last years both from cytologists and zoologists. In effect, from one side spermiohistogenetic events in Clitellata are a good field for the study of some cytological mechanisms of cell differentiation (function of microtubules, fate of the centrioles, organelle symmetry/asymmetry, organelle movements); from the other side, the ultrastructure of the sperm cell can be used as a character for the study of the phylogeny and taxonomy of the Oligochaeta (see Jamieson, 1984).

The spermatozoon

The mature spermatozoon of Oligochaeta is a very narrow (about 0.2 μm) and long (up to 100 μm) filiform cell. At the apex there is an acrosome which, although varying among the various members of the group, is made up of the same parts (Fig. 1): an acrosomal tube, characteristic of the Clitellata spermatozoa, and absent in the other animal groups sofar studied; the acrosomal vesicle, often a minor component of the acrosomal complex, is partly withdrawn into the acrosomal tube, and from its base a thinner secondary tube starts; under the acrosomal vesicle there is a subacrosomal space, where an acrosomal rod can be more or less clearly identified. The function of the acrosomal rod is to be connected with some fertilization event. The perforatorium has been isolated in our laboratory from *Eisenia foetida* and, although its chemical nature has not been proved, its appearance is very similar to that of the actin perforatorium of, for instance, *Limulus* and *Mytilus* (for references, see Baccetti & Afzelius, 1976).

The nucleus is always very condensed and thin. Its shape varies from straight (e.g. in lumbricids and megascolecids: Jamieson, 1978) to variously coiled (*Limnodriloides*: Jamieson & Daddow 1979; *Tubifex*: Braidotti & Ferraguti, 1982). Mitochondria (Fig. 2), tightly packed together, vary in number from 2 (*Tubifex*: Ferraguti & Lanzavecchia, 1971) to 8 (*Sparganophilus*: Jamieson *et al.*, 1982) and in

Hydrobiologia 115, 59–64 (1984).

60

shape: they can be straight, as in *Lumbricus* (Anderson *et al.*, 1967) or coiled, as in *Phreodrilus* (Jamieson, 1981a). Only one centriole is present, highly modified, at the base of the long flagellum. Accessory fibers, always present in spermatozoa of animals with internal fertilization, are replaced in Oligochaeta, by a series of glycogen granules acting as energy accumulators (Ferraguti, 1983).

Spermiogenesis

Some of the events leading to such a complicated sperm model have been elucidated in the last years, although the general scheme of the spermiogenesis was known from the end of the last century (the following description will be based on the recent literature: for references, see Jamieson, 1981b).

Spermatogonia, produced by the testes, undergo a series of mitotic divisions without cytodieresis and give rise to a characteristic structure termed a 'morula', composed of a central nutrient mass (cytophore) and the spermatogonia housed in buds at its periphery. The morulae are released by the testes, and fall into the seminal vesicles where further

maturation stages, including further mitosis, and meiosis, are to be found.

The spermiohistogenetic process which gives rise to the mature sperms is very interesting from the cytological point of view (Fig. 3). Early spermatids appear as buds connected to the cytophore by a collar. The first event of the differentiation is the

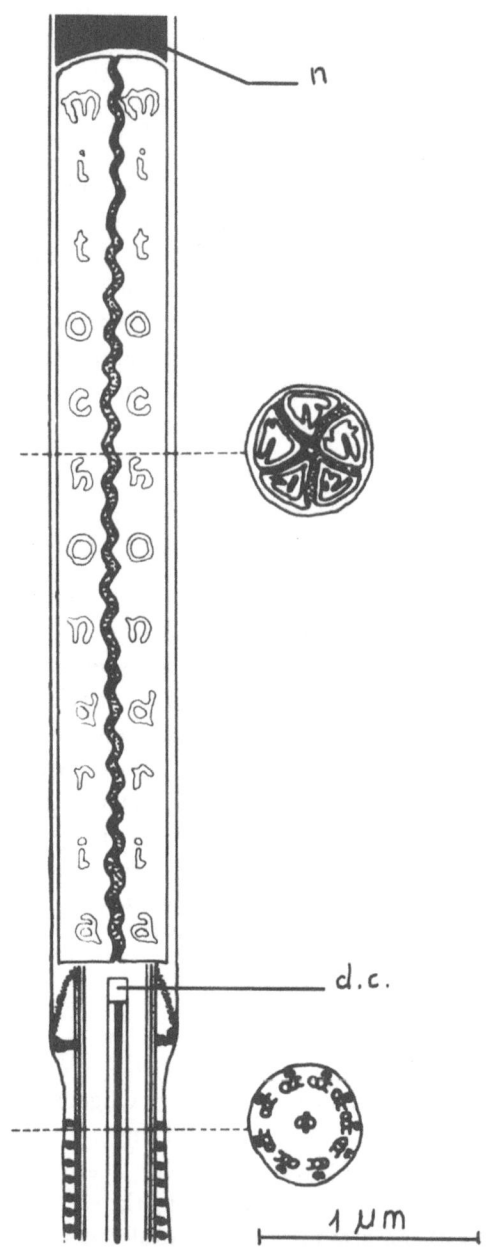

Fig. 2. Scheme of *Glossoscolex paulistus* mid-piece. Note the long mitochondria and the aspect of the flagellum.

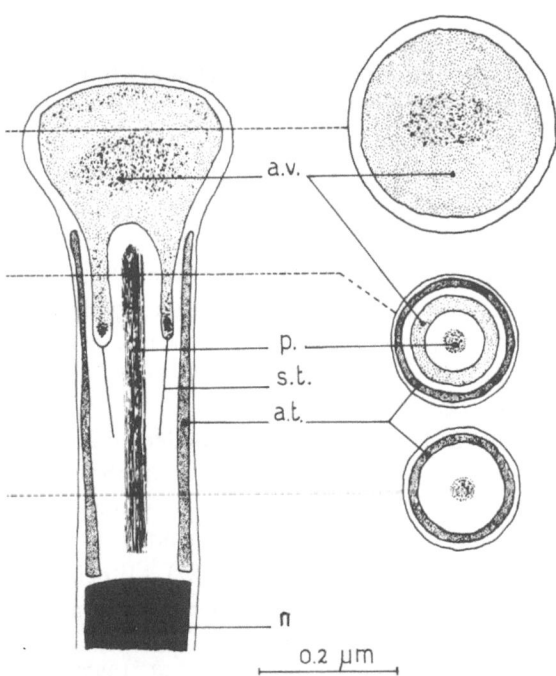

Fig. 1. Scheme of a generalized microdrile acrosome. Dotted lines indicate the level of cross sections showns to the right.

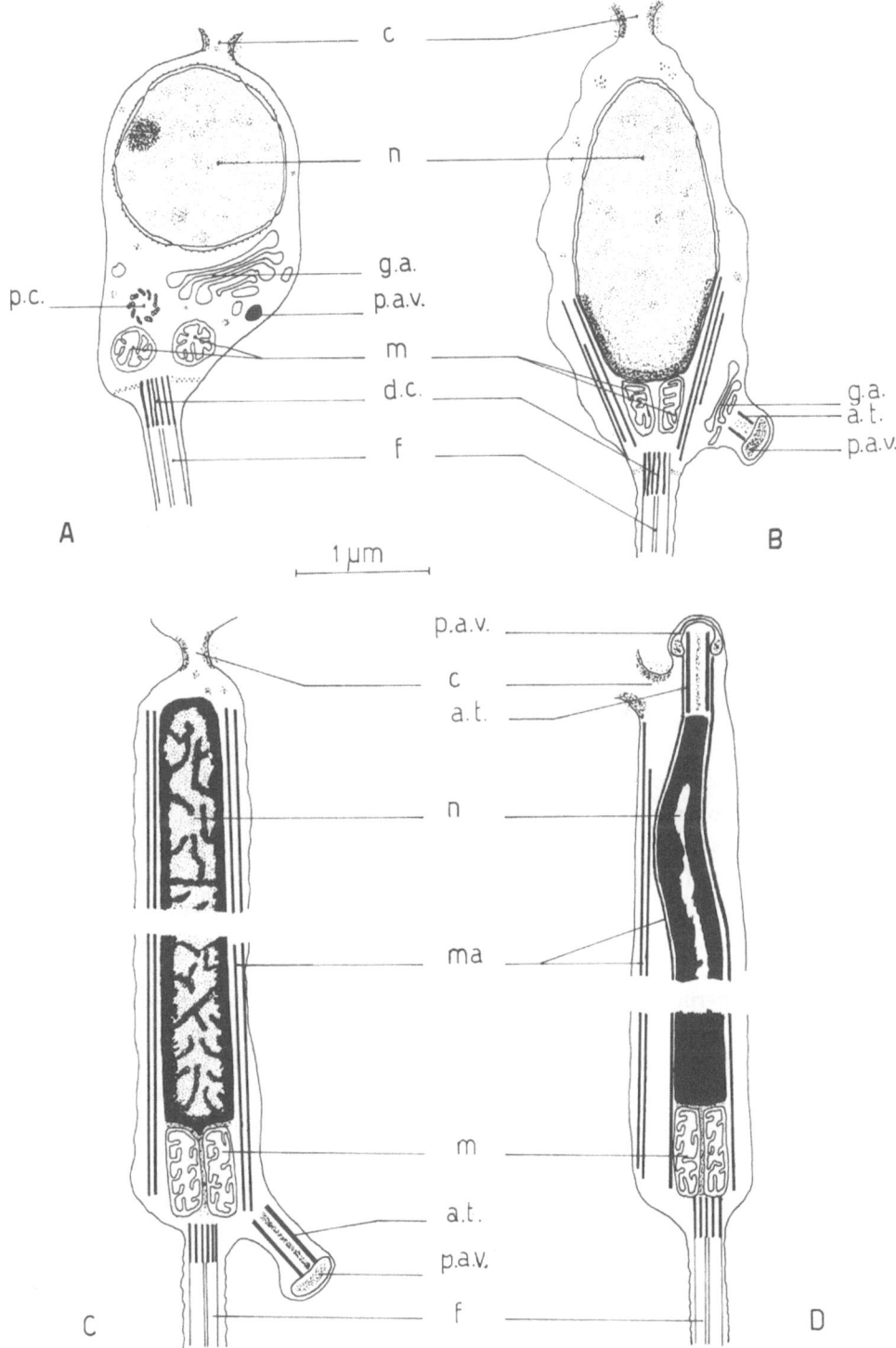

Fig. 3. Generalized sequence of some major events in microdrile spermiohistogenesis. A. Early spermatid. Two centrioles are present, the cell is polarized, and the proacrosomal vesicle has been secreted by the Golgi apparatus. Dotted lines represent the anchoring apparatus of the distal centriole. B. Elongation and chromatin condensation begin contemporaneously with the first appearance of the manchette of microtubules. The acrosomal tube begins its differentiation. C. Acrosomal tube elongates while it is still in the mid-piece area. A continuous layer of condensed chromatin is present at the periphery of the nucleus. D. The acrosome has reached its final position, and chromatin condensation is nearly complete.

emission of the flagellum by the distal centriole (the proximal one disappears soon after). At this stage the cell is already polarized (Fig. 3a): the flagellum is at the opposite pole with respect to the collar; close to the centriole are the mitochondria, and a prominent Golgi apparatus. The centriole is linked to the plasma membrane by an anchoring apparatus similar to the one of 'primitive' marine spermatozoa.

Soon after, the proacrosomal vesicle appears at the top of Golgi apparatus, at first as a small electron-dense vesicle, then as a flattened structure adhering to the plasma membrane. The other structures of the acrosome develop under the proacrosomal vesicle (Fig. 3b). The migration of the acrosome to the top of the nucleus is a later event (Fig. 3d); (Ferragutti & Jamieson, 1984).

The subsequent events are nuclear and mitochondrial morphogenesis. The nucleus is roundish at the beginning of spermiogenesis and mitochondria are grouped in the centriolar area. At maturity the nucleus is long, thin and compact and mitochondria are tightly packed (Fig. 3). This dramatic transformation has been deeply studied in the last years by several groups: it seems that some connection exists between the presence of a microtubular manchette, nuclear morphogenesis and, perhaps, onset of chromatin condensation (for different interpretation of the manchette functions, see Ferraguti, 1983; Fawcett et al., 1971).

Modified sperm line and spermatozeugmata

Although some form of non-conventional sperm and spermiogenesis have been described in the past (see, e.g., Dixon, 1915), the existence of a double line of spermatogenesis in tubificids has been proved only recently by Block & Goodnight (1980) and Braidotti et al. (1980). The modified sperm is different from the conventional model mainly in the shape and size of the nucleus, which is dramatically shorter (3 μm), and in the separation of the axoneme from the plasma membrane. Spermatogenetic events too have been proved to be different (Ferraguti et al., 1983); these observations can throw some light on the interpretation of the spermatozeugma. The modified spermatozoa, in fact, constitute the external portion of it, are helically wound all around, the heads inside and the tails freely floating at the exterior. The central portion of the spermatozeugma, on the contrary, is made up by parallely arranged conventional spermatozoa, which are apparently not as many as the modified ones.

Discussion

I will try to summarize some of the points which remain to be clarified in oligochaete spermiogenesis.

Number of spermatozoa produced by each morula

In the literature there are often different figures concerning the same species. For *Tubifex*, for instance, they vary from 128 in the classical texts, to 1056 in the more recent papers (Block & Goodnight, 1980; Braidotti & Ferraguti, 1982). This is due, I think, to the technical difficulty of the observation, as well as to the ill-defined systematics of the genus. Furthermore, a preliminary study of the early spermatogenetic events in *Tubifex* suggests that the synchrony of the cell divisions whithin a morula is quite far from perfection. Jamieson (1981a) reports in *Phreodrilus* the presence of morulae in which some of the germ cells are grouped and linked to the cytophore by the same collar, suggesting a common origin from a single gonium of the early morula. In the same morulae, other germ cells are not grouped.

Organelle movement and distribution

From the preceding description it is evident that, starting from the original situation of the early spermatid, a profound reorganization of the organelles occurs: the mitochondria, at first scattered, are grouped between nucleus and centriole; the acrosomal vesicle migrates from the original position, close to the centriole, to the top of the nucleus. Troyer & Cameron (1980) distinguish between movements linked to membrane fluidity (like those of the acrosome) and movements of the other cytoplasmic organelles. The absence of direct connection between manchette and moving organelles, as well as the absence of any structure directly supporting the movements, requires a detailed study of the mechanisms of cytoplasmic movements in oligochaete spermatids.

Manchette functions

The precise connection, shown by Ferraguti & Lanzavecchia (1971) in *Tubifex* between presence of a microtubular manchette, onset of chromatin condensation and nuclear morphogenesis has been discussed by many authors working on more or less related Oligochaeta. For instance, it has been shown in *Eisenia* by Martinucci & Felluga (1979) that chromatin condensation begins not only in connection with the manchette, but also with the mitochondria, when they make contact with the perinuclear cisterna. This phenomenon has been seen also in *Tubifex* modified spermiogenesis (Ferraguti *et al.*, 1983). Furthermore, Webster & Richards (1977) have shown that, in *Lumbricillus*, the angle between the manchette microtubules and the nuclear axis differs from the one between nuclear axis and the nuclear flange in spermatids. These and other similar observations, point to a more complicated model of microtubules-chromatin and microtubules-organelles interaction which must be further studied.

The double sperm line

In a recent paper (1983), Ferraguti *et al.* have demonstrated that the differentiation of the two lines commences within the testes. A double population of morulae seems, in fact, to exist as early as at the 4 cells stage, the earliest found in *Tubifex* seminal vesicles. The mechanisms of this differentiation, evidently different from the ones shown by Christensen (1980) in the sub-amphimictic spermiogenesis of *Enchytraeus*, are actually unknown.

The spermatozeugmata

The function of the modified line can be perhaps better understood by studying the function of spermatozeugmata. Is the presence of the spermatozeugmata always linked to the presence of a double line? This problem can be solved by means of a microscopical study of many spermatozeugmata models. My experience is, up to now, confined to *Tubifex*, *Limnodrilus*, and *Psammoryctides*. In these cases, the double population of sperm is present in the spermatozeugmata, but these few data do not allow any general statement. In particular, we do not have any cytological data on the scattered

spermatophores of lumbricids, studied, among others, by Bouché (1975) and Ljungström (1968).

Neither do we know whether the species devoid of spermatozeugmata, like, for instance, all the naidids, are devoid of the double sperm line. This would be obvious if the only function of the modified line is the protection of the conventional spermatozoa. It seems to me that a single mechanical function is nothing much for such an abundant production of differentiated cells. However, it is interesting to note that an identical disposition of the two categories of sperm can be found in some lepidopterans (Friedlander, personal communication).

Abbreviations in the figures

a.t. acrosomal tube
a.v. acrosomal vesicle
c collar
d.c. distal centriole
f flagellum
g.a. Golgi apparatus
m mitochondria
ma manchette
n nucleus
p perforatorium
p.a.v. proacrosomal vesicle
p.c. proximal centriole
s.t. secondary tube

Acknowledgements

This work has been supported by a grant from Consiglio Nazionale delle Ricerche (Rome).

References

Anderson, W. A., A. Weissman & R. A. Ellis, 1967. Cytodifferentiation during spermiogenesis in Lumbricus terrestris. J. Cell Biol. 32: 11–26.
Baccetti, B. & B. Afzelius, 1976. Biology of the sperm cell. Karger. 254 pp.
Block, E. M. & C. J. Goodnight, 1980. Spermatogenesis in Limnodrilus hoffmeisteri (Annelida, Tubificidae): A morphological study of the development of two sperm types. Trans. am. microsc. Soc. 99: 368–384.
Bouché, M. B., 1975. La reproduction de Spermophorodrilus albanianus nov. gen., nov. sp. (Lumbricidae) explique-t-elle la fonction des spermatophores? Zool. Jb. Syst. Okol. Geogr. Tiere 102: 1–11.

Braidotti, P., M. Ferraguti & T. P. Fleming, 1980. Cell junctions between spermatozoa flagella within the spermatozeugmata of Tubifex tubifex (Annelida: Oligochaeta). J. Ultrastruct. Res. 73: 299–309.

Braidotti, P. & M. Ferraguti, 1982. Two sperm types in the spermatozeugmata of Tubifex tubifex (Annelida, Oligochaeta). J. Morphol. 171: 123–136.

Christensen, B., 1980. Annelida. 2 in B. John (ed.), Animal Cytogenetics. Gebrüder Borntraeger, Berl.; Stuttgart, 81 pp.

Dixon, G. C., 1915. Tubifex. L.M.B.C. Mem. typ. Br. mar. Pl. Anim. 23: 1–100.

Fawcett, D. W., W. A. Anderson & D. M. Phillips, 1971. Morphogenetic factors influencing the shape of the sperm heads. Devl. Biol. 26: 220–251.

Ferraguti, M., 1983. Clitellata. In K. G. Adiyodi & R. Adiyodi (eds.), Reproductive Biology of Invertebrates. Vol. II. J. Wiley & Sons, Chichester, pp. 343–376.

Ferraguti, M., P. Braidotti & A. Trigari, 1983. Two different lines in Tubifex tubifex spermiogenesis. In J. André (ed.), The sperm cell. M. Nijhoff, The Hague, pp. 446–449.

Ferraguti, M. & B. G. M. Jamieson, 1984. Spermatogenesis and spermatozoal ultrastructure in Hormogaster (Hormogastridae, Oligochaeta, Annelida). J. submicrosc. Cytol. 16: 307–316.

Ferraguti, M. & G. Lanzavecchia, 1971. Morphogenetic effects of microtubules. I. Spermiogenesis in Annelida Tubificidae. J. submicrosc. Cytol. 3: 121–137.

Jamieson, B. G. M., 1978. A comparison of spermiogenesis and spermatozoal ultrastructure in megascolecid and lumbricid earthworms (Oligochaeta: Annelida). Aust. J. Zool. 26: 225–240.

Jamieson, B. G. M., 1981a. Ultrastructure of spermiogenesis in Phreodrilus (Phreodrilidae, Oligochaeta, Annelida). J. Zool., Lond. 194: 393–408.

Jamieson, B. G. M., 1981b. The Ultrastructure of the Oligochaeta. Academic Press, Lond.; N.Y., 462 pp.

Jamieson, B. G. M., 1984. A phenetic and cladistic study of spermatozoal ultrastructure in the Oligochaeta (Annelida). In G. Bonomi & C. Erséus (eds.), Aquatic Oligochaeta. Dev. Hydrobiol. (this volume).

Jamieson, B. G. M. & L. Daddow, 1979. An ultrastructural study of microtubules and the acrosome in the spermiogenesis of Tubificidae (Oligochaeta). J. Ultrastruct. Res. 67: 209–224.

Jamieson, B. G. M., T. P. Fleming & K. S. Richards, 1982. An ultrastructural study of spermatogenesis and spermatozoal morphology in Sparganophilus tamensis (Sparganophilidae, Oligochaeta, Annelida). J. Zool., Lond. 196: 63–79.

Ljungström, P. O., 1968. Spermatophores in three species of Swedish lumbricids (Oligochaeta). Zool. Anz. 181: 53–60.

Martinucci, G. B. & B. Felluga, 1979. Mitochondria-mediated chromatin condensation and nucleus reshaping during spermiogenesis in Lumbricidae. J. submicrosc. Cytol. 11: 221–228.

Troyer, D. & M. L. Cameron, 1980. Spermiogenesis in lumbricid earthworms revisited: 1. Function and fate of centrioles, fusion of organelles, and organelle movements. Biol. cell. 37: 273–286.

Webster, P. & K. S. Richards, 1977. Spermiogenesis in the enchytraeid Lumbricillus rivalis (Oligochaeta: Annelida). J. Ultrastruct. Res. 61: 62–77.

Life-cycle and karyology of *Branchiura sowerbyi* Beddard (Oligochaeta, Tubificidae)

Sandra Casellato
Istituto di Biologia Animale, Università di Padova, Via Loredan 10, I-35100 Padova, Italy

Keywords: aquatic Oligochaeta, Tubificidae, life-cycle, karyology

Abstract

Data on the life-cycle of a population of *Branchiura sowerbyi* Beddard in a water-lily tank at the Botanical Garden in Padua are reported. The breeding period is from April to July, after which the reproductive system is partially resorbed (August–September) and reformed later in the autumn. The karyology of the species was also studied, and revealed 38 mitotic chromosomes in the gonia, and 19 bivalents in the primary spermatocytes and in the primary oocytes.

Introduction

Branchiura sowerbyi is a widespread tubificid species. According to Michaelsen (1908), Stephenson (1917, 1930) and more recently Timm (1980), it was introduced by man into Europe from the subtropical regions. It is found in a variety of different habitats; it is favoured by an high organic content in the substrate. Carrol & Dorris (1972) demonstrated that its life-cycle adapts to local environmental conditions.

In Italy it is quite common: Brinkhurst (1963) recorded it in Lake Maggiore, Omodeo (pers. commun.) found it in Calabria, and I have collected it along the banks of several rivers in Northeastern areas (rivers Adige, Po, Brenta, Bacchiglione, Tergola).

Branchiura sowerbyi is a cosmopolitan species, but it would be interesting to compare specimens from different geographical areas, not only for their morphology and life-cycle, which can be influenced by the climate, but also from a cytotaxonomical and biochemical point of view. The aim of the present study is to contribute in this direction.

Material and methods

To follow the reproductive cycle of *B. sowerbyi* throughout the year, frequent samplings were made (more numerous during the period of cocoon laying) in a water-lily tank at the Botanical Garden in Padua. Here *B. sowerbyi* occurs in great numbers together with *Limnodrilus hoffmeisteri, Tubifex tubifex, Chironomus plumosus* and *Lymnaea stagnalis*. The tank is filled by a thermal spring (20 °C) and the temperature fluctuates throughout the year between 15 and 23 °C.

The cytological technique was as follows: the testes, ovaries, sperm sacs and ovisacs were removed from mature specimens, just collected, but after a light anaesthesia with 20% alcohol. I carried out hypotonic treatment on this material with frequently changed bidistilled water (for 15–20 minutes). I then stained it with acetic orcein. Because the material was very brittle, I followed up all the phases under the microscope; lastly I made the squash, and the best preparations, observed at the phase contrast microscope, were subsequently stained with Giemsa. Sometimes, before dissecting the worms, I treated them with a 10^{-6} M colchicin

Hydrobiologia 115, 65–69 (1984).

solution for 6-24 hours; this treatment, however, turned out to be of little use.

Life-cycle

During the winter, until the beginning of the spring, the *B. sowerbyi* population studied was formed by specimens, 8-9 cm long (measured alive), with a complete sexual apparatus but without clitellum, together with immature specimens (4-6 cm long).

In the second half of April the first specimens with a developed clitellum appeared. They were maximally 11-12 cm long, and through their transparent skin the eggs appeared as small yellowish bodies. In that period the temperature of the water was 17 °C and the oxygen concentration was 5 mg l^{-1} (it varied from 3 to 6 mg l^{-1} during the year).

Within a few days the clitellate specimens laid cocoons and since not all the individuals in the population were mature at the same time, cocoons continued to be laid until the end of July.

In samples taken in the same period along the banks of the rivers Adige and Brenta, the first clitellate specimens were found from the end of May and were smaller (6-7 cm alive).

In order to check accurately when the worms emerged from the cocoons, I placed several (50-60) newly-laid cocoons in glass containers with silt and lime at a thermostatic temperature of 20 °C. For comparison I also studied cocoons at room temperature (25-27 °C).

It took 12-14 days for the worms to complete their development at 20 °C and 7-8 days at 25-27 °C (Table 1). They subsequently remained a few additional days in the cocoon. Only 70% of the eggs developed completely; the remainder, having passed the stage of division, perished and served as food

Table 1. Influence of temperature on embryonic development in *Branchiura sowerbyi* Beddard.

	20 °C	25-27 °C
Period (in days) of embryonic development	12-14	7-8
Period (in days) of stay in cocoons after embryonic development	5-6	2-3
% worms emerging from cocoons	70%	50%
% cocoons degenerating	20-30%	65-70%

for the developing worms. A certain number of cocoons did not develop, being damaged during the transportation or attacked by fungal-hyphae. At room temperature the number of cocoons that degenerated was higher and the number of worms emerging was lower, as compared to 20 °C.

At the beginning of August the cocoon-laying was finished; I still found one or two but they did not develop in the laboratory. During this period the population consisted of specimens of different sizes. Together with young worms 1-3 months old (2-4 cm long), there were also the adults which had just reproduced. They had no clitellum and were thinner and less numerous than in the spring. Poddubnaya (1958) recorded a mortality of 30% among specimens of *Tubifex albicola* which had just reproduced.

I dissected a certain number of specimens a few weeks after cocoon-laying, and they presented partial resorption of the sexual apparatus. Ovaries and testes were smaller and sometimes spermathecae and sperm sacs were completely resorbed.

Resorption of reproductive organs after the breeding period is known for tubificids. Černosvitov (1930) observed resorption of the sexual apparatus in *T. tubifex,* and Poddubnaya (1971) in her study of *Isochaetides newaensis* observed partial regression and subsequent reformation of the genital organs.

The genital apparatus of *B. sowerbyi* reached its highest degree of involution in the late summer. The ovaries and testes presented few cells, and no mitotic divisions. Spermathecae were often lacking, as were sperm sacs and ovisacs. Sometimes a few eggs in degeneration remained.

The following autumn mitotic divisions began again in the ovaries and testes, both of which increased in size.

Only in spring I could find mature specimens with clitellum and with well developed spermathecae empty of spermatozoa.

Karyological observations

The best karyological results were obtained for the male germinal line from spermatogonial mitosis, in which I could easily observe a diploid number of 38 chromosomes (Fig. 1).

The mean value of the size of these spermatogonia was 3.5×4.2 μm.

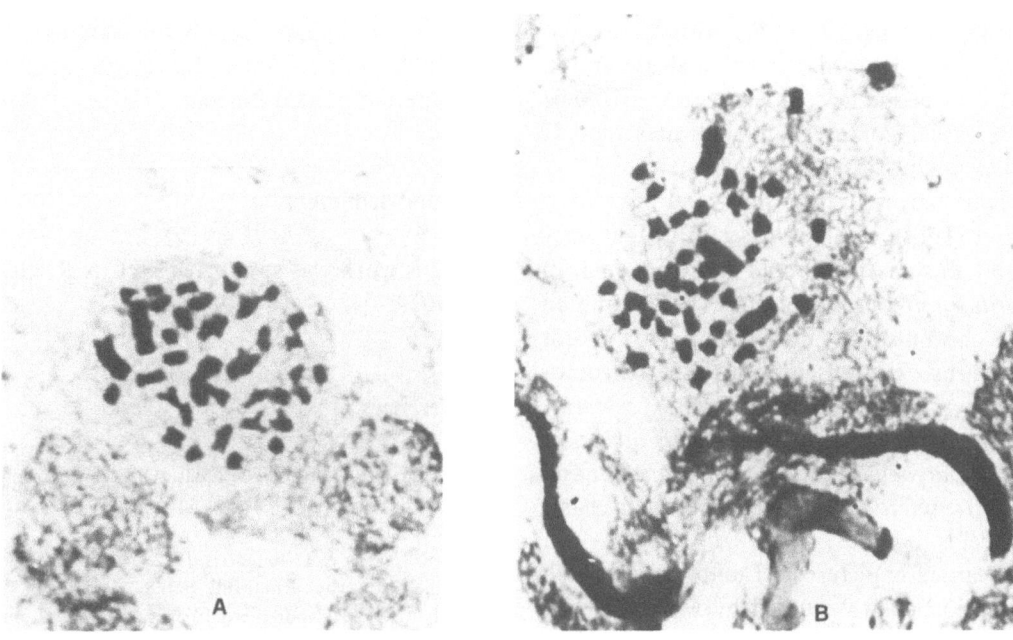

Fig. 1. (A and B) Mitotic metaphases in sperm sacs of *Branchiura sowerbyi.* × 1500.

In the testis of *B. sowerbyi* three mitotic divisions of spermatogonia occur (Hirao, 1973) and the cells deriving from the first remain interconnected by a cytoplasmic bridge to form a morula. This 8 spermatogonia-morula is released from the testis into the sperm sac where another mitotic division occurs before meiosis. The 16 spermatogonia-morulae will contain 32 primary spermatocytes after the first meiotic division and 64 secondary spermatocytes after the second. Maturation continues in the sperm sacs until they become complete spermatozoa.

Counting the bivalents in spermatocytes and oocytes was rather difficult because of a certain degree of stickiness. After numerous counts I decided that 19 was the most probable number of meiotic bivalents, so that the diploid number of 38 was confirmed.

I also used the cocoons that had just been laid in order to find images of bivalents. In fact, in tubificids the primary oocytes released from the ovaries into the ovisacs in metaphase of the first meiotic division are released in this state into the cocoons from the female pores (Hirao, 1968). Here they will be fertilized after completing meiosis in 8–9 hours. I could not, therefore, find good images of the ripe eggs because of the yolk.

Discussion

The population of *B. sowerbyi* studied has a breeding period between April and July. The sexual apparatus undergoes a certain degree of degeneration in the late summer, but regenerates in the following autumn. The specimens of the population in water tanks were bigger than those collected along the banks of the rivers in the Veneto region. Also the breeding period was rather different: the tank population breeded earlier than the river populations. I think that this phenomenon is due to the peculiarity of the environment of the tank where the temperature is stable throughout the year.

The karyological study indicated 38 mitotic chromosomes in the gonia and 19 bivalents in oocytes and spermatocytes. This is an unusual basic number within Oligochaeta and cannot be easily slotted into the karyology of the Tubificidae.

The karyology of this family is little known and confused in the literature. Gathy (1900) counted about 110 chromosomes in oogonia of *Tubifex rivulorum (= T. tubifex);* this number appears approximately triploid in comparison with the data reported by Černosvitov (1927), who counted about 76 chromosomes in the gonia of *T. tubifex.* Dixon

(1915) reported 24 bivalents in oocytes of the same species. In specimens of *T. tubifex* from Lake Suviana (near Bologna) I myself counted about 100 chromosomes in spermatogonia of the testis. The picture is rather confused and indicates that *T. tubifex* is a polyploid species.

In *Tubifex bavaricus (= Potamothrix bavaricus)* Oschmann (1914) observed a set of 20 chromosomes both in somatic cells and in the oogonia. In *Limnodrilus hoffmeisteri* Gavrilov (1935) counted about 96 chromosomes in the gonia. In a recent monograph by Christensen (1980) other chromosome numbers are recorded for *Tubifex costatus, Psammoryctides barbatus, Limnodrilus udekemianus, L. claparedeianus, Peloscolex benedeni* and *Ilyodrilus templetoni*. The most common basic number seems to be 25.

There is a clearer picture for Lumbricidae, which have been particularly studied by Muldal (1949, 1952), Omodeo (1952, 1955) and Vedovini (1973). The most frequent basic number in this family is 18. The numbers 16 and 17 are less frequent, 15 has been proposed for *Eiseniona sineporis,* 11 for *Eisenia foetida,* and 19 was proposed by Muldal (1952) for *Octolasium lacteum* and by Omodeo (1955) for *O. croaticum.* The latter number turned out to be erroneous in subsequent studies (for *O. lacteum,* cf. Vedovini 1973; for *O. croaticum,* cf. Casellato & Rodighiero, 1972). I found a diploid strain of *O. croaticum* from Corfù, with 32 chromosomes in the gonia.

Omodeo (1952) considered n = 18 to be the ancestral basic number of the lumbricid family, from which the others have been derived probably because of centromere fusion.

In the Enchytraeidae, studied particularly by Christensen (1961) and Christensen & Nielsen (1955), the chromosome numbers are more varied. In 71 species studied, many of them polyploids, the cytotype varies between 12 and 150–160. The most frequent basic numbers fall between 12 and 24. Christensen (1961), comparing chromosome numbers within the genera, concluded that n = 16 is the ancestral basic haploid number in the family, from which all others have been derived.

A similar general consideration for the Tubificidae will be possible only when more karyologic data from many species are available.

For *Branchiura sowerbyi,* the only known species of the genus, a better resolution of its karyotype and a comparison with specimens from several different geographical regions will allow us to make an exact hypothesis about the origin, area, manner and period of its diffusion.

Acknowledgments

This work was supported by C.N.R. Grant No. 81.00298.04.

References

Brinkhurst, R. O., 1963. The aquatic Oligochaeta recorded from Lake Maggiore with notes on the species known from Italy. Mem. Ist. ital. Idrobiol. 16: 137–150.

Carrol, J. H. jr. & T. T. Dorris, 1972. The life history of Branchiura sowerbyi. Am. Midl. Nat. 87: 413–422.

Casellato, S. & R. Rodighiero, 1972. Karyology of Lumbricidae. III° Contribution. Caryologia 25: 513–524.

Černosvitov, L., 1927. Die Selbstbefruchtung bei den Oligochaeten. Biol. Zbl. 47: 587–595.

Černosvitov, L., 1930. La régression physiologique des organes génitaux du Tubifex tubifex. Bull. biol. Fr. Belg. 64: 211–248.

Christensen, B., 1961. Studies on cytotaxonomy and reproduction in the Enchytraeidae. Hereditas 47: 387–450.

Christensen, B., 1980. Annelida. In: Animal Cytogenetics, 2., B. John (ed.). Berlin, Stuttgart: Gebrüder Borntraeger, 1–79.

Christensen, B. & C. O. Nielsen, 1955. Studies on Enchytraeidae, 4. Preliminary report on chromosome numbers of 7 Danish genera. Chromosoma 7: 460–468.

Dixon, G. C., 1915. Tubifex. L.M.B.C. Mem. typ. Br. mar. Pl. Anim. 23: 1–100.

Gathy, E., 1900. Contribution à l'étude du développement de l'oeuf et de la fécondation chez les Annélides. Cellule 17: 7–62.

Gavrilov, C., 1935. Contribution à l'étude de l'autofécondation chez les Oligochètes. Acta zool. 16: 21–64.

Hirao, Y., 1968. Cytological study of fertilization in Tubifex egg. Zool. Mag., Tokyo 77: 340–346.

Hirao, Y., 1973. Spermatogenesis in a freshwater oligochaete Branchiura sowerbyi Beddard. Mem. Wakayama prefect. Univ. Med. 3: 55–62.

Michaelsen, W., 1908. Zur Kenntnis der Tubificiden. Arch. Naturgesch. 74: 129–162.

Muldal, S., 1949. Cytotaxonomy of British earthworms. Proc. linn. Soc. Lond. 161: 116–155.

Muldal, S., 1952. The chromosomes of the earthworms, 1. The evolution of polyploidy. Heredity, Lond. 6: 55–76.

Omodeo, P., 1952. Cariologia dei Lumbricidae. 1 Contributo. Caryologia 4: 173–274.

Omodeo, P., 1955. Cariologia dei Lumbricidae. 2 Contributo. Caryologia 8: 138–178.

Oschmann, A., 1914. Die Ovogenese von Tubifex (Ilyodrilus) bavaricus. Arch. Zellforsch. 12: 299–358.

Poddubnaya, T. D., 1958. Some data on the multiplication of the Tubificidae. Dokl. Akad. Nauk. SSSR 120: 422-424 (in Russian).

Poddubnaya, T. D., 1971. Resorption and regeneration of the reproductive system in tubificids using Isochaetides newaensis Mich. (Oligochaeta, Tubificidae) as an example. Trudy Inst. Biol. Vodokhran 22: 81-90 (in Russian).

Stephenson, J., 1917. Aquatic Oligochaeta from Japan and China. Mem. Asiat. Soc. Beng. 6: 83-99.

Stephenson, J., 1930. The Oligochaeta. Clarendon Press, Oxford, 978 pp.

Timm, T., 1980. Distribution of aquatic Oligochaetes. In R. O. Brinkhurst & D. G. Cook (eds.), Aquatic Oligochaete Biology. Plenum Press, New York: 55-78.

Vedovini, A., 1973. Systématique, caryologie et écologie des oligochètes terrestres de la région provençale. Centre de documentation du C.N.R.S. Thèse, Fac. Sci., Univ. Provence, 164 pp.

Diet and histophysiology of the alimentary canal of *Lumbricillus lineatus* (Oligochaeta, Enchytraeidae)

Stuart R. Gelder

Biology Section, Division of Math/Science, University of Maine at Presque Isle, Presque Isle, ME04769, U.S.A. (present address), and University of Maine at Orono, U.S.A.

Keywords: aquatic Oligochaeta, nutrition, sediment, histophysiology

Abstract

Lumbricillus lineatus selectively ingests masses of organic and inorganic interstitial particles from a sand-clay substratum in the upper littoral zone. Particle-masses are ingested, passed along the esophagus and into the anterior intestine where the pH becomes acid. A- and C-esterases, acid β-galactosidase, acid phosphatase and β-N-acetylglucosaminidase are present in the epithelium, while the rotating food masses are surrounded by a membrane of sulphated, acid glycoprotein. These enzymes, with the exception of acid phosphatase and the addition of aminopeptidase M, are also present in the epithelia of the mid and posterior intestinal regions where the pH is alkaline. The cells in the ventral wall of the mid intestinal region contain high concentrations of alkaline phosphatase, acid β-galactosidase and β-N-acetylglucosaminidase. The food consists of absorbed organics and bacteria with absorption and intracellular digestion occurring along the intestine, particularly in the mid ventral region.

Introduction

Information on the feeding mechanisms and digestive physiology in aquatic oligochaetes (Jeuniaux, 1969; Pandian, 1975; Michel & DeVillez, 1978) has largely been ignored as an integral part of studying their productivity and trophic interactions with other organisms of an ecosystem. Attempts to relate an oligochaete's diet with the digestive enzymes present in the gut have so far proved inconclusive (Nielsen, 1962; Kristensen, 1972; Dash *et al.*, 1981). While biochemical methods can identify enzymes accurately, the value of the information is greatly reduced and even suspect when a 'digestive' function is inferred without supporting histo- and cytochemical data. This is especially true when specimens are small and whole worms have to be homogenized. Combined biochemical and cytochemical studies have so far been restricted to *Eisenia foetida* (Van Gansen, 1962) and *Lumbricus terrestris* (Bibighaus *et al.*, 1972). Information on gut structure and feeding mechanisms has usually been reported as part of more general morphological studies (Avel, 1959; Brinkhurst & Jamieson, 1971; Jamieson, 1981). These data are important in understanding the mechanical processing of the food and in determining limiting factors such as maximum ingestible particle-size. It has long been known that not all the organic material ingested by a specimen is digested and absorbed even when selective feeding operates (Brinkhurst & Austin, 1979). Brinkhurst & Chua (1969) reported that although numerous species of bacteria were ingested, certain different species survived passage through the alimentary canal of the respective sympatric species of tubificids. Specimens of *Lumbricillus lineatus* (re-identified as *L. rivalis* by Learner, 1972) were reported to feed on detritus and microorganisms in sewage filter beds (Palka & Spaul, 1970), while Harper *et al.* (1981a, b) demonstrated conclusively that *Nais variabilis* from a similar habitat digested and assimilated bacteria. Giere (1975) re-

Hydrobiologia 115, 71–81 (1984).
© Dr W. Junk Publishers, Dordrecht.

ported and reviewed the information on aquatic oligochaete populations, food relations and ecological roles. He noted a direct correlation in littoral substratum distribution between *Marionina subterranea* and pennate diatoms, its food organisms. Other littoral species, *L. rivalis* (Bülow, 1957) and *L. lineatus* (Giere, 1975), were observed to be intimately associated with macrophytal debris and Giere & Hauschildt (1979) state that in their experiments the nutritive source for *L. lineatus* was algal material and not bacterial films.

In view of the variety of feeding manipulations attributed to the pharyngeal pad (Palka & Spaul, 1970) and the lack of specific information on the digestive physiology in *Lumbricillus* spp., the present study was undertaken. Its aim was to characterize the organic material ingested and demonstrate the site and sequence of certain digestive enzymes in *Lumbricillus lineatus* from an upper littoral habitat. Based upon preliminary observations on the worm's diet it was decided to concentrate on those enzymes most likely to hydrolyze the hexosamine components of glycoproteins and peptidoglycans associated with microorganisms. Inclusion of the alimentary canal description was considered necessary as it differs significantly in places from that given by Palka & Spaul (1970).

Materials and methods

Over 100 specimens of *L. lineatus* were hand-picked from sand-clay samples collected in the upper littoral zone of Lowes Cove, Maine, U.S.A. Live individuals were observed to ingest particles of the substratum and fixed at known intervals following ingestion while other specimens were fixed for histological and non-enzyme histochemical techniques. They were subsequently dehydrated in graded ethanol solutions, cleared in xylene, infiltrated in paraffin wax (m.p. 54 °C Fisher), cut serially (6–8 μm thick sections), and stained by the appropriate protocol. Sections of specimens fixed in Hollande-Bouin fluid (Humason, 1979) were stained by Mallory's 1905 method (Lillie & Fullmer, 1976) for general histology, while acid glycoproteins and 1,2 glycol group containing compounds were demonstrated by the Alcian blue (pH 2.5 or pH 1.0) and periodic acid-Schiff reaction respectively (AB-PAS; Humason, 1979). Basic

proteins were demonstrated with the mercuric bromophenol blue method (MBB; Humason, 1979), while all proteins were stained following ethanol-acetic acid fixation and Coomassie BB R250 staining (CR250; Cawood *et al.*, 1978). These histochemical techniques were also performed on aliquots of the sandy-clay substratum and observed microscopically using polarized light (Gelder, 1983).

Oligochaetes were fixed in Trump's solution (McDowell, 1978) for subsequent enzyme histochemical procedures or 4% formaldehyde buffered in 0.1 M Tris HCl, pH 7.2 when phosphatases were being sought. Fixation proceeded for 1 to 2 h at 4 °C; the specimens were then washed in repeated changes of a 25% sucrose solution for 1 to 5 h also at 4 °C and finally rinsed in distilled water prior to incubation. Whole specimens were incubated with naphthol AS-BI β-N-acetylglucosaminide for β-N-acetylglucosaminidase, naphthol AS-BI β-glucuronide for β-glucuronidase, L-leucyl-4-methoxy-2-naphthylamide for aminopeptidase M, naphthol AS-BI phosphate for acid and alkaline phosphatases, all of which used hexazonium-p-rosaniline for the final reaction product (see Lojda *et al.*, 1979), L-leucyl-β-naphthylamide HCl with Fast Garnet GBC for arylamidase (Burstone & Folk, 1956) and 5-bromo-4-chloro-3-indoxyl-β-galactoside for acid β-galactosidase (Lojda *et al.*, 1979). Esterases were demonstrated by Holt's indigogenic reaction using 5-bromo-3-indoxyl acetate as substrate and characterized into types A, B and C with specific inhibitors (p-chloromercuribenzoate, eserine, and β-phenylpropionic acid respectively) and activator (cysteine) (Pearse, 1972). Negative controls for all of these protocols consisted of heat inactivated specimens or the absence of enzyme substrate from the incubation media; positive controls involved simultaneous incubations of mammalian tissue sections. Following the respective incubation procedure, specimens were rinsed in distilled water, cleared in glycerine jelly (Humason, 1979) or dehydrated and cleared in xylene depending upon the solubility of the final reaction product, mounted on a slide, examined and photographed. Specimens in glycerine jelly were sectioned in a cryostat (Gelder, 1982) while those in xylene were embedded in paraffin wax and the procedure previously described was followed.

Results

Mature *L. lineatus* grow up to 0.6 mm in diameter and 7 to 10 mm in length with 38 to 42 segments (Fig. 1). In the laboratory, specimens were observed to tunnel through the interstices of sand-clay samples collected from the upper littoral zone. Providing interstitial water is present, the oligochaetes force their way through the substratum and line the resulting tunnel with a mucoid substance. The mucus binds the separated grains together and thus maintains a passage. Although the worm can pass freely back and forth in the tunnel, the walls retain their adhesiveness and will seal off the passage should opposite walls touch. The tube consists of sulphated, acid glycoprotein (AB pH 1.0) secreted from the epidermal goblet cells distributed over the surface of the worm. These are referred to as type I acid mucus secreting cells by Richards (1977). The mucoid secretion contains high levels of β-glucuronidase and low levels of acid phosphatase prior to release, while C-esterase and sometimes acid phosphatase are present both in the cytoplasm of these cells and the adjacent epidermal cells.

The prostomium and peristomium have tufts of sensory cilia distributed over their surfaces. In one or two of the epidermal cells adjacent to the cilliary tuft cells, acid β-galactosidase and A-esterase are present. Ciliary tufts and enzymes are similarly located in the epidermis of the pygidium.

Description of the alimentary canal

The alimentary canal (Fig. 1) consists of a mouth, buccal cavity, pharynx, pharyngeal glands, esophagus, intestine and subterminal anus. The gut is ciliated starting with the pharynx through to the end of the intestine; a typhlosole is absent.

The unciliated cuboidal cells which line the mouth and buccal cavity contain high concentrations of C-esterase granules. The epithelium also contains goblet cells which secrete acid glycoprotein (AB pH 2.5) into the lumen through small apertures in the thin, PAS-staining cuticle. The transition from the buccal cavity to the pharynx is marked by the cessation of the cuticle and initiation of ciliated cuboidal and columnar cells.

Three pharyngeal glands lie on either side of the esophagus, a single pair being situated in each of the posterior regions of segments 4, 5 and 6 respective-

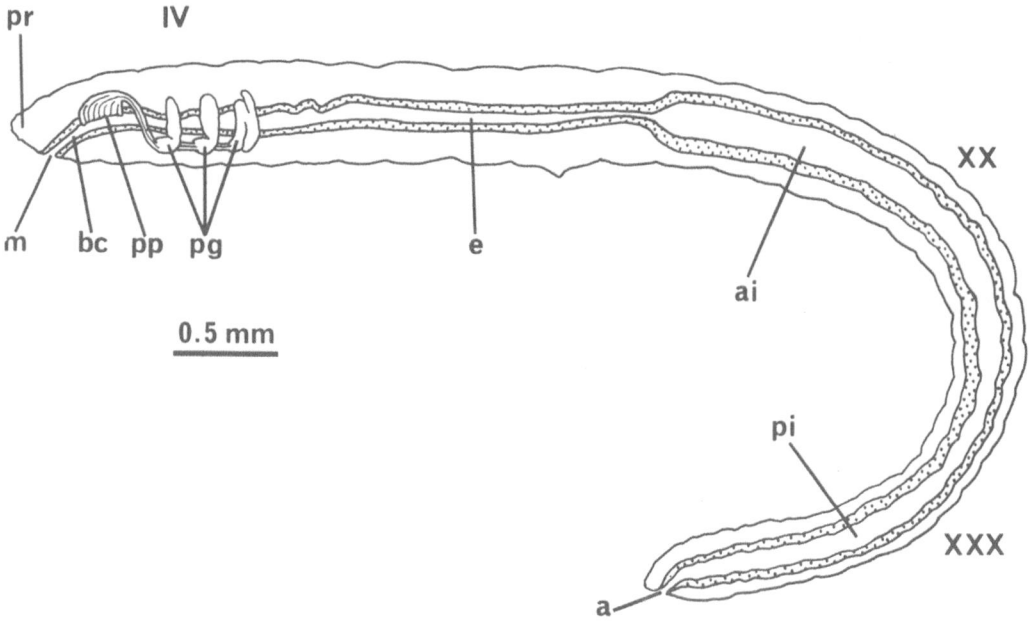

Fig. 1. A diagram of the alimentary canal and pharyngeal glands in *Lumbricillus lineatus* (a = anus, ai = anterior intestine, bc = buccal cavity, e = esophagus, m = mouth, pg = pharyngeal glands, pi = posterior intestine, pp = pharyngeal pad, pr = prostomium; IV, XX, XXX = segment numbers).

74

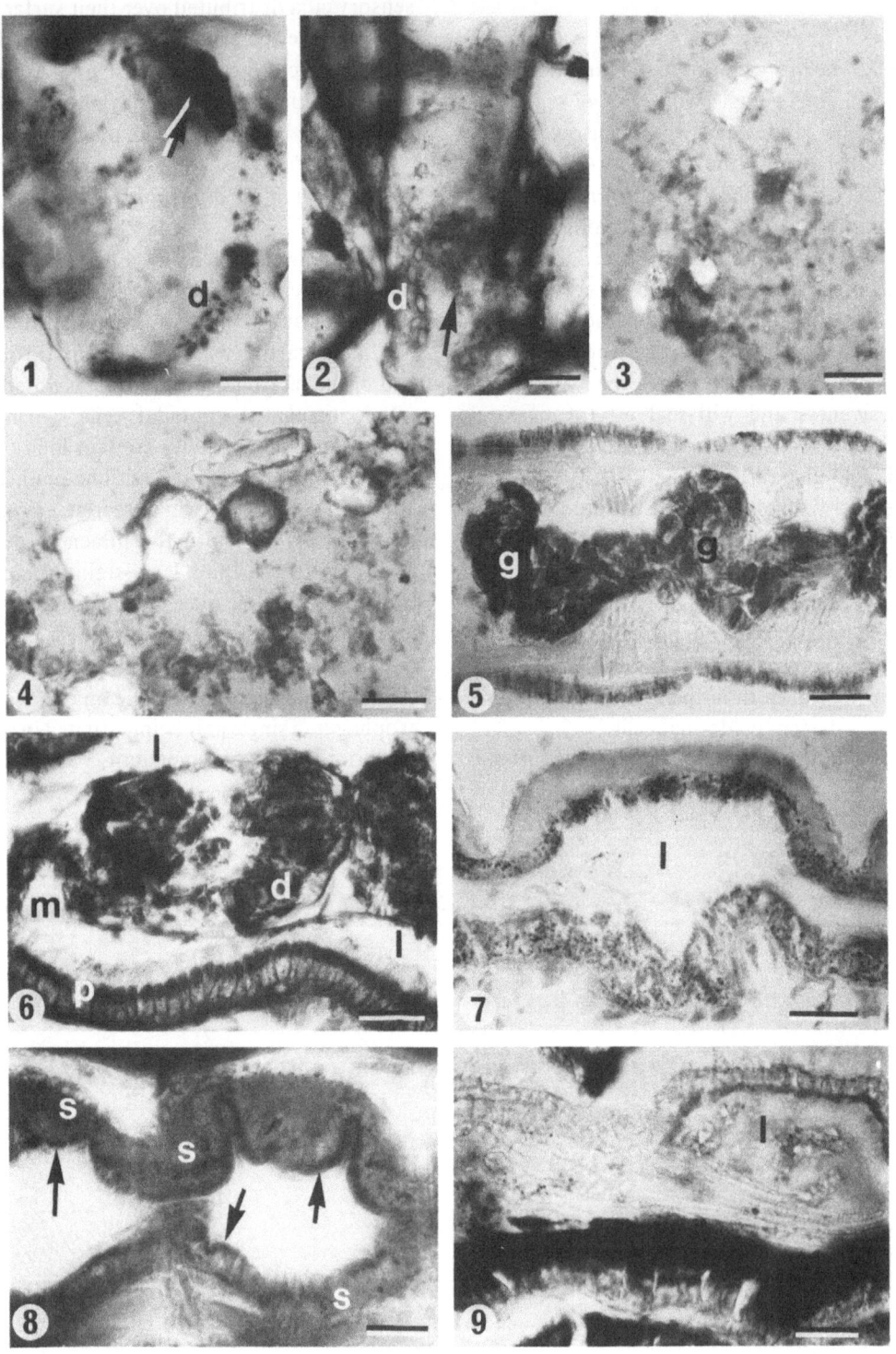

ly. A process from each gland cell passes ventrally then anteriorly along its respective side of the body, around the esophagus to converge dorsally and terminate on the surface of the pharyngeal pad (Fig. 1). The terminal region of the processes often becomes dilated from the build-up of secretion granules. The pharyngeal gland cells show a strong reaction for acid phosphatase (Plate 1.5) in addition to some scattered sites of A-esterase activity. The secretion granules are composed of an acid glycoprotein (AB pH 2.5) with a basic protein (MBB) component; they stain concomitantly with acid fuchsin and aniline blue in the Mallory technique. The secretion processes contain alkaline phosphatase and this enzyme can also be observed in the material covering the surface of the pad.

The esophagus extends from segments III to XIII and its walls consist of basophilic, ciliated cuboidal cells. A few of the epithelial cells contain low levels of acid phosphatase together with granules of C-esterase in the anterior region of the esophagus and A-esterase in the posterior region. The granules in the distal margin of these epithelial cells stain positive for basic protein (MBB) and β-N-acetylglucosaminidase. A glycocalyx of sulphated acid glycoprotein (AB pH 1.0) frequently containing alkaline phosphatase covers the lumenar surface.

Between segments XIII and XIV the narrow esophagus passes into the sligtly wider intestine; the respective lumens are usually separated by partially contracted circular muscle fibers and cilia. The epithelium of the intestine is composed predominantly of ciliated cuboidal and columnar cells with unciliated, conical 'gland cells' scattered among them. The lumenar surface of the ciliated cells is covered by a glycocalyx of sulphated acid glycoprotein (AB pH 1.0), while 2–3 μm particles of the same

material are temporarily attached to the cilia (Plate 1.6). The cells of the anterior intestine in segments XIV to XIX have strongly acidophilic cytoplasm in the mid and basal regions with sulphated acid glycoprotein and acid protein (CR 250) predominating in the distal region. The acid glycoprotein granules measure 2–3 μm in diameter in the intestinal cells of the first segment (XIV); these granules become smaller in size and less numerous in the succeeding segments. The ciliated cells in most of the remaining intestine are also acidophilic but their distal regions contain either acid glycoprotein or basic protein granules. Cells in the ventral walls of segments XXVI to XXVIII have prominent vacuoles basally while acid glycoprotein and basic protein granules fill the distal region (Plate 1.6). Interspersed among these vacuolated cells are ciliated, non-vacuolar cells containing strongly staining basic protein granules. The cytoplasm of these cells appears very similar to that of the unciliated, conical gland cell. The anus is located sub-terminally on the dorsal surface of the pygidium. The anus sometimes appears sunken so that the ciliated intestinal epithelium passes into a nonciliated chamber before the lumen opens to the exterior. This is believed to be caused by the intestinal musculature's contracting during fixation.

The ciliated epithelial cells, with the exception of the ventral vacuolated cells, appeared morphologically identical even though their cytophysiology was demonstrated to vary from cell to cell. Certain ciliated cells in the dorsal wall of the anterior intestine (segments XIV to XXV) contain high concentrations of C-esterase staining granules (Plate 1.7). These and the other cells in the anterior walls also contain A-esterase staining granules with much of this isoenzyme being further characterized as acid

Plate 1. 1. Medium sand grain with diatoms and proteinaceous fragment (arrowed), stained with Coomassie R 250 (scale 30 μm). 2. Medium sand grain showing diatoms surrounded by acid glycoprotein material (arrowed), stained by the AB pH 2.5/PAS procedure (scale 30 μm). 3 and 4. Medium and fine silt grains with associated acid glycoprotein material, stained by the AB pH 2.5/PAS procedure and illuminated by partially polarized light (scale 25 μm). 5. Longitudinal section through the pharyngeal glands, incubated for acid phosphatase (scale 20 μm). 6. Longitudinal section through a food pellet in the intestine of segment XXVI, stained by the AB pH 2.5/PAS procedure (scale 25 μm). 7. Longitudinal section through the anterior intestine in segment XVIII to show the distribution of A-esterase granules; preincubated in 1×10^{-4} M eserine, then incubated in the indoxylesterase medium and counterstained with nuclear fast red (scale 40 μm). 8. Longitudinal section through the intestine in segments XXI and XXII to demonstrate β-N-acetylglucosaminidase (arrowed) (scale 40 μm). 9. Longitudinal section through the intestine in segments XXVI and XXVII to show the distribution of alkaline phosphatase (scale 25 μm).

Abbreviations: d = diatom, g = pharyngeal gland, l = lumen of intestine, m = sulphated acid glycoprotein (? peritrophic) membrane, p = basic protein granules, s = sporozoan.

β-galactosidase (Plate 1.7). Some cells are completely filled with acid phosphatase while in others it is limited to the distal portion. β-N-acetylglucosaminidase staining granules are localized in the distal margin of the ciliated cells along the whole intestine, while high concentrations occur throughout the cytoplasm of the vacuolated cells of the ventral mid intestinal wall. From this region (segment XXIV) on, alkaline phosphatase is present in the glycocalyx, while aminopeptidase M is found scattered among β-N-acetylglucosaminidase in the distal margin. High concentrations of acid β-galactosidase staining granules are localized in some of the ventral vacuolated cells and in lower concentrations in some of the opposite cells in the dorsal wall. C-esterase granules also appear in moderate concentrations in scattered ciliated cells. Neither arylamidase nor β-glucuronidase were demonstrated in the alimentary canal.

The detailed description of the alimentary canal was complicated by the presence of spores and trophozoites of at least one species of sporozoan parasite. The trophozoites were observed in and below the epithelium of both the esophagus and intestine, while spores were localized in chlorogague tissue and gastrodermal cells. The mature trophozoites contained strongly PAS positive staining granules, suggestive of glycogen. β-N-acetylglucosaminidase and acid phosphatase activity appeared in the parasitophorous vacuoles around and sometimes inside the sporozoans located in the esophageal and anterior intestinal epithelium. High levels of these enzymes are present intracellularly in all of the sporozoans in the vacuolated cells of the mid ventral intestine but absent from those sporozoans in the posterior intestinal epithelium. The only observed host response to the parasite involved coelomocytes (granular mucocytes, after Richards, 1980) aggregating around some of the many spores in the coelom. These coelomocytes contained lysosomes in which A- and C-esterases, β-glucuronidase and acid phosphatase were demonstrated.

Diet, feeding mechanisms and digestion

The natural diet of *L. lineatus* was studied by comparing aliquots of substratum taken from around the prostomium immediately after a specimen was observed to feed with material in the alimentary canal.

The aliquots contained mineral grains of aluminosilicates (quarts, feldspar and mica) ranging from sand grain size of 350 μm in diameter to silt-clay (minimum observed diameter being 2 μm). Over half of the material in each aliquot consisted of grains, many of which were birefringent (Plate 1.1–1.4). The remaining portion of the aliquot consisted of organic material that stained PAS positive; this comprised mainly amorphous flocculant material, microorganisms and organic fragments. Blue-green algae, diatoms and small amounts of the amorphous flocculant material (Plate 1.1) stained with MBB and CR250, thus indicating their proteinaceous composition. Those microorganisms which were viable at the time of fixation stained strongly, while those in various stages of cytoplasmic dissociation only stained lightly. Diatoms and blue-green algae were invariably observed attached to the surfaces of sand and some large silt grains (Plate 1.1–1.2); they were surrounded by sulphated and non-sulphated acid glycoproteins (AB pH 1.0 and 2.5 respectively) (Plate 1.2). The interstices contained diatoms, blue-green algae, fungal hyphae, whole spores, bacteria and organic fragments together with chitinous and siliceous skeletons, and silt and clay grains (Plate 1.3–1.4). These interstitial particles were either totally or partially surrounded by a matrix of acid- and some neutral glycoproteins.

An examination of the food material in the alimentary canal showed that *Lumbricillus* ingests silt and clay grains, siliceous debris, macrofloral and chitinous fragments, diatoms, fungal hyphae, filamentous microorganisms and amorphous glycoproteins. The composition of the ingested material reflects the same variety and proportions as that observed in the less than 60 μm long fraction of the aliquot. These observations indicate that an upper particle-size limitation is imposed on the potential food, probably by the maximum dilated aperture of the mouth.

Small, unattached particles of substratum can be sucked through the mouth when the buccal and pharyngeal lumens rapidly dilate. However, the usual method of feeding involves the worm's placing its dilated mouth over a selected mass of substratum. The pharyngeal pad, its surface covered with pharyngeal gland secretions, is protracted until the food mass adheres to the pad's surface; the pad is then retracted into the pharynx and the food mass is thus ingested. Small amounts of acid glycoproteins from the buccal epithelium become attached to the mass as it is ingested, thereby helping to protect the epithelium from abrasion and aiding in the consolidation of loose fragments. These secretions differ from those of the pad as the latter contain alkaline phosphatase. Food is detached from the pad and transported along the esophagus and into the anterior intestine by peristaltic movements and ciliary action. The particles of sulphated acid glycoprotein containing C- or A-esterases attached to the surface of the cilia do not appear to become transferred to the food. The presence of alkaline phosphatase in the glycocalyx of the esophagus with acid phosphatase, esterases and β-N-acetylglucosaminidase in the distal regions of some cells is consistent with the absorption and intracellular digestion of dissolved organics such as hexosamines.

The food-mass enters the anterior intestine and, if succeeded by further masses, is moved along the intestine. Discrete organisms in the food contain intrinsic acid phosphatase while much of the lumenar mucoids and the glycocalyx contain alkaline phosphatase. The food-masses are slowly rotated by actively beating cilia which, like those cilia in the esophagus, also have mucoid-esterase particles attached to their surfaces. After a short time the rotating food-masses are formed into spherical or cigar-shaped pellets and surrounded with a sulphated acid glycoprotein membrane (AB pH 1.0; CR 250) (Plate 1.6). During the formation of the membrane, the pH of the anterior intestinal lumen fell so that alkaline phosphatase activity ceased within the pellet and the lumen. Concomitant with the drop in pH, the concentration of localized intracellular C-esterase granules increased, along with a general rise in A-esterase, including acid β-galactosidase granules in most of the anterior intestinal cells (Plate 1.7). The distal regions of these cells also contain acid phosphatase and low levels of β-N-acetylglucosaminidase (Plate 1.8). As the level of the extracellular esterases does not increase during this period and neither acid β-galactosidase nor β-N-acetylglucosaminidase appear in the lumen, these lysosomal enzymes are deduced to be involved only in the intracellular digestion of absorbed glucosamines.

After 1 to 5 h the pH in the lumen starts to rise until it is alkaline, then alkaline phosphatase reappears in the lumenar mucoids and glycocalyx. By this time the pellets are at least in segment XXVI and adjacent to the ventral vacuolated cells. The levels of intracellular acid β-galactosidase rise and those of alkaline phosphatase greatly increase (Plate 1.9); C-esterase is also present and aminopeptidase M occurs intermittantly. Based on these observations absorption and intracellular digestion continues in the alkaline phase during the pellets' passage through the posterior intestine. The pellets are expelled through the anus where they become attached to the adhesive wall of the worm's mucoid tube. Most of the microphyta and fungi that escaped mechanical damage prior to ingestion appear to be still viable upon egestion from the worm. No examination of bacterial flora was performed on the pellets.

Discussion

The ultrastructure of the pharyngeal glands in the enchytraeids, *Enchytraeus albidus* (Reger, 1967) and *Lumbricillus lineatus* (Ude, 1971, 1977), showed that secretion bodies were transported to the pharyngeal pad from the cell bodies inside their respective cytoplasmic processes. A similar arrangement was found in the present study. Acid phosphatase and A-esterase activity in the pharyngeal gland cells of *L. lineatus* and inosindiphosphate, acid phosphatase, thiaminpyrophosphatase and peroxidase in *L. lineatus* (Ude, 1975) appear to be concerned with secretion synthesis rather than being secreted as digestive enzymes. Reger (1967) and Ude (1971, 1977) both suggest secretion granules are moved along the cellular processes by cytoplasmic flow and this is consistent with the strong reaction obtained for alkaline phosphatase in *L. lineatus*. Although the pharyngeal gland secretion is an acid glycoprotein similar to that in *Eisenia foetida* (Van Gansen, 1962) no amylase was noted; proteases (Keilin, 1920) and trypsin (Bibighaus et al., 1972) demonstrated in *Lumbricus terrestris* were similarly absent.

The primary function of the pharyngeal pad secretion appears to be as an adhesive that is strong enough to hold food particles against the pad while the protruded pharynx is retracted. The pad was never observed to act as a 'sucker' (Palka & Spaul, 1970). Giere & Hauschildt (1979) noted a 'bruising' action, however this was intended to read 'browsing' (Giere, pers. commun.). The possibility of ingested grains being abraded together as a mechanism for dislodging epi-organisms is prevented by the mucoid material which completely or partially surrounds their surfaces. *Lumbricillus lineatus* is a selective feeder (Giere & Hauschildt, 1979); it ingests only mineral grains and detrital masses less than 60 μm in diameter.

Determination of food particle sizes, characterization of the organic material ingested and preferences are among the essential parameters required to understand the nutritional role of benthic deposit-feeders (Levinton, 1980). Whitlatch (1981) characterized the substratum on the basis of inorganic and organic particle morphology and found that most of the organic material contained 1,2 glycol groups (PAS positive). The organic material in the substratum samples in this study proved similar to those previously reported and demonstrated by histochemical techniques by other workers (Frankel & Mead, 1973; Whitlatch & Johnson, 1974), predominantly acid glycoproteins and occasionally proteins adsorbed onto mineral grains. The abundance of acid glycoprotein led Hobbie & Lee (1980) to suggest that such extracellular particulate matter may make up most of the food of detritivores. Any deficiency in nitrogen could be supplemented by the adsorbed proteins and microflora.

β-glucuronidase, β-N-acetylglucosaminidase and acid β-galactosidase were chosen as indicator enzymes as they are known to have key roles in the anabolism and catabolism of glycoproteins (Kornfeld & Kornfeld, 1980; Rodén, 1980; Berger *et al.*, 1982). The most likely effect of β-N-acetylglucosaminidase and β-N-acetylgalactosaminidase, which are indistinguishable by the histochemical technique used (Lojda *et al.*, 1979), is the liberating of the proteoglycan monomers from the hyaluronic acid backbone (Rodén, 1980). Under certain conditions the exposed hyaluronic acid chain can then be hydrolyzed by exo-β-glucuronidase and exo-β-N-acetylglucosaminidase to monosaccharide units (Rodén, 1980). These processes would provide a significant, if variable degree of degradation of most glycoproteins including those surrounding bacteria. Consequently endo-β-N-acetylglucosaminidases, one of which is referred to as lysozyme, could hydrolyze specific sites on the peptidoglycan chain which forms the chief component of the bacterial cell wall (Kimball, 1983). Such a breach in the cell wall would remove the chance of pathogenicity following uptake and concomitantly expose the cytoplasm of the bacteria as a food source.

The intracellular enzymes demonstrated in the glycocalyx and distal region of the esophageal epithelial cells suggest absorption of nutrients for intracellular digestion. However, as food masses usually took 30 to 60 s to pass from the mouth to the anterior intestine, only dissolved, small molecules such as hexosamines would be available for uptake so soon after ingestion.

The constant rotation of the food masses by the epithelial cilia results in the masses' becoming ovoid or elongate and surrounded by a heterogeneous membrane of Alcian blue pH 2.5 and PAS staining material. A comparable membrane in *Enchytraeus albidus*, referred to as the peritrophic membrane, lacks chitin (Peters, 1968) whereas in *Dero obtusa* (Peters, 1968) and *Lumbricus terrestris* the membrane contains chitin. Aminopeptidase was demonstrated in the peritrophic membrane of *Lumbricus* (Vierhaus, 1971) and thought to have a role in digestion.

It is difficult to believe that extracellular digestion does not occur; however no secretion of extracellular or intrinsic enzymes was demonstrated. This suggests that enzymes probably were acting upon the food but were different from the ones sought. It is surprising that β-glucuronidase was entirely absent from the digestive enzyme complement of *L. lineatus*. Haase (1969) reported β-glucuronidase from the typhlosole of lumbricids. A typhlosole is absent in these specimens of *L. lineatus* (reported as *L. rivalis*; Palka & Spaul, 1970), but it is not thought that these two facts are connected. β-glucuronidase was reported in the nematode *Monhystera denticulata* (Jennings & Deutsch, 1975) and the polychaete *Histriobdella homari* (Jennings & Gelder, 1976), and its presence interpreted as being consistent with a bacterial diet. The postcoupling method (Pearse, 1972) used in the cited works was compared with the simultaneous coupling technique which uses naphthol AS-BI β-glucuronide with hexazonium pararosaniline (Lojda *et al.*, 1979) on specimens of *H. homari*. The postcoupling method demonstrates sites of β-glucuronidase in the stomach epithelium (Jennings & Gelder, 1976), while the simultaneous coupling technique visualizes the enzyme in the acid glycoprotein secreting (6th) pair of the salivary glands;

positive controls for this study consisted of mouse kidney sections (Gelder, unpubl. obs.). These observations could suggest that two of the several isoenzymes of β-glucuronidase referred to by Lojda *et al.* (1979) were being demonstrated by the respective substrates. This appears highly unlikely as the same authority reports that 6-bromo-2-naphthol will give positive reactions in enzyme inactivated sections and false-negatives in cells containing β-glucuronidase; the validity of results obtained using this substrate must be questioned.

Sulphate acid glycoprotein and β-glucuronidase occur together in the 6th pair of salivary glands of *H. homari* and in the type I acid mucus secreting cells in the epidermis of *L. lineatus* (nomenclature after Richards 1977). This latter cell-type can be demonstrated in living specimens using neutral red stain (see Kossmagk-Stephan, 1984). It would appear that the β-glucuronidase keeps the mucoid components unpolymerized so that upon secretion, whether into the buccal cavity or onto the tunnel's wall, the enzyme is inactivated and the glycoprotein units polymerize. The converse was reported by Corner *et al.* (1960) where β-glucuronidase was related to the digestion of algal sulphated polysaccharides. The absence of this enzyme from the alimentary canal of *L. lineatus* indicates that hydrolysis of β-glucuronidase units are not present or are not as important as in the study just cited. However, as Hobbie & Lee (1980) note, it is the release of the nitrogen containing sugar units, namely N-acetylglucosamine and N-acetylgalactosamine, that are nutritionally important.

The sites of β-N-acetylglucosaminidase, which according to Lojda *et al.* (1979) probably include β-N-acetylgalactosaminidase, occur along the lumenar margin of the intestinal epithelium and in high concentrations throughout the vacuolated cells in the ventral wall of the mid intestine. Although no microorganisms were seen to be phagocytosed into the mid-ventral intestinal cells, bacteria-size particles could have been taken up unobserved and the high level of β-N-acetylglucosaminidase activity could reflect their intracellular digestion as hypothesized earlier. Such active uptake would certainly be supported by the intense alkaline phosphatase activity demonstrated along the lumenar margin. Acid β-galactosidase was detected in the intestinal epithelium and was also found in a similar location in *Mesenchytraeus glandulosus*

(Nielsen, 1962). β-galactosidase is known to act upon yeast and bacteria (Gottschalk, 1958) and particularly upon galactolipids found in algae (Benson & Shibuya, 1962). The histochemical technique used for A-esterase detection in the present study also demonstrates acid β-galactosidase activity. A comparable visualization of A-esterase, interpreted as 'endopeptidase', was observed on the surface of yeast cells, detrital and other microorganismal food particles in the stomach of *Dinophilus gyrociliatus* (Jennings & Gelder, 1969). In view of the results in the present study and the nature of the food substrate in the stomach of *D. gyrociliatus* it would appear that the A-esterase demonstrated also contained acid β-galactosidase. Evidence of proteases in *L. lineatus* is restricted to very low levels of aminopeptidase M and possibly endopeptidase activity intracellularly. The combination of β-N-acetylglucosaminidase, acid β-galactosidase and A- and C-esterases indicates the importance of intracellular digestion in *L. lineatus* but does not rule out an extracellular phase.

In view of the intricacies demonstrated and those suggested in resource partitioning of microorganisms by hydrobids (Hylleberg, 1975; Lopez & Levinton, 1978) and the review of particle feeding by deposit feeders (Levinton, 1980), further studies are needed on the nutrition of *L. lineatus*. These studies will also have to contend with the possibility of different nutritional regimes in the identifiable genetic populations of *L. lineatus* in the various intertidal levels (Christensen, 1980; see Christensen, 1984). Therefore, in determining the full trophic impact of *L. lineatus* on the intertidal ecosystem, it is important that any trophic variations within the intertidal populations are not lost and data become apparently contradictory.

Acknowledgements

The author would like to acknowledge financial assistance from NSF-EPSCOR awarded to the University of Maine at Orono, Marine Fisheries Research Institute at Orono, Sigma Xi and the University of Maine at Presque Isle with special thanks to Drs S. Tyler, L. Watling, L. M. Mayer and A. M. Gorman.

References

Avel, M., 1959. Classe des annélides oligochaetes. In P.-P. Grassé (ed.), Traité de Zoologie 5. Masson, Paris: 224–470.

Benson, A. A. & I. Shibuya, 1962. Surfactants lipids. In R. A. Lewis (ed.), Physiology and biochemistry of algae. Academic Press, N.Y.: 371–383.

Berger, E. G., E. Buddecke, J. P. Kamerling, A. Kobata, J. C. Paulson & J. F. G. Vliegenthart, 1982. Structure, biosynthesis and functions of glycoprotein glycans. Experientia 38: 1129–1162.

Bibighaus, C. L., E. J. DeVillez & A. L. Allenspach, 1972. Histochemical localization of production and secretion of trypsin in the earthworm Lumbricus terrestris. Z. Zellforsch. 130: 429–439.

Brinkhurst, R. O. & M. J. Austin, 1979. Assimilation by aquatic Oligochaeta. Int. Revue ges. Hydrobiol. 64: 245–250.

Brinkhurst, R. O. & K. E. Chua, 1969. Preliminary investigation of the exploitation of some potential nutritional resources by three sympatric tubificid oligochaetes. J. Fish. Res. Bd. Can. 26: 2659–2667.

Brinkhurst, R. O. & B. G. M. Jamieson, 1971. Aquatic Oligochaeta of the world. Oliver & Boyd, Edinb., 860 pp.

Bülow, T., 1957. Systematisch-autökologische Studien an eulitoralen Oligochaeten der Kimbrischen Halbinsel. Kieler Meeresforsch. 13: 69–116.

Burstone, M. S. & J. E. Folk, 1956. Histochemical demonstration of aminopeptidase. J. Histochem. Cytochem. 4: 217–226.

Cawood, A. H., U. Potter & H. G. Dickinson, 1978. An evaluation of Coomassie brilliant blue as a stain for quantitative microdensitometry of protein sections. J. Histochem. Cytochem. 26: 645–650.

Christensen, B., 1980. Constant differential distribution of genetic variants in polyploid parthenogenetic forms of Lumbricillus lineatus (Enchytraeidae, Oligochaeta). Hereditas 92: 193–198.

Christensen, B., 1984. Asexual propagation and reproductive strategies in aquatic Oligochaeta. In G. Bonomi & C. Erséus (eds.), Aquatic Oligochaeta. Dev. Hydrobiol. (this volume).

Corner, E. D. S., Y. A. Leon & R. D. Bulbrook, 1960. Sulphatases and β-glucuronidase in marine invertebrates. J. mar. biol. Ass. U.K. 39: 51.

Dash, M. C., B. Nanda & P. C. Mishra, 1981. Digestive enzymes in three species of Enchytraeidae (Oligochaeta). Oikos 36: 316–318.

Frankel, L. & D. J. Mead, 1973. Mucilaginous matrix of some estuarine sands in Connecticut. J. sedim. Petrol. 43: 1090–1095.

Gelder, S. R., 1982. Sectioning small invertebrates containing enzyme histochemical azo-dye final reaction products. Proc. r. microsc. Soc. 17: 93–94.

Gelder, S. R., 1983. Enhancement of histochemically demonstrated organic materials on sand-silt grains using polarized light. Tech. Inf. Bull., Leitz, USA 1: 11–12.

Giere, O., 1975. Population structure, food relations and ecological role of marine oligochaetes, with special reference to meiobenthic species. Mar. Biol. 31: 139–156.

Giere, O. & D. Hauschildt, 1979. Experimental studies on the life cycle and production of the littoral oligochaete Lumbricillus lineatus and its response to oil pollution. In E. Naylor & R. G. Hartnoll (eds.), Cyclic phenomena in marine plants and animals. Pergamon Press, Oxford: 113–122.

Gottschalk, A., 1958. The enzymes controlling hydrolytic, phosphorolytic and transfer reactions of the oligosaccharides. Handb. Pfl. Physiol. 6: 87–124.

Haase, E., 1969. Zur Histophysiologie des Regenwurmdarmes. Zool. Anz., Suppl. 33: 535–539.

Harper, R. M., J. C. Fry & M. A. Learner, 1981a. Digestion of bacteria by Nais variabilis (Oligochaeta) as established by autoradiography. Oikos 36: 211–218.

Harper, R. M., J. C. Fry & M. A. Learner, 1981b. A bacteriological investigation to elucidate the feeding biology of Nais variabilis (Oligochaeta: Naididae). Freshwat. Biol. 11: 227–236.

Hobbie, J. E. & C. Lee, 1980. Microbial production of extracellular material: importance in benthic ecology. In K. R. Tenore & B. C. Coull (eds.), Marine benthic dynamics. Belle W. Baruch Library, Univ. S. Carolina: 341–346.

Humason, G. L., 1979. Animal tissue techniques. 4th ed. Freeman & Co., San Francisco, 661 pp.

Hylleberg, J., 1975. Resource partitioning on basis of hydrolytic enzymes in deposit-feeding mud snails (Hydrobiidae). 2. Studies on niche overlap. Oecologia 29: 115–125.

Jamieson, B. G. M., 1981. The ultrastructure of the Oligochaeta. Academic Press, N.Y., 462 pp.

Jennings, J. B. & A. Deutsch, 1975. Occurrence and possible adaptive significance of β-glucuronidase and arylamidase ('leucine aminopeptidase') activity in two species of marine nematodes. Comp. Biochem. Physiol. 52A: 611–614.

Jennings, J. B. & S. R. Gelder, 1969. Feeding and digestion in Dinophilus gyrociliatus (Annelida: Archiannelida). J. Zool., Lond. 158: 441–451.

Jennings, J. B. & S. R. Gelder, 1976. Observations on the feeding mechanism, diet and digestive physiology of Histriobdella homari van Beneden, 1858: an aberrant polychaete symbiotic with North American and European lobsters. Biol. Bull. mar. biol. Lab., Woods Hole 151: 489–517.

Jeuniaux, C., 1969. Nutrition and digestion. In M. Florkin & B. T. Scheer (eds.), Chemical zoology 4. Academic Press, N.Y.: 69–91.

Keilin, D., 1920. On the pharyngeal or salivary gland of the earthworm. Q. J. microsc. Sci. 65: 33–61.

Kimball, J. W., 1983. Biology. 5th ed. Addison-Wesley Publishers, Reading, Mass., 747 pp.

Kornfeld, R. & S. Kornfeld, 1980. Structure of glycoproteins and their oligosaccharide units. In W. J. Lennarz (ed.), The biochemistry of glycoproteins and proteoglycans. Plenum Press, N.Y.: 1–33.

Kossmagk-Stephan, K.-J., 1984. A method of identifying immature specimens of marine Enchytraeidae (Oligochaeta) by vital staining of epidermal glands. In G. Bonomi & C. Erséus (eds.), Aquatic Oligochaeta. Dev. Hydrobiol. (this volume).

Kristensen, J. H., 1972. Carbohydrases of some marine invertebrates with notes on their food and on the natural occurrence of the carbohydrates studied. Mar. Biol. 14: 130–142.

Learner, M. A., 1972. Laboratory studies on the life-histories of four enchytraeid worms (Oligochaeta) which inhabit sewage percolating filters. Ann. appl. Biol. 70: 251–266.

Levinton, J. S., 1980. Particle feeding by deposit-feeders: models, data and a prospectus. In K. R. Tenore & B. C. Coull (eds.), Marine benthic dynamics. Belle W. Baruch Library, Univ. S. Carolina: 423–440.

Lillie, R. D. & H. M. Fullmer, 1976. Histopathologic technic and practical histochemistry. 4th ed. McGraw-Hill, N.Y., 942 pp.

Lojda, Z., R. Gossran & T. H. Schiebler, 1979. Enzyme histochemistry. Springer-Verlag, N.Y. 339 pp.

Lopez, G. R. & J. S. Levinton, 1978. The availability of microorganisms attached to sediment particles as food for Hydrobia ventrosa Montagu (Gastropoda: Prosobranchia). Oecologia 32: 263–275.

McDowell, E. M., 1978. Fixation. In A. Trump & J. Jones (eds.), Diagnostic electron microscopy. J. Wiley & Sons, N.Y.: 113–139.

Michel, C. & E. J. DeVillez, 1978. Digestion. In P. J. Mill (ed.), Physiology of annelids. Academic Press, N.Y.: 509–554.

Nielsen, C. O., 1962. Carbohydrases in soil and litter invertebrates. Oikos 13: 200–215.

Palka, J. & E. A. Spaul, 1970. Studies of feeding and digestion in the enchytraeid worm Lumbricillus lineatus Müll. in relation to its activity in sewage bacteria beds. Proc. Leeds phil. lit. Soc. 10: 45–59.

Pandian, T. J., 1975. Mechanisms of heterotrophy. In O. Kinne (ed.), Marine ecology, 2, 1. J. Wiley & Sons, N.Y.: 61–249.

Pearse, A. G. E., 1972. Histochemistry: theoretical and applied, 2. 3rd ed. Churchill Livingstone, Edinb., 761–1518.

Peters, W., 1968. Verkommen, Zusammensetzung und Feinstruktur peritrophischer Membranen im Tierreich. Z. Morph. Tiere 62: 9–57.

Reger, J. F., 1967. A fine structure study on the organization and innervation of pharyngeal glands and associated ciliated epithelium in the annelid Enchytraeus albidus. J. Ultrastruct. Res. 20: 451–461.

Richards, K. S., 1977. Structure and function in the oligochaete epidermis (Annelida). Symp. zool. Soc. Lond. 39: 171–193.

Richards, K. S., 1980. The histochemical and ultrastructure of the coelomocytes of species of Lumbricillus, and observations on certain other enchytraeid genera (Oligochaeta: Annelida). J. Zool., Lond. 191: 557–577.

Ude, J., 1971. Untersuchungen zur Ultrastruktur des Septaldrüsensystems von Pachydrilus lineatus (Annelida-Oligochaeta). Acta histochem., Suppl. 11: 217–226.

Ude, J., 1975. Elektronenmikroskopisch-enzymhistochemische Untersuchungen zum Sekretionsprozess in den Septaldrüsenzellen von Pachydrilus lineatus. Acta histochem., Suppl. 15: 177–186.

Ude, J., 1977. Licht- und elektronenmikroskopische Untersuchung des Septaldrüsensystems von Pachydrilus lineatus (Annelida-Oligochaeta). Zool. Jb. Allgem. Zool. Physiol. Tiere 81: 42–82.

Van Gansen, P., 1962. Structures et fonctions du tube digestif du lombricien Eisenia foetida Savigny. Ann. Soc. r. zool. Belg. 93: 1–121.

Vierhaus, H., 1971. Über peritrophische Membranen und andere chitinhaltige Strukturen bei Anneliden, unter besonderer Berücksichtigung von Lumbricus terrestris L. Doct. Diss.

Whitlatch, R. B., 1981. Animal-sediment relationships in intertidal marine benthic habitats: Some determinants of deposit-feeding species diversity. J. exp. mar. Biol. Ecol. 53: 31–45.

Whitlatch, R. B. & R. G. Johnson, 1974. Methods for staining organic matter in marine sediments. J. sedim. Petrol. 44: 1310–1312.

The gutless oligochaete *Phallodrilus leukodermatus* Giere, a tubificid of structural, ecological and physiological relevance

O. Giere[1], H. Felbeck[2], R. Dawson[3] & G. Liebezeit[3]
[1] *Zoologisches Institut und Museum der Universität Hamburg, Martin-Luther-King-Platz 3, D-2000 Hamburg 13, Federal Republic of Germany*
[2] *Scripps Institution of Oceanography, A-002, La Jolla, CA 92093, U.S.A.*
[3] *Sonderforschungsbereich 95, Universität Kiel, Olshausen Str. 40/60, D-2300 Kiel 1, Federal Republic of Germany*

Keywords: aquatic Oligochaeta, gutless oligochaetes, structure, ecology, physiology

Abstract

Phallodrilus leukodermatus is not only characterized by the complete absence of mouth, gut, anus and nephridia, but also by an exceptional dermal ultrastructure which is associated with gram-negative bacteria. The vertical distribution of the worms from Bermudian carbonate sands is also unusual in attaining population maximum at oligoxic or anoxic depths around the redox discontinuity (RPD) layer, where extremely high concentrations of amino acids and sugars are to be recorded. Based on results from current ecophysiological and ultrastructural studies, an interpretation of the unique biology of the worms is attempted.

Introduction

Described from calcareous sublittoral sands around Bermuda (Giere, 1979), *Phallodrilus leukodermatus* was one of the first free-living oligochaetes known to lack intestinal structures including mouth and anus. Although other gutless phallodriline tubificids have been described (Jamieson, 1977; Erséus, 1979, 1981; Erséus & Baker, 1982), *P. leukodermatus* is the only species which is intensively investigated in a multidisciplinary approach integrating morphological, ultrastructural, ecological and physiological aspects. This paper will compile results from this project which started in Bermuda in 1980. In addition, some preliminary data from current evaluation are briefly reported here.

Material and methods

The large number of specimens needed particularly for ecological and biochemical studies and uptake experiments was obtained from Flatts Inlet, Bermuda, an area described in detail by Giere *et al.* (1982). The methods used in the various investigations originating from different disciplines are too diverse to be listed here. They have been reported by Giere (1981) for the morphological part, by Giere *et al.* (1982) for the ecological and chemical work, and by Felbeck (1981) and Felbeck *et al.* (1981) for the enzyme physiology.

Results

The outstanding morphological features of *P. leukodermatus* are the absence of the gut, mouth and anus, two well-developed blood vessels and the nerve cord traversing the spacious but septate coelomic cavity, and complete reduction of nephridia (Fig. 1). Fluffy call digitations emerging from the vascular endothelium are best shown in cross sections (see Giere, 1981: Fig. 1). These structures more or less fill the coelom and contain an unusually 'active' cytoplasma with numerous mitochon-

Hydrobiologia 115, 83–89 (1984).

84

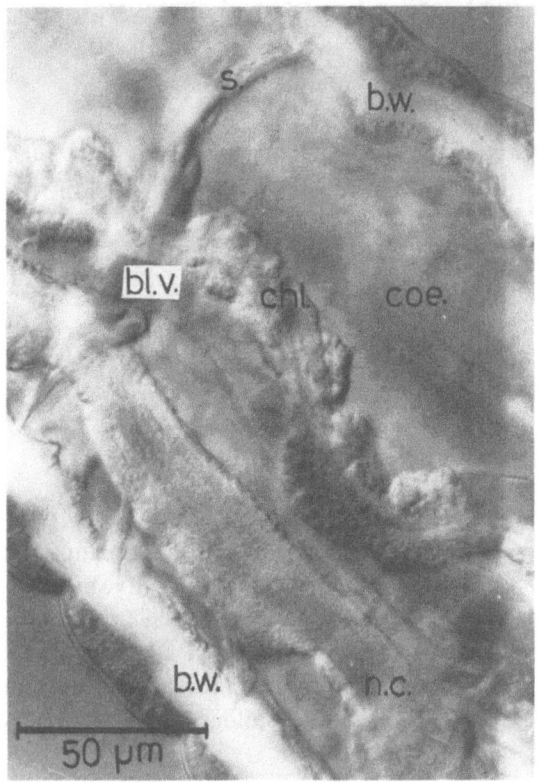

Fig. 1. Lateral view of *Phallodrilus leukodermatus* (bl.v. = blood vessel, b.w. = body wall, chl. = chloragocytes, coe. = coelomic cavity, n.c. = nerve cord, s. = septum); live specimen, microflash, interference contrast.

dria, Golgi-apparatus and a well-developed rough endoplasmatic reticulum.

In the thick body wall of the worms, a tenuous outer ring of circular muscle fibres surrounds a solid inner layer of longitudinal muscle cells. Embedded in this are conical portions of musculature, a feature quite unusual in annelids. They contain many mitochondria. A muscular sheath also encloses dorsally the thick nerve cord.

The integument covering the long, slender, much annulated body of *P. leukodermatus* (ca 20–30 mm long, ca 0.15 mm wide) is a solid, conspicuous 'mantle'. Its surface, often folded into numerous tiny irregular ridges, is covered by countless epicuticular projections (nomenclature according to Richards, 1977) which increase the surface area of the worm by approximately ten times (Giere, 1981: Fig. 4). They emerge from digitated ramifications of the numerous long and irregular microvilli which cross the orthogonal grid of cuticular fibres and the loose matrix of the epicuticle (Fig. 2).

While the cuticular structure is fairly typical for many annelids (Storch & Welsch, 1970; Richards, 1977; Jamieson, 1981), the great distance between the irregular surface of epidermal cells and the outer cuticle is unique to *P. leukodermatus*. Scattered between the cytoplasmatic strands and microvilli extending from the epidermis are abundant large globular membrane-bound bodies (ca 3 μm in length), filled with numerous vesicles (Multi-Vesiculate Bodies, MVBs) (Fig. 3; see also Giere, 1981: Fig. 6). Their content, possibly in a liquid to gelatinous state, is unknown. They lack any organelles typical for animal cells. In some instances their membrane appeared double- or treble-layered, thus resembling that of gram-negative bacteria. However, a Feulgen-test failed to prove the existence of chromatin material (test made by K. S. Richards). Due to their vesiculate content, which often appears strikingly similar to that in the bacteria (see below), these corpuscles are light-refringent and cause the chalkiness of the worm; their distribution along the animal's body (they are absent from the anteclitellar and terminal segments) clearly coincides with the white coloration (compare also *P. albidus;* Richards *et al.,* 1982). The population consists, however, of two forms, co-existing in all samples: the majority are 'whites' resulting from the dense filling of their integument with lucent MVBs, while 10–20% of the worms are pale grey as they contain fewer MVBs with fewer lucent vesicles.

The distal MVBs are regularly surrounded by smaller particles which are particularly frequent directly underneath the cuticle, where they form a distinct layer. Membrane structure, staining characteristics by Gram- and Feulgen-methods and enzyme analyses proved these particles to be gram-negative bacteria with a pronounced enzymatic activity (Figs 4–5; see Giere, 1981: Figs 5 & 7). Although the bacteria occurred in all of the specimens observed, even in the youngest of the juveniles, stages of bacterial cell division were only rarely seen.

Among the array of peculiarities in *P. leukodermatus,* the absence of nephridia is striking. In sections mainly taken though the lateral body walls, numerous vesicles were found aggregated in the epicuticular layer. Membrane-bound bubbles have

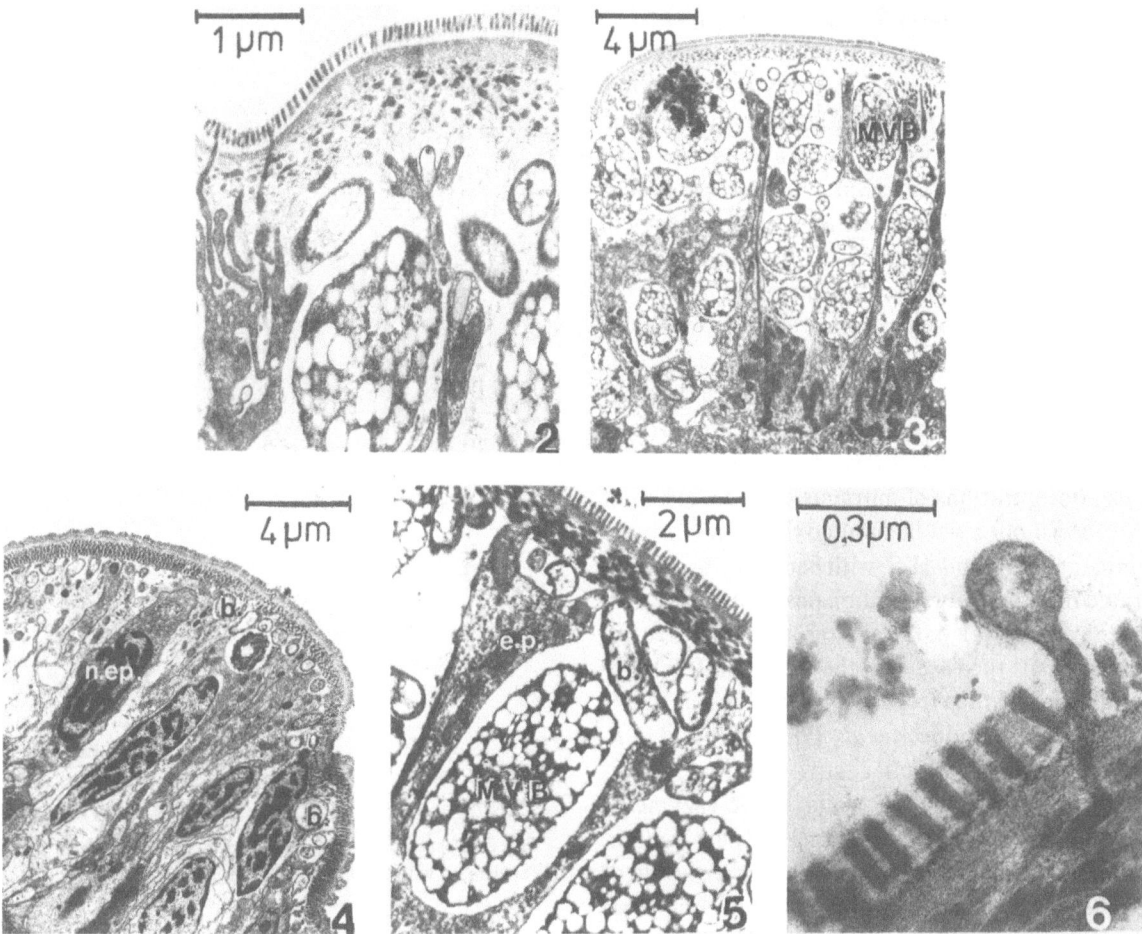

Fig. 2–6. Body wall of *Phallodrilus leukodermatus,* transmission electron microscope micrographs. 2. Microvilli (originating from ramifications of epidermal cell extensions) traversing the cuticular layers; cross-section. 3. Integument with many 'multi-vesiculate bodies' (MVBs) between plasmatic funicles of epidermal cells. 4. Integument with numerous subcuticular bacteria (b. = bacteria, n.ep. = nuclei in epidermal cell); oblique cross-section. 5. Detail of integument with multi-vesicular bodies (MVBs) between epidermal cell extensions (ep.) and subcuticular bacteria (b.); cross-section. 6. 'Bubble' with metabolic wastes (?) excreted through epicuticle; cross-section.

also been encountered between the epicuticular projections (Fig. 6).

Our ecological studies focussed on (1) the question of food uptake in these mouth- and gutless organisms, and (2) their microdistribution, particularly their occurrence in the deeper H_2S-layers. The substratum at the main sampling station consisted of little-sorted carbonate sand(s), median-values 180–240 μm with a sorting coefficient of 2.2. The pore water contained extremely high concentrations of sugars (mostly glucose) and fairly high levels of dissolved, free amino-acids compared to other similar locations (mostly 6–10 mg GLC-equi-

valents total carbohydrates and 150 μmol amino acids l^{-1}; for details see Giere *et al.,* 1982). In this environment with a rich supply of algal debris, up to 85 000 worms m^{-2} could be found, whereas the offshore coralline sands, poorer in organic substances, harboured only a few specimens. Distribution analysis and uptake experiment also demonstrated that *P. leukodermatus* utilizes these organic solutions as important nutrients by transepidermal absorption. The vertical concentration pattern of these substances, with a maximum usually around a depth of 5 cm, agrees fairly well with the distribution of the worms in the sediment column (Giere *et al.*, 1982: Fig. 3). With the H_2S gradient also increasing to high concentrations at a depth of 5 cm, the occurrence and activity of these worms in these often anoxic layers are ecologically as exceptional as their morphology: while other interstitial annelids are known to be very demanding of oxygen, the viability of *P. leukodermatus* in survival experiments was not restricted either by several days of complete anoxia or by strong H_2S-concentrations. This apparent independence from oxygen and hydrogen sulphide concentrations was the case in both juvenile and mature worms, in 'whites' as well as 'pales'. This explains why, despite anoxia and the presence of H_2S, the aggregations of worms around or underneath the RPD-layer can be parallelled to high concentrations of nutrients at these horizons.

Life without a gut in an oligoxic or anoxic environment containing H_2S, with bacteria closely integrated into the body structure, poses questions as to the physiological pathways the animals have invented. As is the case with the vestimentiferan pogonophore *Riftia* from deep-sea thermal vents (Felbeck, 1981; Felbeck *et al.*, 1981) we found in *P. leukodermatus* substantial activity of ribulose-1,5-biphosphate-carboxylase, a diagnostic enzyme of the Calvin-Benson cycle existing only in bacteria and plants. The presence of high activity levels of ATP-sulphurylase (10–20 International Units) and of sulphite oxidases also indicates that enzymatic sulphur metabolism is performed by the bacteria. Better understanding of the ecophysiology of *P. leukodermatus* emerges from uptake experiments. From presently evaluated data it can be stated that, among hexoses, mainly glucose is absorbed (ca 150 μg GLC l^{-1} mg^{-2} dry wt h^{-1}; Fig. 7), while pentoses seem to be neglected. Among the free amino-acids, aspartate is preferred with the highest uptake rates in combination with glutamic acid. Solutions of these 'favored' substances were immediately incorporated, whereas other compounds required an initial lag-phase of some hours before uptake started. It is probable that here, prior to uptake, a transport mechanism has to be activated, e.g. by special proteins. Thereafter, uptake rates (unpubl. data) are well in the order of magnitude found in pogonophorans (Southward & Southward, 1981).

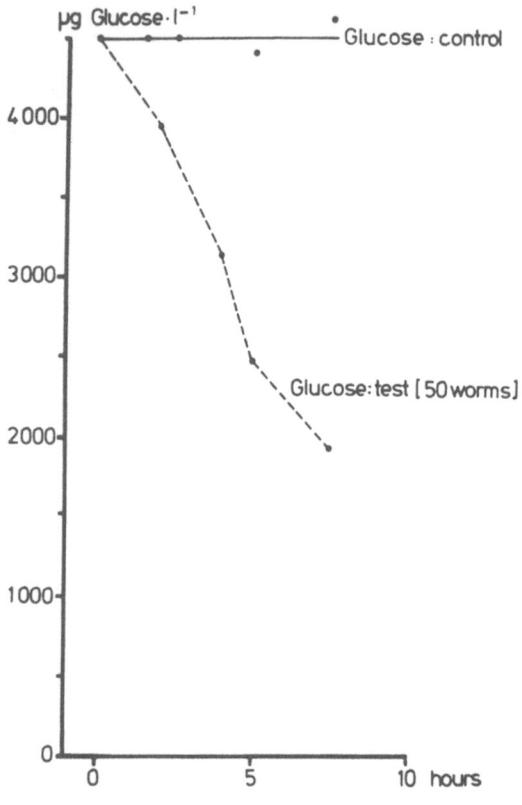

Fig. 7. Decrease of external glucose concentration in uptake experiment with *Phallodrilus leukodermatus*.

Discussion

Since many of the morphological, ecophysiological and chemical data given here are derived from current, not yet fully evaluated investigations, their interpretation has to be of a tentative nature.

Structural relations of Phallodrilus leukodermatus *to* P. albidus

Among the other numerous gutless oligochaetes so far described (see Erséus, 1979, 1981; Erséus & Baker, 1982), only *P. albidus* has been histologically studied (Richards *et al.*, 1982). There are apparent coincidences between this Australian species and *P. leukodermatus:* subcuticular gram-negative bacteria, absence of nephridia, increased density of epicuticular projections, even similar problems in resolving the tri-lamellar structure of the cell wall in gram-negative bacteria (see also Heckmann, 1975). However, a closer look reveals many differences between *P. albidus* and *P. leukodermatus.* The first species is characterized by different bacteria (based on cell size and cytoplasmatic structure) bearing 'filiform extensions', and by a diverging arrangement of body musculature and blood vessels. Fluffy cell structures filling the coelomic cavity seem to be restricted to *P. leukodermatus* (K. S. Richards, pers. commun.). In the latter species, a wide epidermis/cuticle interface, traversed by long cytoplasmic extensions and filled with numerous MVBs and bacteria, develops a thick 'mantle' underneath the cuticle. The MVBs in *P. leukodermatus* are larger ($2 \times 3.2 \mu$m) than the 'multigranular bodies' (1.4–1.9 μm in diameter) in *P. albidus.*

MVBs are of very heterogeneous electron-density and contrast in micrographs, often even visible in the same EM-section. Their varying number and aspects certainly indicate a considerable metabolic relevance for *P. leukodermatus* (further details on their origin, fate and function will be published elsewhere).

The reduction of nephridia in *P. albidus* has already led Richards *et al.* (1982) to speculate about some physiological consequences. In *P. leukodermatus,* production of metabolic wastes seems to include urea, since analyzes showed a distinct increase in urea concentration in the ambient test water (15 ng mg^{-1} dry wt h^{-1}). Thus, the amount of liquid metabolic wastes could be considerably reduced, enabling the worms to eliminate them by the observed numerous epidermal vesicles and epicuticular 'bubbles' (see Fig. 6).

Ecophysiological aspects

Enzyme studies suggest that *P. leukodermatus,* through its bacteria, is able to use ATP and the reducing capacity originating from sulphur oxidation for the fixation of inorganic CO_2 from the ambient water. This process corresponds with the recently found metabolic pattern in the huge vent worm *Riftia* (Felbeck, 1981; Williams *et al.,* 1981). Whether sulphide oxidation in *P. leukodermatus* should be interpreted mainly as a means of detoxifying the poisonous H_2S in the body tissues (see Powell *et al.,* 1979, 1980) or of utilizing this energy source for its metabolic needs (Felbeck, 1981), remains as yet undecided. In any case, it could explain why *P. leukodermatus* can actively live in sediments strongly smelling of H_2S, an amazing capacity, un-

parallelled, to our knowledge, by most other annelids (Giere & Pfannkuche, 1982).

Thus, the population can benefit from the rich supply of dissolved organic nutrients at or below the RPD-layer (Giere *et al.,* 1982), thereby exploring an ecological niche so far unoccupied by other 'higher' interstitial animals. No ecological remarks on the biome of the gutless annelids from Pacific reef sands were given by Jouin (1979) or Richards *et al.* (1982). Neither is pertinent information about the microhabitat available from the few ecological comments in Jamieson (1977).

It is questionable that *P. leukodermatus* is highly dependant on its bacterial metabolism, closely connected to its autotrophic capacity. This is concluded from (1) the small area that the bacteria occupy in their host's body compared to the massive 'throphosome' full of prokaryonts in the Vestimentifera, (2) the relatively small portion of subcuticular bacteria observed by INT-staining (Iodo-Nitro-Tetrazolium-Violet; see Zimmermann *et al.,* 1978) to be metabolically active at a time, and (3) the fact that the ability to absorb dissolved organics via the integument is also well developed in 'normal' aquatic oligochaetes with a gut (Siebers, 1976; Giere & Pfannkuche, 1982). Therefore, in the overall metabolism of the worms, the bacterial contribution is held to be of subordinate rank.

Considerations on a thiobiotic life

One of the ecologically most relevant points, the association of *P. leukodermatus* to a thiobios, a debated community in the meiobenthos, needs to be discussed. If we define the thiobiotic environment from its etymological root primarily as a biome with an ample supply of sulphides and only secondarily as a habitat with little or no access to oxygen, i.e. if we stress the repellant effect or the energetic value H_2S may have, *P. leukodermatus* certainly is a thiobiotic animal capable of surviving for weeks in highly concentrated H_2S, independent of oxygen. Obligatory anaerobiosis as a criterion for the thiobios is ecologically not very meaningful since at the RPD-chemocline it is the sulphur metabolism itself which makes possible a rich microbial life. These bacteria are, in turn, the trophic basis of all thiobionts. Moreover, the strict separation of aerobic and thiobiotic life stems from the former concept that oxygen and H_2S are mutually exclusive compounds, an approach hardly tenable in the light of results from sediment microchemistry (see Boaden, 1977; Fenchel & Riedl, 1970; Jørgensen, 1977; Kepkay & Novitsky, 1980; Powell & Bright, 1981).

The thiobios, in the above sense, belongs to the biotas with 'extraordinary energy sources' (Reid, 1980), where many inhabitants seem to benefit energetically from H_2S and other sulphides. Here, gutless animals play a characteristic role: pogonophorans in mud or around hydrothermal vents, the bivalve *Solemya* from sewage sludge and, last but not least, *P. leukodermatus* from sediments rich in reduced detritus. Powell & Bright (1981) claim that 'thiobiotic communities probably depend on prokaryotic, sulphur-dependent organisms'. This would be confirmed by the presence of bacteria in all these gutless forms, including *P. albidus* and *P. planus* where activity of ribulose-diphosphate-carboxylase first revealed occurrence of bacteria, meanwhile confirmed by EM-micrographs (unpublished data). Histological and ecological information is still too scanty to allow further conclusions to be drawn on other gutless forms.

Phylogenetic views

Following Erséus (1980, 1981), the numerous gutless tubificids are attributed to the wide range of phallodriline organisations with genital features grouping them in several more or less independent and allopatric species-flocks closely allied to 'normal' relatives. Hence, within this subfamily, a convergent and quite recent trend seems to have released food uptake from intestinal structures with a subsequent degeneration of the gut. This would stand in contrast to the situation in the Pogonophora as a well-defined anenteric phylum, but would correspond to the position of the gutless archiannelid *Astomus* (Jouin, 1979) and the bivalve *Solemya* (Reid & Bernard, 1980).

The 'invention' of physiological aid from bacteria might have been the trigger for this degenerative evolution in some specialists. As a result of this metabolic cooperation between bacteria and animals, formerly adverse habitats may have been invaded and new food sources utilized. In oligochaetes, one may conclude that these habitats lie in subsurface horizons of sublittoral carbonate sands in warm climates. They ought to contain high

amounts of dissolved organic substances and much H_2S.

We expect, from further studies on *P. leukodermatus* and other gutless species like *P. planus, P. longissimus* and 'normal' relatives, to provide more evidence for solving the problems and forming hypotheses and to answer a number of other unanswered questions, such as: do the worms depend on H_2S? What favours their occurrence in carbonate sands in warm-water areas? What are the steps in intestinal degeneration (studies on embryos and on species with intestinal vestiges)? How are the bacteria transferred through the generations? What is the significance of the reduction of nephridia?

The basic agreement in some enzymatic reactions between *P. leukodermatus* and the pogonophoran *Riftia,* proven by our working group, demonstrates that even investigation of other, non-related forms can contribute substantial information to these questions. This underlines the importance of a joint approach by various disciplines which is the only way of coping with the scientific problems posed by the anenteric way of life in higher animals.

Acknowledgements

O. G. would like to thank Dr. K. Sylvia Richards, Keele (England), for much help and fruitful discussions. Contribution No. 917, Bermuda Biological Station for Research, Bermuda: Report No. 406, Sonderforschungsbereich 95, Universität Kiel, Kiel, Federal Republic of Germany.

References

Boaden, C., 1977. Thiobiotic facts and fancies (aspects of the distribution and evolution of anaerobic meiofauna). Mikrofauna Meeresboden 61: 45–63.

Erséus, C., 1979. Taxonomic revision of the marine genus Phallodrilus Pierantoni (Oligochaeta, Tubificidae), with descriptions of thirteen new species. Zool. Scr. 8: 187–208.

Erséus, C., 1980. Morphology and taxonomy of the marine tubificid subfamily Phallodrilinae Brinkhurst, 1971 (Oligochaeta). Ph.D. Thesis, Univ. Göteborg, Sweden, 83 pp.

Erséus, C., 1981. Taxonomic studies of Phallodrilinae (Oligochaeta, Tubificidae) from the Great Barrier Reef and the Comoro Islands with descriptions of ten new species and one new genus. Zool. Scr. 10: 15–31.

Erséus, C. & H. R. Baker, 1982. New species of the gutless marine genus Inanidrilus (Oligochaeta, Tubificidae) from the Gulf of Mexico and Barbados. Can. J. Zool. 60: 3063-3067.

Felbeck, H., 1981. Chemoautotrophic potential of the hydrothermal vent tube worm, Riftia pachyptila Jones (Vestimentifera). Science 213: 336-338.

Felbeck, H., J. J. Childress & G. N. Somero, 1981. Calvin-Benson cycle and sulphide oxidation enzymes in animals from sulphide-rich habitats. Nature 293: 291-293.

Fenchel, T. M. & R. J. Riedl, 1970. The sulphide system: a new biotic community underneath the oxidized layer of marine sand bottoms. Mar. Biol. 7: 255-268.

Giere, O., 1979. Studies on marine Oligochaeta from Bermuda, with emphasis on new Phallodrilus-species (Tubificidae). Cah. Biol. mar. 20: 301-314.

Giere, O., 1981. The gutless marine oligochaete Phallodrilus leukodermatus. Structural studies on an aberrant tubificid associated with bacteria. Mar. Ecol. Prog. Ser. 5: 353-357.

Giere, O., G. Liebezeit & R. Dawson, 1982. Habitat conditions and distribution pattern of the gutless oligochaete Phallodrilus leukodermatus. Mar. Ecol. Prog. Ser. 8: 291-299.

Giere, O. & O. Pfannkuche, 1982. Ecology and biology of marine Oligochaeta, a review. Oceanogr. mar. Biol. ann. Rev. 20: 173-308.

Heckmann, K., 1975. Omikron, ein essentieller Endosymbiont von Euplotes aediculatus. J. Protozool. 22: 97-104.

Jamieson, B. G. M., 1977. Marine meiobenthic Oligochaeta from Heron and Wistari Reefs (Great Barrier Reef) of the genera Clitellio, Limnodriloides and Phallodrilus (Tubificidae) and Grania (Enchytraeidae). Zool. J. linn. Soc. 61: 329-349.

Jamieson, B. G. M., 1981. The ultrastructure of the Oligochaeta. Academic Press, N.Y. 462 pp.

Jørgensen, B. B., 1977. The sulphur cycle of a coastal marine sediment (Limfjorden, Denmark). Limnol. Oceanogr. 22: 814-832.

Jouin, C., 1979. Description of a free-living polychaete without gut: Astomus taenioides n. gen., n. sp. (Protodrilidae, Archiannelida). Can. J. Zool. 57: 2448-2456.

Kepkay, P. E. & J. A. Novitsky, 1980. Microbial control of organic carbon in marine sediments: Coupled chemoautotrophy and heterotrophy. Mar. Biol. 55: 261-266.

Powell, E. N. & T. J. Bright, 1981. A thiobios does exist - gnathostomulid domination of the Canyon community at the East Flower Garden brine seep. Int. Revue ges. Hydrobiol. 66: 675-683.

Powell, E. N., M. A. Crenshaw & R. M. Rieger, 1979. Adaptations to sulfide in the meiofauna of the sulfide system. 1. ^{35}S-sulfide accumulation and the presence of a sulfide detoxification system. J. exp. mar. Biol. Ecol. 37: 57-76.

Powell, E. N., M. A. Crenshaw & R. M. Rieger, 1980. Adaptations to sulfide in sulfide-system meiofauna. Endproducts of sulfide detoxification in three turbellarians and a gastrotrich. Mar. Ecol. Prog. Ser. 2: 169-177.

Reid, R. G. B., 1980. Aspects of the biology of a gutless species of Solemya (Bivalvia: Protobranchia). Can. J. Zool. 58: 386-393.

Reid, R. G. & F. R. Bernard, 1980. Gutless bivalves. Science 208: 609-610.

Richards, K. S., 1977. Structure and function in the oligochaete epidermis (Annelida). Symp. zool. Soc. Lond. 39: 171-193.

Richards, K. S., T. P. Fleming & B. G. M. Jamieson, 1982. An ultrastructural study of the distal epidermis and the occurrence of subcuticular bacteria in the gutless tubificid Phallodrilus albidus (Oligochaeta: Annelida). Aust. J. Zool. 30: 327-336.

Siebers, D., 1976. Absorption of neutral and basic amino acids across the body surface of two annelid species. Helgoländer wiss. Meeresunters. 28: 456-466.

Southward, A. J. & E. C. Southward, 1981. Dissolved organic matter and the nutrition of the Pogonophora: a reassessment based on recent studies of their morphology and biology. Kieler Meeresforsch., Sonderh. 5: 445-453.

Storch, V. & U. Welsch, 1970. Über die Feinstruktur der Polychaeten-Epidermis (Annelida). Z. Morph. Tiere 66: 310-322.

Williams, P. M., K. L. Smith, E. M. Druffel & T. W. Linick, 1981. Dietary carbon sources of mussels and tubeworms from Galápagos hydrothermal vents determined from tissue ^{14}C activity. Nature 292: 448-449.

Zimmermann, R., R. Iturriaga & J. Becker-Birck, 1978. Simultaneous determination of the total number of aquatic bacteria and the number thereof involved in respiration. Appl. envir. Microbiol. 36: 926-935.

Asexual propagation and reproductive strategies in aquatic Oligochaeta

Bent Christensen
Institute of Population Biology, Universitetsparken 15, DK-2100 Copenhagen, Denmark

Keywords: aquatic Oligochaeta, asexual reproduction, genetical-ecological mechanisms

Abstract

A great variety of asexual reproductive modes are known among aquatic oligochaetes. The main types of these modes are shortly described. Based upon observations on natural populations, the possible genetical and ecological implications of asexual reproduction are discussed. The following points are emphasized: (1) The often expressed expectations of a strong predominance of one particularly adaptive genotype is not born out. (2) In most cases, a number of genetically distinct clones are present in each population, and they show a strong differential distribution in heterogeneous environments, indicating an effective exploitation of the available resources. (3) Most cases of asexual propagation are reproductive strategies of their own and not escape mechanisms. (4) The mechanisms underlying asexual propagation are complex and involve many aspects of the life history. The great variety of types among aquatic oligochaetes offer particularly useful models for the study of these problems.

Introduction

In the animal kingdom as a whole, asexual reproduction is rare compared to amphimixis. A possible reason for this is the disadvantages thought to be associated with non-recombinational genetic systems. Having forsaken segregation and recombination, asexual organisms have also sacrificed the genetic plasticity thought to be necessary for further evolutionary change. As a consequence, asexual organisms have earned the label of evolutionary dead ends.

Despite the above-mentioned disadvantages, there are many asexually reproducing species that appear to be remarkably successful by most standards. Among aquatic oligochaetes in particular, several instances of extremely common and widespread species or even higher taxons, which have adopted some form of asexual reproduction are known. It is the aim of this contribution to present a short survey of the known reproductive strategies and to discuss their possible genetic and ecological implications.

Survey of findings

The various forms of asexual and parasexual reproduction among aquatic oligochaetes are outlined in the following four sections. They represent an entirely artificial classification with no phylogenetic implications, but an attempt has been made to present a sequence of increasing deviation from normal sexual reproduction.

A. Sub-amphimictic reproduction

The characteristic feature of this mechanism is that some of the chromosomes regularly form bivalents while others appear as univalents at meiosis. Thus one part of the chromosome complement follows the normal meiotic pattern of sexual reproduc-

Hydrobiologia 115, 91–95 (1984).

tion, while the other chromosomes follow a mitotic pathway characteristic of ameiotic parthenogenesis. The cytology of a polyploid form of the enchytraeid *Enchytraeus lacteus* is described by Christensen & Jensen (1964).

A similar reproductive behaviour is found in the tubificid *Aktedrilus* (*Phallodrilus*) *monospermathecus* (Christensen, 1980b), and probably also in the megascolid *Ilyogenia santixavieri* (Černosvitov, 1931).

B. Alternating asexual and sexual reproduction

The alternation of different phases in the life cycle comprising sexual reproduction (amphimixis) and asexual proliferation either through parthenogenetically developing eggs or through fission is known in several groups of animals. In aquatic oligochaetes with different reproductive phases the asexual proliferation is through transverse fission. In the well-known cases of naidids and aelosomatids the separation of new individuals is preceded by the formation of a budding or fission zone where a new anterior end is formed before the individuals separate (Paratomy). Organogenesis in the fission zone has been studied by several authors; for reviews see Jamiesen (1981) and Christensen (in press). In the above mentioned families fission is apparently universal and this type of reproduction has undoubtedly arisen before, or very early, in the evolution of the taxa in question. But other instances of asexual fission among aquatic oligochaetes have clearly arisen as a secondary phenomenon in groups with normal sexual reproduction and it is a remarkable coincidence that in these cases a different type of transverse fission is found. The individual suddenly divides into fragments caused by vigorous contractions of the body wall, and such fragments are completed through a subsequent formation of new anterior and posterior ends (Architomy). Such independently arising cases of asexual fission are known in several families, notably Lumbriculidae, Enchytraeidae and Tubificidae, see Christensen (in press). Organogenesis has been studied by Stephan-Dubois (1956) and Christensen (1964).

C. Parthenogenetic reproduction

Parthenogenetic reproduction probably occurs within all major groups of aquatic oligochaetes, but cytologically verified cases are only known within Tubificidae and Enchytraeidae. Conventionally, a distinction is made between meiotic and ameiotic parthenogenesis. In the former, meiosis still occurs in the developing oocyte, but is compensated for by a doubling of the chromosome number at some stage. In the latter, meiosis is in various ways suppressed, the maturation division(s) being mitotic in character.

Several types of meiotic parthenogenesis can be distinguished. In the least specialized case meiosis is essentially normal and the somatic chromosome number is restored by fusion of the female pronucleus and the second polar body. This mode of restoration is not yet recorded in aquatic oligochaetes, but has been described in a few terrestrial enchytraeid species (Christensen, 1961). The most characteristic type of meiotic parthenogenesis among aquatic oligochaetes, and for that matter among oligochaetes in general, is one involving a premeiotic doubling of the chromosome number. Several cases have been demonstrated among earthworms of the family Lumbricidae (Omodeo, 1952, 1955) and within the aquatic family Tubificidae, notably in the genera *Limnodrilus* and *Tubifex* (Christensen, 1980b).

The cytology of ameiotic parthenogenesis is in most cases very simple. There is no synapsis, no bivalents are formed, and only a single maturation division, which is an ordinary mitosis, occurs in the oocyte. Polyploid forms of the earthworm *Dendrobaena octaedra* (Omodeo, 1955) are known to follow this simple pattern, while the polyploid forms of the littoral enchytraeid *Lumbricillus lineatus* show a much more complicated mode of ameiotic parthenogenesis (Christensen, 1960, 1980b).

D. Propagation exclusively through fission

In a sense, species which reproduce by asexual fission only represent the greatest deviation from normal sexual reproduction because they have entirely abolished propagation through gametes. However, apart from that, the group as such is not very well demarcated. The modes of fission are the same as those found under group B, and in addition, the abolishing of sexual reproduction would seem to be a gradual phenomenon. In the typical case, the alternation of asexual and sexual reproduction follows a regular annual cycle in naidids and aelosomatids, but in some species sexual reproduction is very rare and apparently somewhat erratic in its occurrence. In another instance, differences in the expression of sexual reproduction may be found even within the same nominal species. The lumbriculid *Lumbriculus variegatus* includes a diploid, as well as various polyploid biotypes (Christensen, 1980a). In the diploid form, gametogenesis is normal, and even though the frequency of sexually mature individuals is low, sexual reproduction definitely takes place. However, in some odd-numbered polyploids, meiosis is so abnormal that asexual fission must be the only possible reproductive mechanism. Future studies will undoubtedly reveal additional cases of this nature.

Discussion

In the introduction, some possible disadvantages associated with asexual reproduction were mentioned. However, since most (all?) of the above cases represent evolutionary novelties that have managed to establish themselves in competition with pre-existing sexually reproducing species, certain positive consequences of asexual reproduction may also be assumed.

One such feature is that asexual reproduction can be expected to fix particular adaptive combinations of genes and perpetuate them *ad infinitum*. Such genotypes are not 'broken down' each generation through segregation and recombination as is the case in normal sexual organisms, with the production of adaptively inferior individuals as an inevitable result. Species with an alternation of reproductive modes can be expected to exploit this possibility during the asexual phase. Is this the case? Does selection during the asexual phase enhance the frequency of the better adapted genotypes and depress the frequency of, or even eliminate, those less well-adapted? This possibility was studied in the naidid, *Stylaria lacustris*, see Christensen (in press). In the offspring of the sexual generation, the various genotypes are expected to occur in Hardy-Weinberg proportions, but if the above mechanism operates, selection would, during the asexual phase, favour and propagate the genotypes especially suitable for the environment in question. Consequently, strong deviations from Hardy-Weinberg proportions are

expected among the individuals entering a new sexual phase. However, the actual observations showed that the frequencies of genotypes at this stage in the life cycle agreed with those expected according to the Hardy-Weinberg law. Thus, no changes in frequencies occurred in the preceding asexual phase, indicating that *Stylaria*, with respect to the two loci investigated, have not experienced radical increases in particularly adaptive genotypes during the asexual propagation.

The same question applies to the species that have abolished sexuality altogether. Observations on *Lumbricillus lineatus* are relevant in this context. Parthenogenesis is here of the ameiotic type; and since segregation does not occur, mutations will tend to accumulate indefinitely. The expectation is therefore a strong dominance of one (or very few) highly heterozygous genotype(s). Actual observations do not bear out this expectation. In the polyploid parthenogenetic forms of *L. lineatus,* the degree of heterozygosity is not very high and in most habitats a number of different genotypes are present (Christensen, 1980a; Christensen *et al.*, 1976, 1978). Thus, a genetic explanation is not readily available in this case either.

The arguments presented above rest on the assumption that the adoption of asexual propagation is an advantage to the organisms in question. However, the opposite view has also been taken: that it is a defensive measure. The strong association between asexual reproduction and polyploidy seen in many animal groups has led to the view that asexual reproduction is merely an 'escape' mechanism, because polyploidy often causes a breakdown of normal gametogenesis. Accordingly, polyploidy should arise first and asexual reproduction next. In organisms with alternating reproductive phases this idea is of course not valid. At least in naidids, gametogenesis is completely normal, and furthermore this family is unique among oligochaetes in the absence of known cases of polyploidy (Christensen, 1980b). But in other aquatic oligochaetes, the association between polyploidy and asexual reproduction is definitely strong. Is asexual reproduction an escape from sterility in these cases? Apparently not. The most common types of parthenogenesis among oligochaetes are also known to occur in diploids, and even in the diploid form of *Lumbriculus variegatus* asexual fission is the principal reproductive method. Thus most cases of both parthenogenesis and asexual fission are apparently reproductive strategies of their own and not 'escape' mechanisms.

The onset of asexual reproduction thus seems to be a complicated phenomenon apparently involving other aspects of the life history than purely genetical ones. Christensen (1973) has shown that, for the enchytraeid *Enchytraeus bigeminus,* sexual reproduction is suppressed in dense cultures of this species so that asexual fission becomes the only reproductive method. Furthermore, dense populations of *E. bigeminus* also suppress sexual reproduction in some potential competitors which are obligatory sexual breeders. This indicates that the adoption of asexual reproduction in this particular case may be an element in the competitive interactions with related oligochaete species, but this is probably a unique situation rather than a universal explanation. In particular, it would seem very difficult to explain the regular, yearly cycles of reproductive modes seen as a fundamental feature in naidids and aelosomatids in this way.

The above mentioned observations on polyploid parthenogenetic *Lumbricillus* (Christensen, 1980a) also involved studies on the micro-distribution of different cyto- and genotypes in a heterogeneous environment. The habitat in question was a short transect (approx. 4.5 m in length) extending from the lowest to the highest tide livel at a particular locality on the west coast of Jutland. Tri-, tetra- and pentaploid cytotypes showed a highly differential distribution, and, within tri- and tetraploids, different electrophoretic variants also showed a differential distribution over very short distances. Since this pattern was constant both in space and time, it would seem as though particular genetic variants within this species constantly exploit particular subniches in this heterogeneous environment. The extremely sharp partitioning observed may have its background in the parthenogenetic mode of reproduction, as combinations of genes that may be adaptations to particular subniches are not broken down by segregation and recombination as is the case in sexual organisms. This ability to develop constantly breeding genetic variants specially adapted to particular subniches may give asexual reproduction an advantage over sexuality, since, in heterogeneous environments, this may result in a more efficient exploitation of the available resources. Again, this is not a universal mechanism, but it may form an important element in a final explanation.

From the presentation above one might gain the impression that subamphimixis represents a first step from sexuality towards asexual reproduction. The situation is probably the opposite. *Enchytraeus lacteus*, at least, is originally an odd-numbered polyploid (and as such an obligatory parthenogen), in which the chromosomes that follow the normal sexual pattern have arisen *ex novo* as a secondary phenomenon. To date no genetic studies have been made on this unique reproductive method, which represents a reversal from asexual reproduction towards sexuality. Hopefully, future research will remedy this; further illumination is awaited.

In conclusion: aquatic oligochaetes have adopted a variety of different reproductive strategies. This makes them extremely useful models for studying the mechanisms underlying the evolution of genetic systems. They will also provide insight into many other biological problems that are best studied in clonally reproducing organisms. The scarcity of general conclusions reached here shows that this work has merely begun.

References

Černosvitov, L., 1931. Zoologische Ergebnisse der Reise des Dr. Storkan nach Mexiko 2. Über die Entwicklung der dimorphen Spermien und die erste Reifungsteilung der Eier bei Ilyogenia santixavieri Eisen. Z. Zellforsch. 12: 53–65.

Christensen, B., 1960. A comparative cytological investigation of the reproductive cycle of an amphimictic diploid and a parthenogenetic triploid form of Lumbricillus lineatus (O.F.M.) (Oligochaeta, Enchytraeidae). Chromosoma 11: 365–379.

Christensen, B., 1961. Studies on cyto-taxonomy and reproduction in the Enchytraeidae. Hereditas 47: 387–450.

Christensen, B., 1964. Regeneration of a new anterior end in Enchytraeus bigeminus (Enchytraeidae, Oligochaeta). Vidensk. Meddr. dansk naturh. Foren. 127: 259–273.

Christensen, B., 1973. Density dependence of sexual reproduction in Enchytraeus bigeminus (Enchytraeidae). Oikos 24: 287–294.

Christensen, B., 1980a. Constant differential distribution of genetic variants in polyploid, parthenogenetic forms of Lumbricillus lineatus (Enchytraeidae, Oligochaeta). Hereditas 92: 193–198.

Christensen, B., 1980b. Annelida. In B. John (ed.), Animal cytogenetics 2. Gebrüder Borntraeger, Berl., Stuttgart: 1–79.

Christensen, B., in press. Asexual propagation and reproductive strategies in Annelida-Clitellata. In K. G. Adiyodi & R. G. Adiyodi (eds.), Reproductive Biology of Invertebrates, J. Wiley & Sons, N.Y.

Christensen, B., J. Jelnes & U. Berg, 1976. A comparative study on enzyme polymorphisms in sympatric diploid and polyploid populations of Lumbricillus lineatus (Enchytraeidae, Oligochaeta). Hereditas 84: 41–48.

Christensen, B., J. Jelnes & U. Berg, 1978. Long-term isozyme variation in parthenogenetic polyploid forms of Lumbricillus lineatus (Enchytraeidae, Oligochaeta) in recently established environments. Hereditas 88: 65–73.

Christensen, B. & J. Jensen, 1964. Sub-amphimictic reproduction in a polyploid cytotype of Enchytraeus lacteus Nielsen and Christensen (Oligochaeta, Enchytraeidae). Hereditas 52: 106–118.

Jamieson, B. G. M., 1981. The ultrastructure of the Oligochaeta. Academic Press, Lond., N.Y., 462 pp.

Omodeo, P., 1952. Cariologia dei Lumbricidae. Caryologia 4: 178–275.

Omodeo, P., 1955. Cariologia dei Lumbricidae 2 Contributo. Caryologia 8: 135–178.

Stephan-Dubois, F., 1956. Migration et différenciation des neoblastes dans la régénération anterieure de Lumbriculus variegatus. C.r. Séanc. Soc. Biol. 150: 1239–1242.

Life cycles and reproductive adaptations of marine Oligochaeta from European boreal shores

Olaf Pfannkuche
Institut für Hydrobiologie und Fischereiwissenschaft, Univ. Hamburg, Zeiseweg 9, D-2000 Hamburg 50, Federal Republic of Germany

Keywords: aquatic Oligochaeta, reproductive strategy, life cycle

Summary

Although European boreal shores are inhabited by only a relatively small number of oligochaete species, the life histories and reproductive strategies of these species vary considerably.

Growth in naidid populations mainly depends on asexual reproduction (paratomy, fragmentation) within a short period of individual life spans, whereas sexual outbreeding plays only a minor role. Naidid populations are characterized by considerable fluctuations of abundance.

Aquatic and mesopsammic tubificids represent a 'conservative' type of sexual reproduction within well defined breeding periods. Population structure is fairly stable.

Enchytraeids show a variety of improvements of the sexual outbreeding scheme which lead to an amazing plasticity. Other forms of reproduction such as parthenogenesis and fragmentation seem to be of greater importance in enchytraeids.

The flexible adaptability of development and reproductive capacity in naidids and enchytraeids is mainly controlled by habitat conditions and population size.

A complete review of the biology and ecology of marine Oligochaeta is published in: Giere, O. & O. Pfannkuche, 1982. Oceanogr. mar. Biol. ann. Rev. 20: 173–308.

Hydrobiologia 115, 96 (1984).
© Dr W. Junk Publishers, Dordrecht.

Parthenogenesis in Tubificidae

Tamara L. Poddubnaya
Institute of Biology of Inland Waters, U.S.S.R. Academy of Sciences, Borok, Nekouf, Jaroslavl 152742, U.S.S.R.

Keywords: aquatic Oligochaeta, Tubificidae, parthenogenesis, experiment

Abstract

The ability of tubificids to reproduce parthenogenetically following or in place of bisexual reproduction has been proved. During parthenogenesis the spermatogenesis ceases at an early stage of sexual cell development, which, together with some peculiarities of the structure of the sexual system, rules out the possibility of self-fertilization in the worms. For two years of cultivation the life cycles of 3 parthenogenetic generations in *Tubifex tubifex* and 2 in *Limnodrilus hoffmeisteri* were traced.

Parthenogenesis in oligochaetes has been studied in detail in the families Lumbricidae and Enchytraeidae. It has been proved as a result that there are two ways of reproduction: 1) bisexual with cross fertilization, found in diploid and, in certain cases, polyploid individuals; 2) parthenogenetic, always associated with polyploidy.

Černosvitov (1927) described self-fertilization in *Limnodrilus udekemianus* Clap. and *Tubifex tubifex* (Müller), and Gavrilov (1935), observing isolated specimens of *L. hoffmeisteri* Clap., *L. claparedeianus* Ratzel and *T. tubifex*, noted the ability of these species to lay cocoons with normally developing eggs without sexual partners. Basing his research on cytological studies of eggs of *L. hoffmeisteri*, Gavrilov could not draw a final conclusion whether eggs develop parthenogenetically or by way of self-fertilization. He never succeeded in finding spermatozoids in the cocoons. Thus, the problem of parthenogenesis in tubificids remained unsolved.

Our investigations on reproduction in tubificids, carried out over many years, have shown too that isolated individuals can reproduce for a long time and give viable offspring. The question arose as to whether this is bisexual reproduction with self-fertilization or parthenogenesis. According to the literature we could expect either. In our case we could assume parthenogenesis. Since Christensen & O'Connor (1958) pointed out that, in a parthenogenetic triploid form of *Lumbricillus lineatus* (Müller) grown in isolation, the process of spermatogenesis is not completed and ceases at the spermatid stage, we paid particular attention to spermatogenesis. In order to cast more light on this process, we analysed the development and functioning of the sexual apparatus of the worms grown alone or within a group. At the same time, we considered the fecundity of the worms reproducing parthenogenetically and assessed the viability and capacity for reproduction of their offspring.

Primary sexual cells appear in worms as early as during the embryonic period of ontogenesis. The gonads appear by the end of the first month of life (at 20 °C). The testes develop first and the ovaries after 3–5 days. At this time the testes increase considerably in size and begin to function. From the testes, groups of cells consisting of spermatogonia separate and pass into the seminal vesicles, where they continue to develop. Spermatogenesis begins before the complete formation of the sexual system. In the seminal vesicles one can see all the stages of development of male gametes, from 4–16 cell spermatogonia to mature spermatozoids irregularly scattered over the cytophores.

By the time of complete development of the sexual system, the function of the testes diminishes but that of the ovaries becomes active. The maximum number of spermatozeugma found in the spermatheca in *L. hoffmeisteri* and *T. tubifex* is 12 (usually 10), i.e., for one period of reproduction the worms can lay not more than 10–12 cocoons. However, according to our observations, each individual can lay a greater number of cocoons (up to 30) during the reproductive period.

From the analysis of spermatogenesis the obvious conclusion may be drawn that a repeated coupling of the worms and replenishment of the spermatozeugma stock is impossible, since the function of the testes is completed and there are no reserves of spermatozoids in the sperm sac. However, after the first part of the cocoons have been laid and after all spermatozeugma from the spermathecae have been used up, there is still a significant number of eggs ready for laying and they continue to pass into the cocoons for a long time. Under natural conditions, if there is a reserve of eggs, it is likely that the worms reproduce parthenogenetically after bisexual reproduction.

Having completed the first period of reproduction, the individual worms either die or show resorption and regeneration of the sexual system. This process does not take a long time, only 20–25 days. The sexual apparatus recovers, intensive spermatogenesis begins and the whole cycle is repeated. All the above features are characteristic of tubificids living in groups.

The development of gonads in isolated individuals shows deviations from the norm. The deviations are found in both *T. tubifex* and *L. hoffmeisteri*, and they are peculiar to the species. It is necessary to emphasize that changes occur only in the course of spermatogenesis and practically do not involve oogenesis.

In *T. tubifex* formation of the gonads proceeds normally in the first parthenogenetic generation. During further development, the testes do not reach their normal dimensions (700–1000 μm) and spermatogenesis proceeds slowly. In the seminal vesicles there are few gametes, most of them at the stage of spermatogonium or spermatocyte. Development stops before reaching spermatid stages. There are practically no mature spermatozoids separated from the cytophore, spermatozeugmata are not formed, spermathecae are always empty. In the second parthenogenetic generation (offspring of the first), the condition of the gonads is the same as in the first one, but the number of spermatogonia in

Table 1. Characteristics of parthenogenetic generations of *T. tubifex* and *L. hoffmeisteri*.

| | Tubifex tubifex | | | | | Limnodrilus hoffmeisteri | | | |
| | Short | | | Long | | Short | | Long | |
Generation	1	2	3	1	2	1	2	1	2
Duration of life cycle, days	95 ± 5.6	106 ± 3.8	–	245 ± 10.0	380 ± 9.9	150 ± 10.1	115 ± 7.0	349 ± 8.7	181 ± 6.4
Duration of maturing, days	52 ± 1.9	58 ± 1.5	94 ± 4.0	60 ± 0.8	72 ± 3.0	93 ± 3.9	62 ± 1.2	90 ± 7.5	66 ± 1.5
Number of reproductive periods	1	1	1	2–3	2–6	1	1	2–5	2–3
Number of eggs	42 ± 4.3	30 ± 3.3	–	101 ± 4.0	92 ± 6.4	55 ± 5.6	75 ± 7.5	223 ± 11.5	88 ± 8.3
Number of individuals in population, %	60	40	10*	40	60	22	31	78	69

*10% of initial number of individuals in experiment.

Table 2. Fecundity in sexual and parthenogenetic reproduction.

Type of reproduction	Tubifex tubifex			Limnodrilus hoffmeisteri		
	Sexual	Parthenogenetic		Sexual	Parthenogenetic	
Generation		1	2		1	2
Number of eggs/cocoon	4 ± 0.2	4 ± 0.5	3 ± 0.8	6 ± 0.5	5 ± 0.3	4 ± 0.4
Number of eggs/month	25 ± 0.8	24 ± 1.2	18 ± 1.5	30 ± 1.0	30 ± 0.9	26 ± 0.5
Number of eggs/whole life of individual	297 ± 3.6	120 ± 4.5	110 ± 3.7	300 ± 5.0	233 ± 2.5	140 ± 4.2

the sperm sacs is still less. In the third parthenogenetic generation (offspring of the second), formation of the gonads is rapidly inhibited, and the spermatogonia are absent.

In *L. hoffmeisteri* a considerable delay in gonad formation was already revealed in the first generation (by the 50th day). Spermatogenesis proceeds slowly and spermatid stage is reached by a small number of gametocytes. In the second parthenogenetic generation of *L. hoffmeisteri* practically everything is repeated.

Our experiments have shown that the life span of reproducing parthenogenetic individuals of *T. tubifex* varies within 70–530 days. During its lifetime in isolation, one specimen can lay from 28 to 128 eggs. It has been established that all the individuals may be clearly divided into two groups: 1) those with a short life cycle, dying after the first period of reproduction, and 2) those with a prolonged life cycle reproducing several times with intervals.

The longevity of *L. hoffmeisteri* individuals which reproduced parthenogenetically, varied within 36–546 days; fecundity was from 15 to 236 eggs. In the population, two groups of individuals were also discovered.

On the whole, the individuals which have a short life span in both the first and the second generation, differ from the rest in that they possess a somewhat faster rate of sexual maturation (in *T. tubifex* a

fecundity three times higher and a life span half as long) (Table 1).

Comparing the fecundities of individuals which reproduce sexually and parthenogenetically, it appeared that in *T. tubifex* and *L. hoffmeisteri* the individual fecundity (number of eggs laid by one individual per month) is practically the same. The number of eggs laid by an individual for the whole life cycle in parthenogenesis is only half the total number of eggs in both species (Table 2).

Parthenogenesis in tubificids is classed as facultative. The occurrence of facultative parthenogenetic reproduction in the investigated mass species of tubificids in water-bodies of various types may serve as an argument in favour of their high adaptability to exist under dynamic conditions and to maintain their abundance in the population; the heterogeneity of the population (short and long life span individuals) secures its stability.

References

Černosvitov, L., 1927. Die Selbstbefruchtung bei den Oligocheten. Biol. Zbl. 47: 587–595.

Christensen, B. & F. B. O'Connor, 1958. Pseudofertilization in the genus Lumbricillus (Enchytraeidae). Nature 181: 1085–1086.

Gavrilov, K., 1935. Contribution à l'etude de l'autofécondation chez les Oligochaetes. Acta zool., Stockh. 16: 21–64.

Potential age of aquatic Oligochaeta

Tarmo Timm
Vörtsjärv Limnological Station of the Institute of Zoology and Botany of the Estonian S.S.R. Academy of Sciences, Rannu 202454, Tartu District, Estonian S.S.R., U.S.S.R.

Keywords: aquatic Oligochaeta, life-cycle, aquarial cultures

Abstract

The maximum lifetime of species with exclusively sexual reproduction covers 5–15 years or even more in aquaria, although most specimens die consecutively earlier. The zooids of paratomic species live some weeks or months, their clones usually less than a year, rarely for some years. In the species with facultative architomy (fragmentation) the clonal age also exceeds that of the individual, both lasting for several years. The actual lifetime of worms in nature may be shorter than in cultures.

It is very difficult to determine the natural lifetime of aquatic oligochaetes, since they have almost no clear age differences; even their sexual maturation is reversible. For a long time researchers assumed mostly a one-year or even shorter life-cycle with obligatory post-reproductive mortality. In recent decades some authors, starting with Brinkhurst (1964), have supposed a two-year life-cycle for various species of Tubificidae and Lumbriculidae, but also with one single reproduction period. At the same time, it is well known that earthworms used for regeneration experiments can live 4–10 years in captivity (Korschelt, 1906). As early as 1913 the idea of a multi-year life-cycle in *Lumbriculus variegatus* and *Rhynchelmis limosella*, with several reproduction periods, was expressed by Mrázek (1914). Repeated reproduction periods in Tubificidae were first described by Hrabě (1939).

The problem can be clarified by means of long-term aquarial observations. For this purpose the author has kept various aquatic oligochaetes in the aquaria of the Vörtsjärv Limnological Station for about 20 years. Small aquaria usually contain about 300 ml of sieved lake mud under a thin layer of water. Ten tiny, new-born worms of a known species or some adult specimens brought from nature are placed in an aquarium. The aquaria are examined 4–6 times a year; the mud is sieved and the old worms from sieving residue are returned to their aquaria with a new portion of mud, but their descendants and cocoons are removed. Even when both are immature and of the same size, old generations differ from young ones by the darker chloragogen tissue on the intestine. Some aquaria are kept in particular temperature conditions (see Fig. 1). These observations continue down to the natural death of the last of the initial worms in the aquarium. Some further details of the methods are described by Timm (1980). Not all species prosper under such conditions. Among the 36 species of aquatic oligochaetes tested, only about a third, mostly Tubificidae, succeeded in sexual reproduction, while some other underwent asexual reproduction. Many species, especially those of Naididae, did not survive long in mud cultures.

It is reasonable to discuss the results separately for 3 groups of oligochaetes with different types of life-cycle.

1. The Naididae and Aeolosomatidae (probably also the Opistocystidae) are characterized by a regular, mostly seasonal alternating of sexual and

Hydrobiologia 115, 101–104 (1984).
© Dr W. Junk Publishers, Dordrecht.

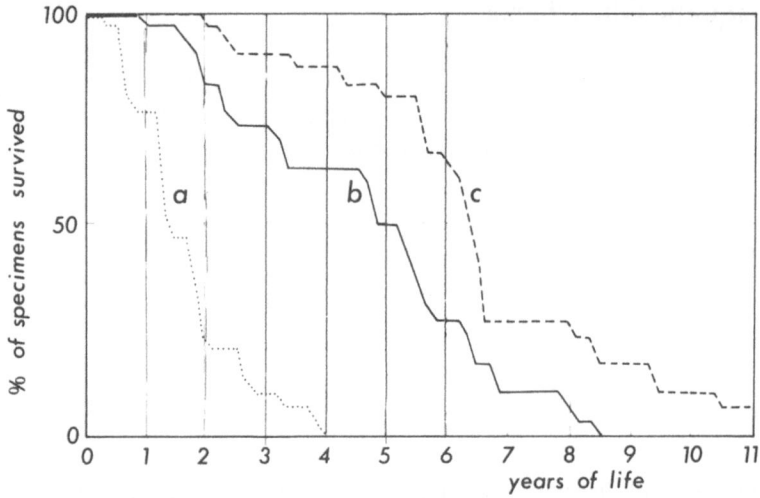

Fig. 1. Survival of the limnophilous form of *Tubifex tubifex* from the spring of Roosna-Alliku under various temperature conditions. a) constant temperature +25 °C, b) room temperature, c) seasonal temperature (about +5 °C in winter and up to +15 °C in midsummer); 100% corresponds to 30 specimens.

asexual reproduction. The sexual generation appears in natural waters, as a rule, before winter or drought and survive by diapausing cocoons only. Old worms always die after the egg-laying period. These species are not suitable for long-term cultivation, since it is difficult to create favourable conditions for their maturation and successful sexual reproduction.

The age of a single specimen (zooid) and the age of a clone must be distinguished here. The zooids of *Stylaria lacustris* and *Chaetogaster diaphanus* die due to decrepitude or after egg-laying at the age of 1–2 months (Stolte, 1962; Poddubnaya, 1966). Specimens of *Aeolosoma hemprichi* can live up to 8–8.5 months and die only after 63–94 (Štolc, 1903) or 130 paratomic buddings (Hämmerling, 1924).

Clones of Naididae live less than a year in natural conditions, in Estonia mostly from spring till autumn. Clones die even when not all specimens mature, e.g. in *Nais elinguis* in springs. In cultures, where stable conditions are not conducive to maturation, paratomy can sometimes continue for several years, slowing down gradually. Thus, a clone of *Nais communis* survived six years in my aquaria. According to Štolc (1903), clones of some *Aeolosoma* can live at least 3 years.

2. Species with the sexual mode of reproduction are the commonest. About 20 of such species, be-

longing to the families Tubificidae, Lumbriculidae and Criodrilidae, were kept in aquaria (see Table 1). They can live and often also repeatedly mature there for many years. At the same time the lifetimes of individual worms vary to a great extent, with the first dying already during the first year (see Fig. 1). The survival time for half of an aquarial population (LT_{50}) is roughly equal to half of the maximum age. Age does not apparently depend on reproduction. The lifetime of worms decreases with the rise of temperature (Fig. 1). Specimens brought from open waters often live longer than those born in captivity. This can be explained with a broader distribution of parthenogenesis in aquarial conditions (Poddubnaya, 1980a).

The maximum age of *Criodrilus lacuum* in aquaria can reach at least 17 years, that of *Isochaetides newaensis, Spirosperma ferox, Rhyacodrilus coccineus* and *Stylodrilus heringianus* more than 15 years, *Psammoryctides barbatus* more than 12 years, some forms of *Tubifex tubifex* and *Limnodrilus hoffmeisteri* more than 11 years. The two latter species demonstrate a great variability among the strains originating from different water bodies.

The potential age achieved in aquaria is probably very seldom realized under natural conditions, where predators and other harmful factors eliminate most of the population as early as during the first year of life. Nevertheless, in natural conditions

Table 1. Maximum lifetimes of some oligochaetes with exclusively sexual reproduction in aquaria, years. Numbers in brackets refer to specimens brought as adults from natural water bodies.

Species, form	Seasonal temperature	Room temperature	20–25 °C
Tubifex tubifex, rheophilous form	$8\frac{1}{2}$	$6\frac{1}{3}$	$4\frac{1}{3}$
Tubifex tubifex, limnophilous form from Lake Peipsi-Pihkva	$6\frac{1}{2}$	$5\frac{2}{3}$	$4\frac{1}{6}$
Tubifex tubifex, limnophilous form from the spring of Roosna-Alliku	>11	$9\frac{1}{2}$	$3\frac{5}{6}$
Tubifex tubifex, limnophilous form from the spring of Norra		$1\frac{1}{4}$	
Tubifex ignotus		<1 (>5)	
Ilyodrilus templetoni	8	5	3
Potamothrix hammoniensis	3–6	$2\frac{1}{2}$–4	$3\frac{1}{2}$
Potamothrix moldaviensis	$3\frac{1}{2}$	$4\frac{1}{2}$	$2\frac{1}{2}$
Limnodrilus hoffmeisteri from the River Suur-Emajõgi	>11	8	
Limnodrilus hoffmeisteri from the sublittoral of Lake Peipsi-Pihkva		$1\frac{1}{2}$ (>$7\frac{1}{2}$)	
Limnodrilus udekemianus		9	
Isochaetides newaensis	8 (>15)	7	1
Isocheatides michaelseni		(>10)	
Psammoryctides barbatus	>12	8 (>10)	3–5
Psammoryctides albicola		(>1->$2\frac{1}{2}$)	
Spirosperma ferox	>12 (>15)	>11 (>14)	8
Rhyacodrilus coccineus	>11 (>15)		
Stylodrilus heringianus	>12	7 (>15)	$3\frac{1}{2}$
Lamprodrilus isoporus	<1 (>1)		
Rhynchelmis limosella	(>$6\frac{1}{2}$)	(>3)	
Rhynchelmis tetratheca	$9\frac{1}{2}$ (>11)		
Criodrilus lacuum		>14 (>17)	

several age-classes take part simultaneously in every reproduction period with gradually decreasing abundance. For example, Poddubnaya (1980b) distinguished at least 3 age-classes in a population of *Isochaetides newaensis*.

The author has no experience with Enchytraeidae in aquaria. Various literature sources refer to their dying after the first reproduction period, with a maximum lifetime in cultures of less than a year (Reynoldson, 1943; Ivleva, 1953; Learner, 1972). However, Springett (1970) gives a lifetime of a year or even more for terrestrial Enchytraeidae in natural conditions. I have observed two immature size-classes in a summer population of *Marionina* sp. in the Baltic Sea, which indicates a possible multiannual life-cycle of this species.

3. Single species of Tubificidae, Lumbriculidae and Enchytraeidae have independently acquired the asexual mode of reproduction by fragmentation. Sexual reproduction occurs seldom and irregularly, particularly in aquaria, and the whole population never acquires maturity simultaneously. Unlike the Naididae, these worms do not necessarily die after egg-laying but resorb their genital organs and continue with fragmentation. *Potamothrix bedoti*, *Bothrioneurum vejdovskyanum* and *Lumbriculus variegatus*, the latter during a shorter period, have been kept in aquaria as representatives of this group.

It is rather difficult to determine individual age in fragmenting species. The author has observed some specimens of *Potamothrix bedoti* with incidental teratological marking for at least five years, and some anterior ends of *Lumbriculus variegatus* for three years.

The clonal age of the third group species is comparable to the individual age in the second group. Some clones of *Potamothrix bedoti* lived up to 15 years in aquaria, but fragmentation strongly decelerates in the last years and, in some aquaria, worms die out in the end. A clone of *Bothrioneurum vejdovskyanum* has flourished for eight years without any sign of degeneration so far.

We can conclude that the maximum age of various species of Tubificidae, Lumbriculidae and Criodrilidae in cultures can reach 5–15 years or even more. The lifetime of fragmenting clones of Tubificidae and Lumbriculidae species with domi-

104

nant asexual reproduction is of the same order. On the other hand, clones of Naididae and Aeolosomatidae usually live less than a year; only in the case of impossibility of sexual maturation can some of them survive up to six years. Among the Enchytraeidae, both one-year and longer life-cycles can be assumed, according to the literature data.

References

Brinkhurst, R. O., 1964. Observations on the biology of the marine oligochaete Tubifex costatus. J. mar. biol. Ass. U.K. 44: 11–16.

Hämmerling, J., 1924. Die ungeschlechtliche Fortpflanzung und Regeneration bei Aeolosoma hemprichii. Zool. Jb. Allgem. Zool. Physiol. Tiere 41: 581–656.

Hrabě, S., 1939. O vyvoji samčiho vyvodniho aparatu u některych niteňek a žižalic. Sb. Přír. Klubu Třebiči 3: 56–65.

Ivleva, I. V., 1953. Growth and reproduction of Enchytraeus albidus Henle. Zool. Zh. 32: 394–404 (in Russian).

Korschelt, E., 1906. Versuche an Lumbriciden und deren Lebensdauer in Vergleich mit anderen wirbellosen Tieren. Verh. dt. zool. Ges. 16: 113–127.

Learner, M. A., 1972. Laboratory studies on the life-histories of four enchytraeid worms (Oligochaeta) which inhabit sewage percolating filters. Ann. appl. Biol. 70: 251–266.

Mrázek, A., 1914. Beiträge zur Naturgeschichte von Lumbriculus. S.B.k. böhm, Ges. Wiss. (II) 1913, 14: 1–54.

Poddubnaya, T. L., 1966. On the systematics of Chaetogaster diaphanus Gruith. (Oligochaeta, Naididae). Trudy Inst. Biol. Vodokhran 12: 120–124 (in Russian).

Poddubnaya, T. L., 1980a. On the parthenogenesis of Tubificidae (Oligochaeta). Trudy Inst. Biol. vnutr. vod 44: 3–13 (in Russian).

Poddubnaya, T. L., 1980b. Life cycles of mass species of Tubificidae. In R. O. Brinkhurst & D. G. Cook (eds.), Aquatic Oligochaete Biology. Plenum Press, N.Y.: 175–184.

Reynoldson, J. B., 1943. A comparative account of the life cycles of Lumbricillus lineatus Mull. and Enchytraeus albidus Henle in relation to temperature. Ann. appl. Biol. 30: 60–66.

Springett, J. A., 1970. The distribution and life histories of some moorland Enchytraeidae (Oligochaeta). J. anim. Ecol. 39: 725–737.

Štolc, A., 1903. Über den Lebenszyclus der nidrigen Süsswasserannulaten und über einige anschliessende biologische Fragen (Aeolosoma-Arten). Bull. int. Acad. tchéque Sci. 7: 74–130.

Stolte, H.-A., 1962. Oligochaeta. In H. G. Bronns (ed.), Klassen und Ordnungen des Tierreichs, IV, 3,3. Leipzig: 891–1141.

Timm, T..E., 1980. Methods of culturing aquatic Oligochaeta. In Aquatic Oligochaeta worms (Proc. Symp. aquat. Oligochaeta Tartu 1967). Amerind Publishing Co., Pvt. Ltd., New Delhi: 119–131 (translated from Russian).

Growth of *Stylaria lacustris* (L.) (Oligochaeta, Naididae)

N. P. Finogenova

Zoological Institute, USSR Academy of Sciences, University Embankment, Leningrad 199034, U.S.S.R.

Keywords: aquatic Oligochaeta, Naididae, growth, experimental

Abstract

The peculiarities of the growth in weight and length of *Stylaria lacustris* (L.) on the basis of observations in experimental vessels are considered. The growth of this species fits a parabolic curve. The equations relating weight to absolute growth rate as well as weight to duration of life are given.

Introduction

The oligochaetes are of great importance in production processes of fresh-water bodies. The peculiarities of their growth have been insufficiently studied. There are some data about the character of length and weight growth of *Stylaria lacustris* L. in the paper of Kamlyuck & Kovaltchuk (1972): in particular they suggested a formula for the relation between weight and length of body for this species, that I adopted for my calculations:

$$W = 0.0035L^{2.1} \qquad (1)$$

where: W = wet weight in mg and L = length, in mm.

These authors have also noticed a rectilinear character of growth of body of *S. lacustris* in length and a high instantaneous growth rate.

Material and methods

The observations on growth of this species were carried out in summer in the Onega-lake. A single worm was placed in each vessel with a volume of 100 ml. Pieces of dead water plants preliminarily gone through a binocular for removing animals living on them were used for food. Length of *S.*

lacustris was measured under a binocular with precision to 1 mm. The vessels were covered with dense gaze and placed into the lake at a depth of 20–30 cm. The animals for experiments came from the same spot. Every day the vessels were gone through, the length of the animals was measured, food and water were substituted. The water temperature in the experimental vessels in June–July was 18–20.5 °C in the morning, and 18.3–22° in the evening; in August 16–19° in the morning, 17.2–20.5° in the evening. The animals in the experimental vessels (only aclitellate individuals) were 3–16 mm long. From the beginning of observations (end of June) to the last decade of August, *S. lacustris* in the vessels and in the field reproduced asexually by division: at first forming chains with 2, rarely 3, zooids, then dividing into 2–3 individuals. Daughter individuals were selected, placed separately and observed. The period of division was on the average 4.2 days, the mean length of the dividing chain was 14 mm; the mother individual in the moment of division had a length of 8 mm, the daughter individual of 6 mm. On the first day after division a significant (average: 30%) increase of body length of both the mother and daughter individuals was observed, which may be explained by change of form of the body after disintegration of the chain. For the analysis of the character of

106

growth of *S. lacustris* these data were not taken into account. Some preliminary results about growth of this species have already been published (Finogenova, 1981).

Results and discussion

It is well known that growth of the different animals may be described by means of the exponential, parabolic or S-forming curves. The cases of parabolic growth are sufficiently frequent (Winberg, 1975). For parabolic growth a linear relation between the logarithm of growth rate and the logarithm of weight is typical:

$$\lg dW/dt = \lg N + (1 - n)\lg W \qquad (2)$$

where dW/dt = absolute growth rate (defined as $\dfrac{W_2 - W_1}{t_2 - t_1}$), N = absolute growth rate at $W = 1$, n = constant.

This condition for *S. lacustris* is fulfilled as it is shown in Fig. 1. Growth in weight of aclitellate individuals of *S. lacustris* fits a parabolic curve and the relation between growth rate and weight in the process of growth may be expressed by the equation:

$$dW/dt = (0.112 \pm 0.025)W^{0.628 \pm 0.145} \qquad (3)$$

[$r = 0.81$, criterion of Fisher 6.9 (number of pairs: 39)] where W is the wet weight in mg.

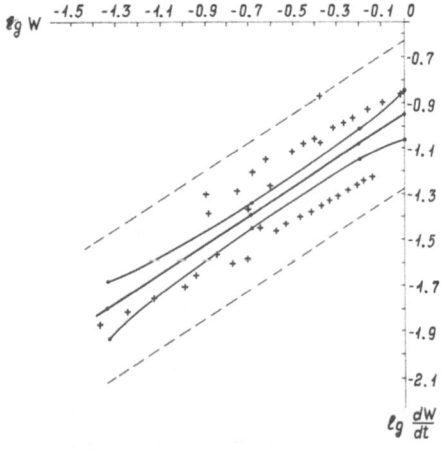

Fig. 1. The relation between the logarithm of growth rate and the logarithm of weight in *Stylaria lacustris*.

From this equation if $1 - n = 0.628$, $n = 0.372$. Knowing the main parameters of growth N and n, one can present weight of an organism as the function of duration of development (Winberg, 1968, 1975):

$$W_t = (Nnt)^{1/n} = W_1 t^{1/n} \qquad (4)$$

where W = weight attained at time t from the start of development; $W_1 = (Nn)^{1/n}$ = weight at $t = 1$; t = calculated value of growth duration from the start of development; $1/n$ = constant of growth. For *S. lacustris* this equation is

$$W_t = 1.93 \cdot 10^{-4} \cdot t^{2.69} \qquad (5)$$

where W = mg wet weight, t = days.

The observations as a rule are carried out from hatching of animals but not from 'beginning' of development and it is necessary to know the value t_o (duration of development of animal till hatching). According to the ideas of parabolic growth (Winberg, 1968):

$$t_o = \frac{W_o^n}{Nn} \qquad (6)$$

Knowing the t_o value one can describe formula (4) as:

$$W_t = W_1 (\tau + t_o)^{1/n} \qquad (7)$$

where τ = growth duration of animal from hatching ($\tau = t - t_o$); t_o = duration of development of animal till hatching. If the length of just hatched *S. lacustris* is being taken as 2.5 mm (for oligochaetes of different sizes hatched animals are 1.5–3 mm in length) and weight as 0.0237 mg, then t_o will be equal to 6 days and the formula (7) will read:

$$W_t = 1.93 \cdot 10^{-4} \cdot (\tau + 6)^{2.69} \qquad (8)$$

The growth of an animal for any interval of time $t_2 - t_1$ may be defined from the equation:

$$t_2 - t_1 = \frac{W_2^n - W_1^n}{Nn} \qquad (9)$$

For example, one can find out the time between divisions at asexual reproduction of *S. lacustris*. As at the time of division of the chains the mother individuals had a mean length of 8 mm, and during the first day it increased 30% by length that may be explained by change of the body form (as a rule this process is accompanied by increase of weight), the initial length of the mother individuals was taken as equal to 10.4 mm and weight from (1) as 0.473 mg. Then from (9) the period of time between divisions is equal to 4.75 days, which is nearly the same as the result from direct observations in the vessels (4.2 days). As it is well known that weight and length of animals are related to each other by the function $W = qL^b$, then W in (4) may be changed with qL^b. After this the formula of linear growth will be described as:

$$L_t = \frac{(Nn)^{1/nb}}{q^{1/b}} \cdot t^{1/nb} \tag{10}$$

For *S. lacustris,* using formula (1), it has the following parameters:

$$L_t = 0.254t^{1.28} \tag{11}$$

where length is in mm, time in days.

The linear growth of *S. lacustris* is therefore expressed by a parabolic curve but with more gentle a slope (nearly straight) then the weight growth.

In the literature I managed to find one more example of parabolic growth for oligochaetes. Thorhauge's (1976) data on growth of *Potamothrix hammoniensis* (Mich.) may be expressed with a parabolic curve (Fig. 2). For this species the relation between growth rate and weight is given by the equation (at 13 °C):

$$dW/dt = (0.014 \pm 0.002)W^{0.488 \pm 0.141} \tag{12}$$

[$r = 0.91$; Fisher's criterion 5.07 (number of pairs 14)] where W is the dry weight in mg.

The equation of relation between weight and duration of development has the following parameters:

$$W_t = 6.86 \cdot 10^{-5} \cdot t^{1.95} \tag{13}$$

where W = mg dry weight, t = time in days.

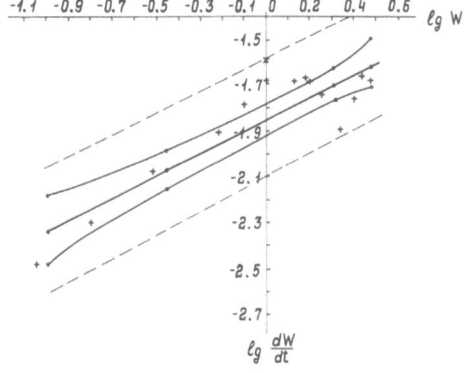

Fig. 2. The relation between the logarithm of growth rate and the logarithm of weight in *Potamothrix hammoniensis.*

References

Finogenova, N. P., 1981. Growth of Stylaria lacustris (L.) (Oligochaeta, Naididae). In: The Principles of Studying of Fresh Water Ecosystems. Leningrad: 125–130 (in Russian).

Kamlyuck, L. V. & M. M. Kovaltchuk, 1972. Some data about number, growth and production of oligochaete Stylaria lacustris (Naididae, Oligochaeta) in littoral of lake Narotch. In: Aquatic Oligochaeta. Proc. 2 All-Union Symposium, Borok 27–30 June 1972, Jaroslawl: 148–151 (in Russian).

Thorhauge, F., 1976. Growth and life cycle of Potamothrix hammoniensis (Tubificidae, Oligochaeta) in the profundal of eutrophic Lake Esrom. A field and laboratory study. Arch. Hydrobiol. 78: 71–85.

Winberg, G. G., 1968. Main regularities of animal growth. In: Methods of Determining of Production of Water Animals. Minsk: 45–58 (in Russian).

Winberg, G. G., 1975. Interrelation between growth and energetic exchange at poikilotherm animals. In: Quantitative Aspects of Organism's Growth. Nauka, Moscow: 7–23 (in Russian).

The biology of *Psammoryctides barbatus* (Grube) in English chalk streams

M. Ladle & G. J. Bird

Freshwater Biological Association, River Laboratory, East Stoke, Wareham, Dorset BH20 6BB, U.K.

Keywords: aquatic Oligochaeta, *Psammoryctides barbatus*, growth, cocoon production

Abstract

The tubificid *Psammoryctides barbatus* is common in the sediments of English chalk streams. The species is primarily a spring and summer breeder with the most intense period of reproduction from April to June throughout its distribution in the river. Recruitment of juveniles takes place chiefly from June to August and most of the worms attain maturity in the following spring. In culture at 15 °C worms had attained 38 mm within 90 days of hatching and showed early signs of maturity. Growth rates and rates of cocoon production in culture are presented.

Introduction

Psammoryctides barbatus (Grube) is abundant in the streams and rivers of southern England (Ladle & Bird, 1980) including the acid streams of the New Forest (Ladle, 1971a). It is a large tubificid which is known to contribute to the diets of cyprinid fishes and to act as an intermediate host of fish parasites (Kennedy, 1969). Unlike many other tubificid species *P. barbatus* is easily recognisable when immature, by the characteristic palmate dorsal setae.

The present paper consists of information derived from several studies which took place from 1966 to 1981. The main sites studied were a longitudinal series of reaches on the River Frome, Dorset, which were sampled by coring of both gravel and fine sediments in 1978–79 and two regions of the Bere Stream, a smaller chalk stream (NGR SY 858929 and NGR SY 835957) which were sampled in 1979–80 by quantitative coring of three habitats, gravel, soft sediment associated with *Ranunculus* plants and marginal silt. A lattice distribution of core samples on a 250 mm (total area 1.25 m × 2.50 m) grid was taken on one occasion to study microdistribution on a gravel bed.

Results

In the gravel bed sediments of the River Frome, *P. barbatus* had a pattern of longitudinal distribution which contrasted with that of the associated *R. coccineus* (Ladle & Bird, 1980), the former being more abundant and occurring more frequently in the fauna of downstream sites and being relatively less numerous in most of the tributaries. In fine sandy sediments the pattern was similar although *P. barbatus* was common at Frampton (8 km from the source) (51.7%) and Bradford Peverell (11 km from the source) (26.7%).

In the Bere Stream, in fine sandy sediment associated with *Ranunculus* where *P. barbatus* ranged from 2–5% of the oligochaete communities (Ladle 1971b), the estimated mean population densities never fell below 1 000 m^{-2} and rose from 1 299 $\overset{x}{\div}$ 1.68 m^{-2} on 24 January 1979 to peaks in late August and late September (4 280 $\overset{x}{\div}$ 1.65 m^{-2} and 4 208 $\overset{x}{\div}$

Hydrobiologia 115, 109—112 (1984).

© Dr W. Junk Publishers, Dordrecht.

1.86 m^{-2} respectively). In the gravel substrata *P. barbatus* was present through 1979 at densities less than 1 000 m^{-2}, population densities were maximal at 524 ± 468 m^{-2} on 24 January 1979 and in the following winter at 993 ± 900 m^{-2} on 2 January 1980. In marginal silt densities were similar to those in gravel, only on one occasion exceeding 1 000 m^{-2} but in this case there was a distinct peak in numbers from late August to late September (max. 28 September 987 $\stackrel{\times}{\div}$ 1.87 m^{-2}).

For analysis of maturity and size structure, data from all substrata were pooled to provide a month comparison. Mature individuals (bearing at least modified spermathecal setae) were recorded at Bere Stream throughout the year with the exception of September and October. The main peaks in relative frequency of mature individuals of *P. barbatus* were recorded in June (31.4%) and May 1980 (28.8%) (Fig. 1). Similar timing of peaks was observed at Moreton and East Stoke on the R. Frome in 1978–79.

Supplementing the above, towards the end of the main period of maturity were records of individual worms devoid of ventral setae on the spermathecal segment (X). Such animals accounted for 4% of the population on 4 August 1980 and were probably post-breeding individuals. In addition, at the commencement of the period of maturity, some worms were noted to have only slightly modified spermathecal setae. A peak of relative abundance of

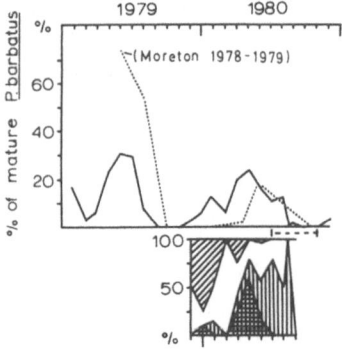

Fig. 1. Percentage maturity of *Psammoryctides barbatus* in the Bere Stream, the dotted line gives data from the Moreton site on the R. Frome for comparison. The lower figure indicates the proportion of worms in different states of maturity. Oblique shading – spermathecal chaetae only, vertical shading – clitellate ovigerous worms, cross hatching – worms with everted penes. The broken line between the two figures indicates the presence of worms with no ventral chaetae in X.

individuals having everted penes, presumably indicating recent copulation, was observed on 2 May 1980.

Clitellate and/or ovigerous specimens formed the highest proportion of the mature class from 2 May until 4 August 1980, which is probably the most intensive period of breeding although both clitellate animals and cocoons were collected as early as 24 January 1980.

The size frequency distributions (Fig. 2) show that from 24 January 1979 until 27 June 1979 there was a trend of increase in the mean body width of worms coincident with the increasing relative frequency of mature worms and diminution of population density. On 31 July 1979 a group of small immature worms was present in the population and from this date until 2 January 1980 an increase in the body width of the group was apparent. There was a concurrent decrease in the frequency of larger individuals. In 1980 the population structure followed a progression almost identical to that observed in 1979.

The more extensive data from microdistribution samples very clearly demonstrate the recruitment of small worms in the form of a bimodal size distribution on 19 August 1980. At this time the small group of larger individuals included many which lacked ventral setae on segment X. On 30 June 1980 juvenile worms with yolk in their guts were numerous. The high population densities observed in July and September were clearly attributable to recruitment of young worms.

In the River Frome large numbers of small worms appeared in samples collected from the gravel in September but the maximum number was recorded (at a slightly larger size) in February 1979 (Fig. 2), probably because of less efficient sampling of the smaller animals than in the Bere Stream study.

Observations on growth and fecundity of this species were made in laboratory cultures. Three large, ovigerous specimens were maintained in river sediment (particle sizes < 250 μm) at natural River Frome temperatures. The culture was set up on 5 March 1980 and one of the three worms died on the following day. The culture was re-examined on 30 May 1980 and contained one large clitellate worm, 54 small free living juveniles, 5 empty cocoons and 4 cocoons containing embryos in varying stages of development.

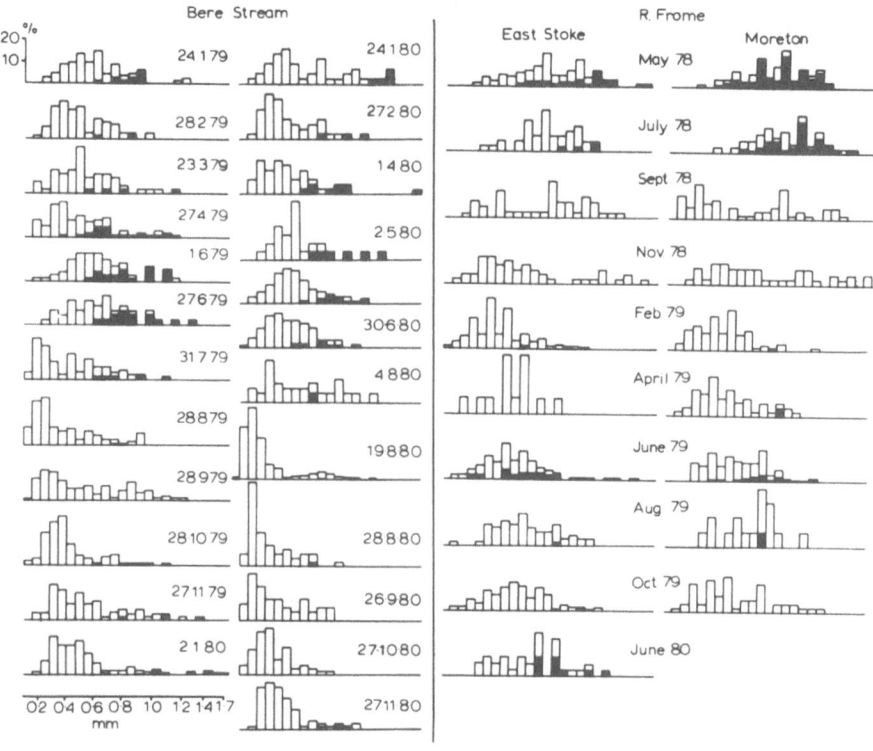

Fig. 2. Size (width of segment VIII) % frequency of samples of *Psammoryctides barbatus.* Bere Stream January 1979–November 1980. R. Frome May 1978–June 1980, at the Moreton and East Stoke sites. Black areas indicate mature worms.

The total progeny from one or (probably) two adult worms over a period of 86 days was thus 25.7–77.0 embryos/juveniles per adult and the mean number of offspring per observed cocoon was 8.56.

The size distribution of free living juveniles from the culture was similar to that of juveniles recorded in the field (Bere Stream 19 August 1980) and 48% of the juveniles from laboratory culture contained yolk in the guts.

A further series of cultures undertaken at a constant temperature of 15 °C during June and July provided additional information on fecundity. The observed rates of cocoon production ranged from 0.10–0.48 cocoons per individual per week ($\bar{x}$ = 0.31 ± 0.05), a level of reproduction over 91 days equivalent to 11.1–53.4 juveniles per adult worm.

To determine the time to hatching for this species, cocoons collected from the River Frome and found to be in early stages of development (containing more or less spherical embryos) were placed individually in small vials containing river water and incubated at 10°, 15° or 20°C. At the two higher temperatures, fungi frequently infected and killed the developing embryos. In subsequent cultures a proprietary aquarium fungicide was added to the water.

Rates of development were related to temperature. These experiments, although they were clearly inadequate in terms of replication, suggest that at the temperatures encountered in the field the development time of embryos within cocoons is between 14 and 30 days.

The growth of animals at constant temperatures of 10°, 15° and 20°C was also examined in culture. To avoid possible limitation of growth due to deficiencies of food only the early stages of growth were examined. A series of containers of fine sediment each containing 50 recently hatched juveniles were initiated at a density equivalent to ca. 6 000 animals m^{-2}. At intervals animals were removed from culture, fixed and their lengths measured. The instan-

112

taneous growth rates G_i and specific growth rates G_s were determined and the former was plotted against the geometric mean body length (Kaufmann, 1981). A growth curve was fitted (Fig. 3).

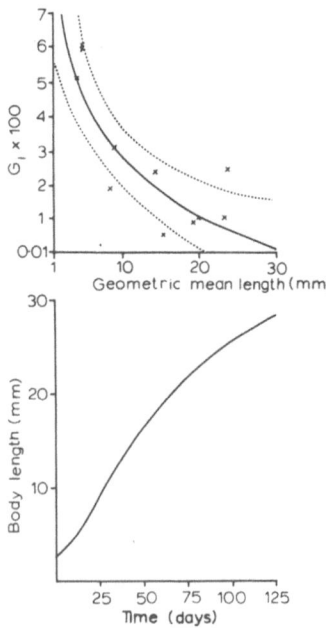

Fig. 3. Growth rate of *Psammoryctides barbatus* in culture. The upper curve represents instantaneous growth rate (ordinate) plotted against geometric mean body length (dotted lines indicate 95% confidence interval for the line). The lower curve is a Gompertz curve derived from the same data.

A dry weight: relaxed length relationship was determined on groups of 4 to 7 selected similar worms. Determinations ranged from ⊅ 0.01 mg at 3 mm length to 1.27 mg at 40 mm length and provided the regression

$$\log_e W = 1.94 \log_e L - 6.973 \quad r^2 = 0.91$$

which, when applied to length values used for growth determinations, gives specific growth (in weight) rates of 3.7–8.6% weight per day.

W = dry weight (mg), L = relaxed length of body (mm)

The conclusions derived from this work are:

1. *P. barbatus* is more abundant in silted sands than in gravel and detritus-rich silts.
2. In the streams examined *P. barbatus* breeds mainly from April–June.
3. The major period of juvenile recruitment is June–August.
4. Maturation is achieved by the spring of the year after hatching.
5. There is high mortality by the end of the breeding season.
6. Survivors of the summer breeding season may mature in November–December and breed for a second time in January–February.
7. Each adult worm can produce at least six cocoons (about 50 young worms) over a three month breeding period.
8. The observations of Brinkhurst (1964) and Timm (1970) are essentially consistent with the present assertion of summer recruitment.
9. Poddubnaya (1973) quotes a development time of 9–12 months. In culture Timm (1970) showed animals maturing at the age of about four months. In the present study at 15 °C two worms showed signs of maturity at the age of 3 months.
10. Growth rates of young worms are high and decrease with size/age.

Acknowledgements

The authors wish to thank their colleagues of the F.B.A. River Laboratory, in particular Jon Bass, Stewart Welton and Dave Cooling, for advice and help with this work.

References

Brinkhurst, R. O., 1964. Observations on the biology of lake dwelling Tubificidae. Arch. Hydrobiol. 60: 385–413.

Kaufmann, K. W., 1981. Fitting and using growth curves. Oecologia (Berl.) 49: 293–299.

Kennedy, C. R., 1969. Tubificid oligochaetes as food of Dace Leuciscus leuciscus (L.). J. Fish Biol. 1: 11–15.

Ladle, M., 1971a. Studies on the biology of oligochaetes from the Phreatic water of an exposed gravel bed. Int. J. Speleol. 3: 311–316.

Ladle, M., 1971b. The biology of Oligochaeta from Dorset chalk streams. Freshwat. Biol. 1: 83–97.

Ladle, M. & G. J. Bird, 1980. Aquatic Oligochaeta of southern England. In R. O. Brinkhurst & D. G. Cook (eds.), Aquatic Oligochaete Biology. Plenum Press, New York: 165–174.

Poddubnaya, T. L., 1973. Characteristics of tubificid and naidid life cycles. In Aquatic Oligochaeta (Systematics, ecology and studies of the Soviet fauna). Fish. Res. Bd Can. Trans. Ser. 2721: 135–146.

Timm, T. E., 1970. On the fauna of the Estonian Oligochaeta. Pedobiologia 10: 52–78.

Cohort cultures of *Psammoryctides barbatus* (Grube) and *Spirosperma ferox* Eisen: a tool for a better understanding of demographic strategies in Tubificidae

L. Adreani, C. Bonacina, G. Bonomi & C. Monti
C.N.R. – Instituto Italiano di Idrobiologia, Largo V. Tonolli 50/52, I-28048 Pallanza NO, Italy

Keywords: aquatic Oligochaeta, Tubificidae, experimental, density regulation, bionomic strategies

Abstract

Laboratory research on the tubificids *Psammoryctides barbatus* and *Spirosperma ferox* was done. Embryonic development time, growth, time required to attain the first cocoon laying and egg production were estimated at different temperatures and population densities. The results allow us to demonstrate some intrinsic density regulation mechanisms in the profundal tubificid communities and substantiate the hypothesis that the succession from oligotrophic to eutrophic species in the profundal of lakes undergoing eutrophication is mainly based on their biotic characteristics.

Introduction

It is well-known that many difficulties are encountered in the study of production and population dynamics of Tubificidae (Johnson & Brinkhurst, 1971; Jónasson & Thorhauge, 1976; Bonomi & Di Cola, 1980; Giere & Pfannkuche, 1982). It may be useful to remind ourselves of some of the main problems:

1) very rarely can the cocoons be classified as to species.

2) the same difficulty is presented by the young individuals that have not yet attained sexual maturation.

3) the complicated life-cycle, with a repeated alternation of maturation → cocoon laying → dematuration with regression of the reproductive system, makes it very difficult to split the field material into biological compartments that give a true reflection of biological stages. The above mentioned difficulties are reflected in the scarcity of bibliographic data not only on population statistics, but even simply on production estimates (see, e.g., Waters, 1977).

Our group was able to avoid some of the usual problems by studying the profundal Oligochaeta of reservoirs. In these man-made water bodies the bottom communities are structurally very simple; for example, in the reservoirs we worked on, the profundal tubificids are represented only by *Tubifex tubifex* Müller and *Limnodrilus hoffmeisteri* Claparède. In this case it was very easy to separate the cocoons, the young, the mature and the ovigerous (individuals with eggs in the ovisac) for the two species; this suggested the idea of starting a programme of culturing cohorts of individuals of the two species, with the aim of estimating – under controlled temperature and density conditions – the time required to cover the different successive biological stages. Some preliminary results on *T. tubifex* were presented at the first Symposium in British Columbia (Bonomi & Di Cola, 1980); then when the results referring to *L. hoffmeisteri* were obtained, a comparison of the two species was shown at one of the benthos sessions of the International Congress of Limnology in Kyoto (Adreani *et al.*, 1981).

More recently we decided to start culturing individuals of *Spirosperma ferox* Eisen and *Psammoryctides barbatus* (Grube). The reason why we se-

Hydrobiologia 115, 113–119 (1984).
© Dr W. Junk Publishers, Dordrecht.

lected these two species was that they were dominant in the profundal benthos of the central part of Lake Maggiore up to about twenty years ago, but are now being replaced by *T. tubifex, L. hoffmeisteri* and *Euilyodrilus hammoniensis* (Michaelsen) (Bonomi, 1967). This ecological succession runs parallel to the increasing eutrophication of this large water body.

The culturing of *S. ferox* and *P. barbatus* turned out to be more difficult than that of *T. tubifex* and *L. hoffmeisteri*: not only do they require better culture conditions (that is, more frequent inspecting and washing of cultures), but make the initial stage of starting their cohort cultures rather laborious. In fact, the selection of definitely identified mature specimens from mixed field material requires microscopic observation; from these identified 'producers' we obtained cohorts of cocoons, which are the starting material for our cohort cultures. Furthermore, as *S. ferox* and *P. barbatus* require a longer time-span to cover their biological cycle, it is evident that we can produce fewer data in the same time.

Embryonic development time

The experiments were performed at four temperatures (5, 10, 15, 18 °C), using small glass jars (diameter: 4 cm), half filled with fine sand covered by filtered lake water. The cohorts of cocoons were buried deep in the sand (eggs not well covered often do not develop) and periodically inspected in order to follow the development of embryos and the hatching process. The results are shown in Table 1, where all data relating to our experiments are reported.

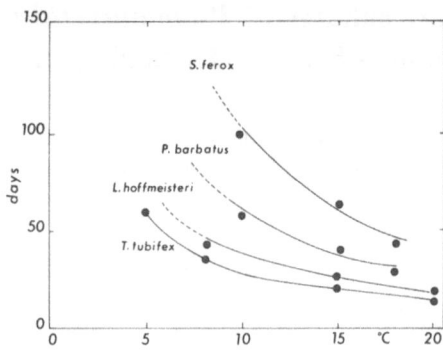

Fig. 1. Relationship between temperature and the time spent inside the cocoon.

Note that we have no data for the 5 °C experiments: indeed, at the lowest temperature the cocoons of the two species are invaded by epiphytic colonies that apparently inhibit the embryos' development. There is a clear correlation between the hatching times (measured as times from the beginning of cocoon cohort cultures and the hatching peak) and the experimental temperatures for both species: *S. ferox* requires a hatching time that is 30–40% longer in the temperature range of our cultures (see also Fig. 1).

The threshold temperatures have also been calculated for the two species, as well as the degree · days requirement for the embryonic development at 10, 15 and 18 °C.

Time required to attain maturation

As was done for *T. tubifex* and *L. hoffmeisteri* (Adreani & Bonomi, 1980; Bonomi, 1979; Bonomi

Table 1. Hatching times.

T °C	No. of cohorts	No. of cocoons	Mean hatching time (d) ± c.l.	Threshold temp. °C	Degree · days
Psammoryctides barbatus					
18	5	31	28.4 ± 2.47		423
15	7	62	43.5 ± 7.10	3.1	518
10	4	23	60.2 ± 14.04		415
Spirosperma ferox					
18	4	64	47.0 ± 11.90		724
15	6	29	62.7 ± 12.34	2.6	778
10	7	36	98.3 ± 15.80		727

& Di Cola, 1980; Adreani *et al.*, 1981) the maturation process of *P. barbatus* and *S. ferox* was observed in cohorts cultured at different temperatures and population densities. The culture method was that described by Kosiorek (1974): glass vessels (6 cm in diameter), half filled with fine sterilized sand covered by filtered lake water, were put in large trays kept in thermostat to maintain the chosen temperature. Pieces of limp lettuce (*Lactuca sativa*) were dug into the sand as food for the newly hatched worms. The cohorts were periodically inspected (more frequently at the highest temperatures), washed and new food added. The quantity of the added lettuce was clearly excessive so that we may reasonably presume that there was no food limitation. During the inspections, we checked the number of the animals and, at the same time, we counted separately the immature, mature and ovigerous individuals; from the moment when the worms had grown enough to be handled without damaging them, we began to measure their individual wet weights. In this way, it was possible to follow the growing process and to construct the frequency distribution diagrams for the individual wet weights. Figure 2a is an example for a cohort of *P. barbatus*. An important feature appears to be the wide weight range presented not only by the immature, but also by the mature and ovigerous individuals. Furthermore, it appears that the faster growing worms mature and become ovigerous before the slower ones, and over a longer period of time. In previous experiments we were able to show that cohorts of *T. tubifex* and *L. hoffmeisteri*, cultured the same way, show a quite similar mechanism, which causes the maturation time to depend upon the individual growth rate (Bonomi, 1979; Adreani *et al.*, 1981). But the maturation time is also density dependent: for instance, in the six cohorts of *P. barbatus* reared at 15 °C, a significant correlation ($r = 0.77$) was found between the cohorts' initial density and the maturation time, roughly measured as the time required for 50% of the individuals to attain maturation.

The growth and development of *P. barbatus*, as seen through the results of our cohorts, can be described as follows: after requiring 460 °C × days for the embryonic development inside the cocoon (e.g.: 67 d at 10 °C), the fastest growing individuals require 1600 °C × days to grow to a weight greater than 1 mg, and a further 2500 °C × days to attain

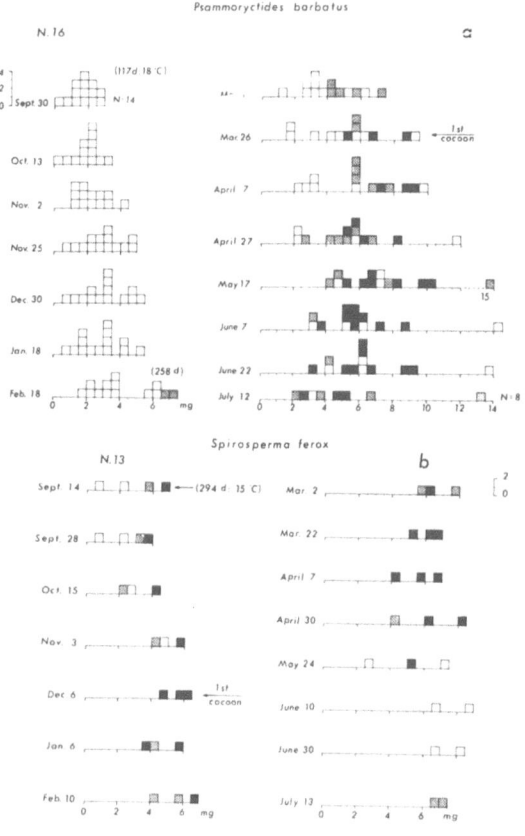

Fig. 2. Weight frequency distribution for cohorts of *P. barbatus* (*a*) and *S. ferox* (*b*). White: young worms, dashed: mature, black: ovigerous.

maturation; the slow-growing individuals, however, require 460, 1600 and 3700 °C × days, respectively.

We have much fewer data from cohorts of *S. ferox* (an example is given in Fig. 2b), but a similar scheme, even if preliminary, may be supplied also for this species: 690 °C × days (embryos) + 3000 °C × days + 2700 °C × days are required, by the individuals with the highest growth rate, to attain maturation.

An attempt to compare the above two species with *T. tubifex* and *L. hoffmeisteri* is given in Table 2, where the compartment y includes young individuals in the range 0–1 mg, Y the following compartment at the next youngest stage, and M the mature compartment before the cocoon deposition (following the scheme of Bonomi & Di Cola, 1980).

It is therefore possible to compare the generation

Table 2. Degree · days requirements of *T. tubifex*, *L. hoffmeisteri*, *P. barbatus* and *S. ferox*, before reaching the first ovigerous stage. Note that the fastest development follows: E + y + Y3 + M3; the slowest: E + y + Y1 + M1. For further explanations, see the text.

Compartment		Weight (mg)	Species			
			T. tubifex	*L. hoffmeisteri*	*P. barbatus*	*S. ferox*
Eggs and embryos	E	'0'	250	310	460	740
young	y	0– 1	800	1200	1600	3000
	Y_1	1– 3	1200	1440	3700	2750
	Y_2	3– 6	900	1110	2800	2700
	Y_3	6–10	600	840	2500	–
mature	M_1	1– 3	550	600	1000	700
	M_2	3– 6	430	600	900	600
	M_3	6–10	310	600	850	–

time in the four tubificids studied. A simple way is to compute the minimum generation time in the four species; some data are shown in Table 3, for a temperature of 10 °C. It appears that, while *S. ferox* may have only 1 generation in 2 years, *P. barbatus* may have 1.3, *L. hoffmeisteri* 2.5 and *T. tubifex* almost 4 in the same time. A comprehensive scheme of the duration of the different life stages, again at 10 °C, a temperature which is close enough to that found in the profundal zone of lakes, is shown in Fig. 3.

A rough proportionality seems to hold (for the four species) between the time spent as egg + embryo and that required to attain the mature or egg-laying stage. This is a point that deserve further investigation.

Several cohorts have been cultured long enough to allow an acceptable measure of fecundity. The results are relatively easy to obtain for *P. barbatus*, but only a couple of cohorts of *S. ferox* could be cultured long enough to allow such measurements. So it is now possible to compare the fecundity of *T. tubifex* (Bonomi & Di Cola, 1980) with that of *L. hoffmeisteri* (Adreani *et al.*, 1981) and *P. barbatus*.

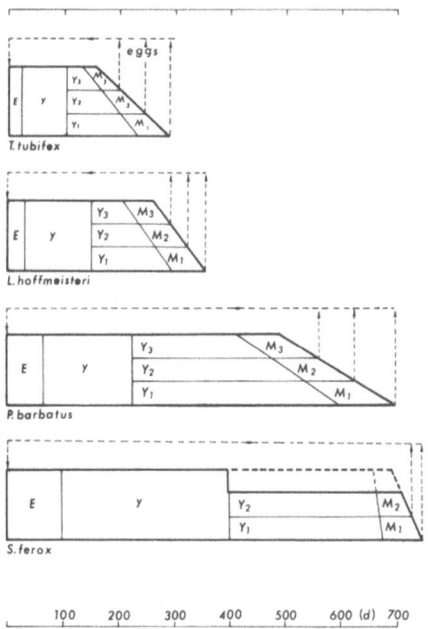

Fig. 3. Estimated residence times for the compartments: egg + embryo (E), young (y + Y) and mature (M) individuals at 10 °C (first generation).

Table 3. Duration of the egg and embryo (E), young (y and Y) and mature (M) stages. See also the text.

(10 °C)	E (d)	y (d)	Y (d)	M (d)	Σ (d)	Max. number of generations in 2 a
Tubifex tubifex	25	80	60	31	196	3.7
Limnodrilus hoffmeisteri	31	120	84	60	295	2.5
Psammoryctides barbatus	67	160	250	85	562	1.3
Spirosperma ferox	100	300	270	60	730	1.0

The results are shown in Fig. 4, where the initial experimental abundance (N_0) is plotted against the total egg production per ovigerous individual at 15 °C. The correlation is significant for all three

species: *T. tubifex*, as expected, reveals the highest fecundity, *L. hoffmeisteri* appears as the most sensitive to population density; *P. barbatus* and *L. hoffmeisteri* seem to have comparable fecundities at low densities, but the egg production of the former is less density dependent. Our observations on experimental egg-laying in the cohorts, range as follows: *T. tubifex*: 155–294 d, *L. hoffmeisteri*: 244–277d , *P. barbatus*: 132–174 d. An attempt to compensate for the different mortalities in the cohorts was made, by plotting the integral of the survival curve (days × individuals) at the time of the last observed egg-laying against the specific egg production. The plot is given in Fig. 4.

One should remember that during an egg-laying period the clutch size (number of egg:cocoon) undergoes a variation: low at the very beginning, it increases and rapidly reaches a peak, slowly decreasing again down to the lowest clutch size at the end of a deposition period (Bonomi & Adreani, 1978). If we measure the cocoon and egg production over a very long time (that is: several production 'waves'), we may allow for this variation and have a significant measure of the mean clutch size for the different species under different population densities. Figure 5 illustrates the density dependence of clutch size for *T. tubifex*, *L. hoffmeisteri* and *P. barbatus*. We may therefore conclude that population density negatively affects not only the egg production, but also the corresponding clutch size.

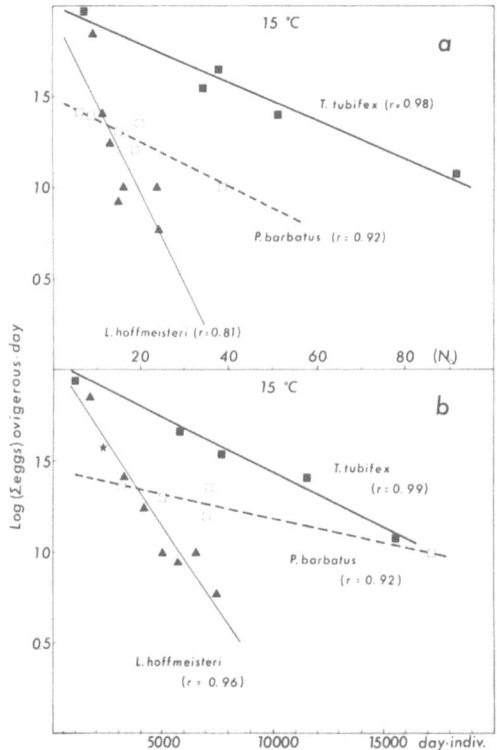

Fig. 4. Plot of the initial density (N_0) of cohorts (panel *a*) and the integral of the survival curve (panel *b*), against the logarithm of specific fecundity.

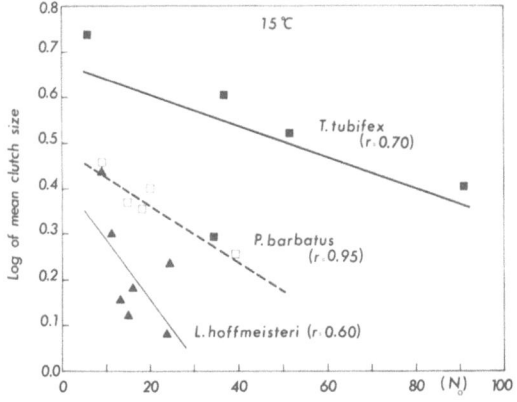

Fig. 5. Plot of the initial density (N_0) of cohorts against logarithm of the mean clutch size (log (Σeggs/Σcocoons)).

Conclusions

The experimental investigations on the tubificids from the profundal zone of lakes and reservoirs have proved fruitful. The measurement of time spent inside the cocoon (egg + embryo compartment) is a contribution towards understanding the comparative population biology in tubificids, and supplies some tools for more refined population studies: allowing the selection of an adequate time-lapse between our periodical field population counts; making it more rewarding to adopt the time-consuming procedure of counting the cocoons and their egg content in the field material (where not too severely limited by taxonomical difficulties); permitting the estimation of field mortality of the youngest compartment (Bonomi & Di Cola, 1980; Adreani *et al.*, 1981).

The results on the measurements of the time required to attain maturation and the cocoon laying, together with those on fecundity, give us a wider view of the bionomic strategies of the different species. Maturation time depends on the individual growth rate, therefore expressing it as the time required to allow a given percentage of the population to attain maturation, is a rather simplistic description of complex population biology. Therefore, for *P. barbatus* and *S. ferox*, as was done before for *T. tubifex* and *L. hoffmeisteri* (Bonomi & Di Cola, 1980; Adreani et al., 1981), we estimated maturation time separately for the different weight-classes (Fig. 3). Maturation time is revealed as density-dependent; density also affects negatively the percentage of the population that reaches maturity (Fig. 6), and the specific fecundity (Figs. 4 and 5). We may therefore expect – and indeed we find – field situations where species with intrinsic low fecundity lay only a limited number of cocoons, so making difficult a numerical description of the egg compartment. On the other hand it is very common

to come across situations where cocoons are extremely abundant, as in the case of *T. tubifex* in many reservoirs (Adreani & Bonomi, 1980; Adreani et al., 1981). In one reservoir in Northern Italy (Lago di Suviana) the egg/ovigerous ratio in *T. tubifex* was found to be comparatively very low and thought to be dependent on the extremely high population abundance ($100-200 \times 10^3$ ind. $\cdot$ m^{-2}; Comini, 1982). It is evident that the quantitative description of the dependence of the maturation rate and of specific egg production on density is a basis for improving the population dynamic models, which have been conceived for the estimation of population statistics for field populations (e.g. Bonomi & Di Cola, 1980).

Generation time and fecundity differentiate the bionomic strategies of the four species studied. *T. tubifex* has the shortest generation time and highest fecundity which explain its dominance in eutrophic environments, not only where oxygen depletion favours this species due to its ability to maintain a high respiration rate at very low oxygen tension (Berg et al., 1962); but also its appearance in the profundal of those lakes undergoing eutrophication whose hypolimnion, due to the enormous volume, does not yet display a decrease in oxygen content (e.g. Lake Maggiore; Bonomi, 1967, 1969; Bonomi et al., 1970, 1979). Its presence in young and unstable environments (reservoirs) and in mountain lakes is now better understood.

The sequence *T. tubifex* → *L. hoffmeisteri* → *P. barbatus* → *S. ferox* is that of an increasing generation time and, at least beyond certain population abundances, that of a decreasing fecundity. But the sequence leads also from 'eutrophic' species to 'oligo-mesotrophic' (*P. barbatus*) and 'oligotrophic' species (*S. ferox*); therefore we may provisionally conclude that, at least in the tubificids of the profundal zone of lakes and reservoirs, oligotrophic species are closer to the so called 'K strategist' species, while eutrophic species are close to the 'r strategist' species.

Finally, our data seem to indicate that some density regulation mechanisms in the tubificids may operate as follows: a decrease in population abundance (e.g. by predation) causes both the percentage maturation and the specific fecundity to increase, so leading to an overall numerical growth; the new situation in its turn will bring about, through the decreased maturation and fecundity, a decline in the population size.

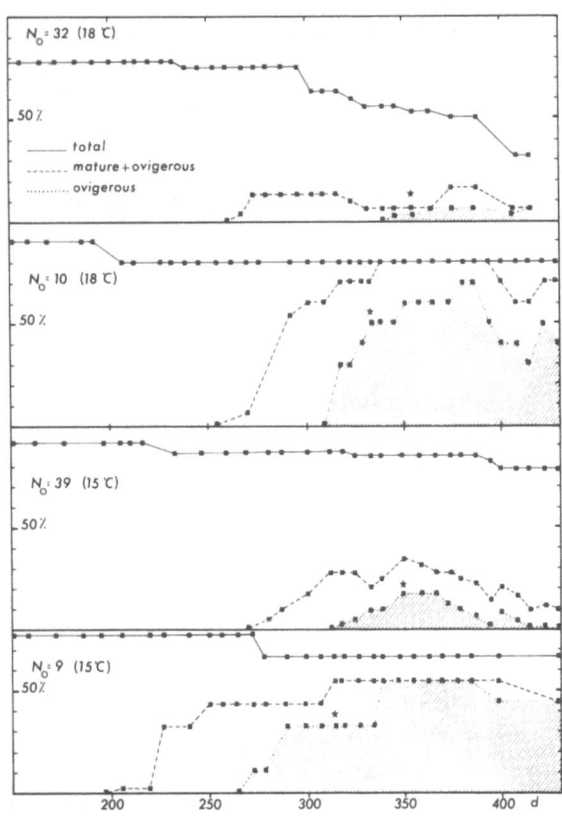

Fig. 6. Percent survival of cohorts of *P. barbatus*; the asterisk indicates the time of laying of the first cocoon; N_0 = initial number.

References

Adreani, L. & G. Bonomi, 1980. Elementi per una dinamica di popolazione dei tubificidi (Annelida, Oligochaeta). In Atti 3° Congresso Associazione Italiana Oceanologia Limnologia, Sorrento, 18-20 Dicembre 1978: 355-366.

Adreani, L., C. Bonacina & G. Bonomi, 1981. Production and population dynamics in profundal lacustrine Oligochaeta. Verh. int. Ver. Limnol. 21: 967-974.

Berg, K., P. Jónasson & K. W. Ockelman, 1962. The respiration of some animals from the profundal zone of a lake. Hydrobiologia 19: 1-39.

Bonomi, G., 1967. L'evoluzione recente del Lago Maggiore rivelata dalle cospicue modificazioni del macrobenton profondo. Mem. Ist. ital. Idrobiol. 21: 196-212.

Bonomi, G., 1969. The use of a new version of the Tonolli mud-burrower for sampling low density benthonic populations. Verh. int. Ver. Limnol. 17: 511-515.

Bonomi, G., 1979. Ponderal production of Tubifex tubifex Müller and Limnodrilus hoffmeisteri Claparède (Oligochaeta, Tubificidae), benthic cohabitants of an artificial lake. Boll. zool. 46: 153-161.

Bonomi, G. & L. Adreani, 1978. Significato adattativo della struttura comunitaria e della dinamica di popolazione nel macrobenton profondo di un lago artificiale. In Il Lago di Pietra del Pertusillo: definizione delle sue caratteristiche limno-ecologische. Ist. ital. Idrobiol.: 133-201.

Bonomi, G. & G. Di Cola, 1980. Population dynamics of Tubifex tubifex, studied by means of a new model. In R. O. Brinkhurst & D. G. Cook (eds.), Aquatic Oligochaete Biology. Plenum Press, New York: 185-203.

Bonomi, G., A. Calderoni & R. Mosello, 1979. Some remarks on the recent evolution of the deep Italian subalpine lakes. Symp. Biol. Hung. 19: 87-111.

Bonomi, G., M. Gerletti, E. Indri & L. Tonolli, 1970. Report on Lake Maggiore. In. O.C.S.E. Symposium on Large Lakes and Impoundments. Uppsala, 13-16 May 1968: 299-341.

Comini, P., 1982. Dinamica di popolazione di Tubifex tubifex Müller (Oligochaeta, Tubificidae) nel Lago di Suviana (Appennino Tosco-Emiliano). Thesis, Univ. Bologna, 65 pp.

Giere, O. & O. Pfannkuche, 1982. Biology and ecology of marine Oligochaeta. A review. Oceanogr. mar. Biol. annu. Rev. 20: 173-308.

Johnson, M. J. & R. O. Brinkhurst, 1971. Production of benthic macroinvertebrates of Bay of Quinte and Lake Ontario. J. Fish. Res. Bd Can. 28: 1699-1714.

Jónasson, P. M. & F. Thorhauge, 1976. Population dynamics of Potamothrix hammoniensis in the profundal of Lake Esrom with special reference to environmental and competitive factors. Oikos 27: 193-203.

Kosiorek, D., 1974. Development cycle of Tubifex tubifex Müller in experimental culture. Pol. Arch. Hydrobiol. 21: 411-422.

Waters, T. F., 1977. Secondary production in inland waters. Adv. ecol. Res. 10: 91-164.

Oxygen demand and long term changes of profundal zoobenthos*

Petur M. Jónasson
Freshwater Biological Laboratory, University of Copenhagen, 51 Elsingørsgade, DK-3400 Hillerød, Denmark

Keywords: aquatic Oligochaeta, eutrophication, oxygen demand, population dynamics, zoobenthos of lakes

Abstract

The paper attempts to combine the low oxygen content of the hypolimnion during stratification and the oxygen uptake of zoobenthos. Data of declining oxygen content in the hypolimnion and critical limits of respiration are combined for *Chironomus anthracinus*, *Potamothrix hammoniensis* and three species of *Pisidium*, *P. casertanum*, *P. subtruncatum* and *P. henslowanum*. The respiratory adaptation to low oxygen content influences both growth and population dynamics of the different species. The results have important bearing on eutrophication of the Lake Esrom ecosystem and temperate eutrophic lakes in general as well as the composition of profundal zoobenthos and its population dynamics.

Introduction

Increasing urbanization contributes to the eutrophication of our lakes: a process in which an increase in nutrient content, primary production and sedimentation influences in various ways the conditions of life of the profundal zoobenthos. In a dimictic, eutrophic lake in its classical sense, i.e. a lake with a small hypolimnion volume, a considerable seasonal variation occurs in physical and chemical parameters. This paper discusses the eutrophication of the Lake Esrom ecosystem with special reference to the declining oxygen content of the hypolimnion and its effect upon the dynamics and composition of the profundal zoobenthos. This paper focuses especially on oxygen and its uptake by the zoobenthos. Oxygen, when present in limited amounts, caused retardation of growth, increases length of life, and lowers production.

The investigation was carried out in Lake Esrom,

which is situated in a moraine landscape in northern Zealand, Denmark (56° N, 12° E). The lake is 8–9 km long and 2–3 km wide. The surface area is 17.3 km², maximum depth 22 m and mean depth 12.3 m. The bottom at a depth of 10–22 m (61.3% of total area) is covered with a thick mud layer of high organic content. The drainage basin of Lake Esrom, including the lake area (17.3 km²) is only 76 km² at the outlet. The lake volume is 212.7×10^6 m^{-3} (Berg, 1938). The area surrounding the lake is calcareous moraine with marlpits in the farmland areas on the eastern side of the lake, sandy beaches at the north and south end, stony beaches are found all around the open lake, and humic ditches in the forest on the western side provide a varied environment. The water supply to the lake is from many small brooks and ditches, of which Fønstrup Bæk is the biggest. Sewage inlets used to be present, but now only minor inlets are still present. An inflow of groundwater is likely. These environmental conditions produce a eutrophic calcareous lake of high stability (Berg, 1938; Jónasson, 1972; Jónasson *et al.*, 1974).

* Publication No. 389 from Freshwater-Biological Laboratory, University of Copenhagen.

Hydrobiologia 115, 121–126 (1984).

Pollution and population dynamics 1953–1973

The first cleaning plant was established in 1914. The effect of agricultural fertilizers began to be felt apparently at the same time. Draining of tributaries through pipes resulted in increased eutrophication of the lake from 1956 onwards. A total loading of 9 000 person equivalents polluted the lake. Untreated sewage, corresponding to 4 500 p.e., polluted the lake in 1961–71, after which the sewage was diverted to another watershed. Further diversions have increased the retention time from 9.6 to 18.1 years. In 1968 28 000 m³ of lumber also polluted the lake.

The progress of oxygen depletion in the profundal at a depth of 18–20 m was as follows: in August 1908 the oxygen content was 7.4 mg l⁻¹, in August 1933 the range was 1.4–3.3 mg l⁻¹, in August 1955–1960 ca 0.2 mg l⁻¹, but in 1969–1970 it had declined to 0 (range 0–0.2 mg l⁻¹). The oxygen depletion period increased from 1 week in 1933 to 3 months in 1970 (Fig. 1).

The main production of organic matter is by phytoplankton, 240 g C m⁻² y⁻¹, based on 15 years' average (1954– 1975). Nitrogen is still limiting for production, while PO₄-P doubled from 1930 to 1970 (Jónasson, et al., 1974). Therefore, production

did not increase during the period, but became more irregular (Jónasson, 1972; Fig. 17; Jónasson et al., 1974: Fig. 1). Large amounts of organic material were added, causing an increase in BOD and COD in hypolimnion water.

The profundal zoobenthos consists mostly of monocultures of *Potamothrix hammoniensis, Chironomus anthracinus, Chaoborus flavicans* and *Procladius pectinatus*. In contrast, 3 species of *Pisidium* occur. *P. hammoniensis* and the *Pisidium* spp. are detritivorous subsurface feeders, *C. anthracinus* a detritivorous surface feeder. *C. flavicans* and *P. pectinatus* are carnivorous.

Density and population dynamics of zoobenthic detritivores were studied in 1953–1973, i.e. through 20 years, in relation to the declining oxygen content of the profundal (Fig. 2). The Pisidia (*P. casertanum, P. subtruncatum,* and *P. henslowanum*) are seriously affected by oxygen decline, because their respiration declines rapidly at two critical limits, 30 and 15% air saturation (Fig. 4), respectively. Correspondingly, their numbers declined from 6 000 m⁻² in 1955–1957 to 690 m⁻² or almost 1/10, i.e. from 19.5 to 2.5% of the total fauna in 1971–1973. The midge *C. anthracinus* had no decline in respiration from 100 to 24% air saturation. A decline occurred between 24 and 2.5% air saturation. At

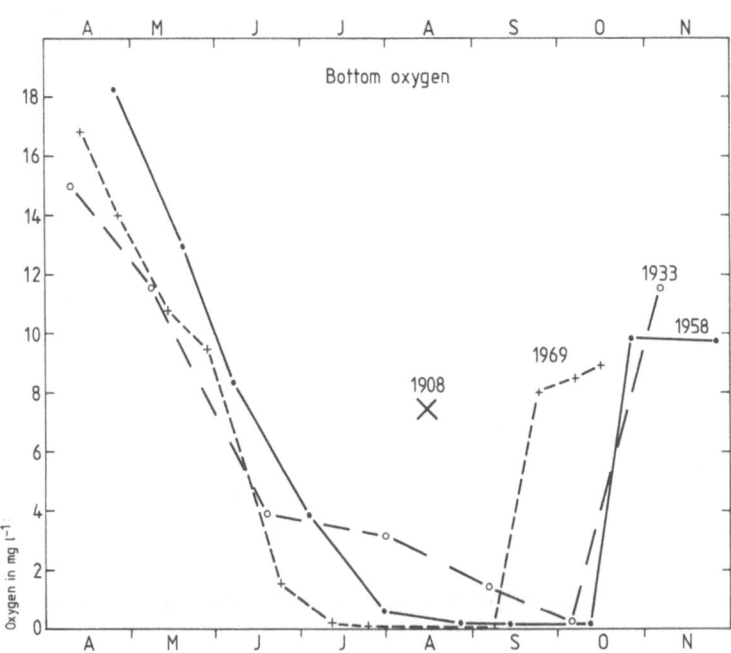

Fig. 1. Increase in length of oxygen depletion period from 1908 to 1979.

this lower point it respires at 75% of the normal respiration rate (Fig. 3). This species is extremely well adapted to low oxygen tensions, which was shown by its importance in the profundal community. It had an average of 9000 m^{-2} in 1958–1960 (range 70 000 to 2 000 m^{-2}) and in 1971–1973 8 580 m^{-2} (range 30 000 to 400 m^{-2}), but its percentage of the total fauna was lower, i.e. 31.5 versus 43.8%, and in 1973 it declined to 10% of the total fauna, and this is permanent. Most likely this species cannot survive the 3 months' period without oxygen. Population dynamics change from a 2-year cycle to a 1-year cycle at different depths. This was also the case at a depth of 20 m in 1962–63 due to low densities (Jónasson, 1972: Figs. 45, 54–57).

Potamothrix hammoniensis is extremely well adapted to low oxygen tensions with a critical limit at 9% air saturation, in spite of a decline of 25% in oxygen uptake between 100 and 9% saturation. The fluctuations in population size seem natural and not influenced by oxygen depletion. Their average increased from 26.8% to 57.7% of total and three-fold in numbers. By 1973 they had reached 22 000 m^{-2} and 80% of the total.

Oxygen demand and growth

Figures 3–5 show that critical limits of respiration coincide with declining oxygen content in the hypolimnion during July, when the hypolimnion temperature was 9 °C. The critical limits are as follows:

Chironomus anthracinus (Fig. 3). Upper critical limit, 24% air sat. equals 2.9 mg O$_2$ l^{-1}, 28 June.

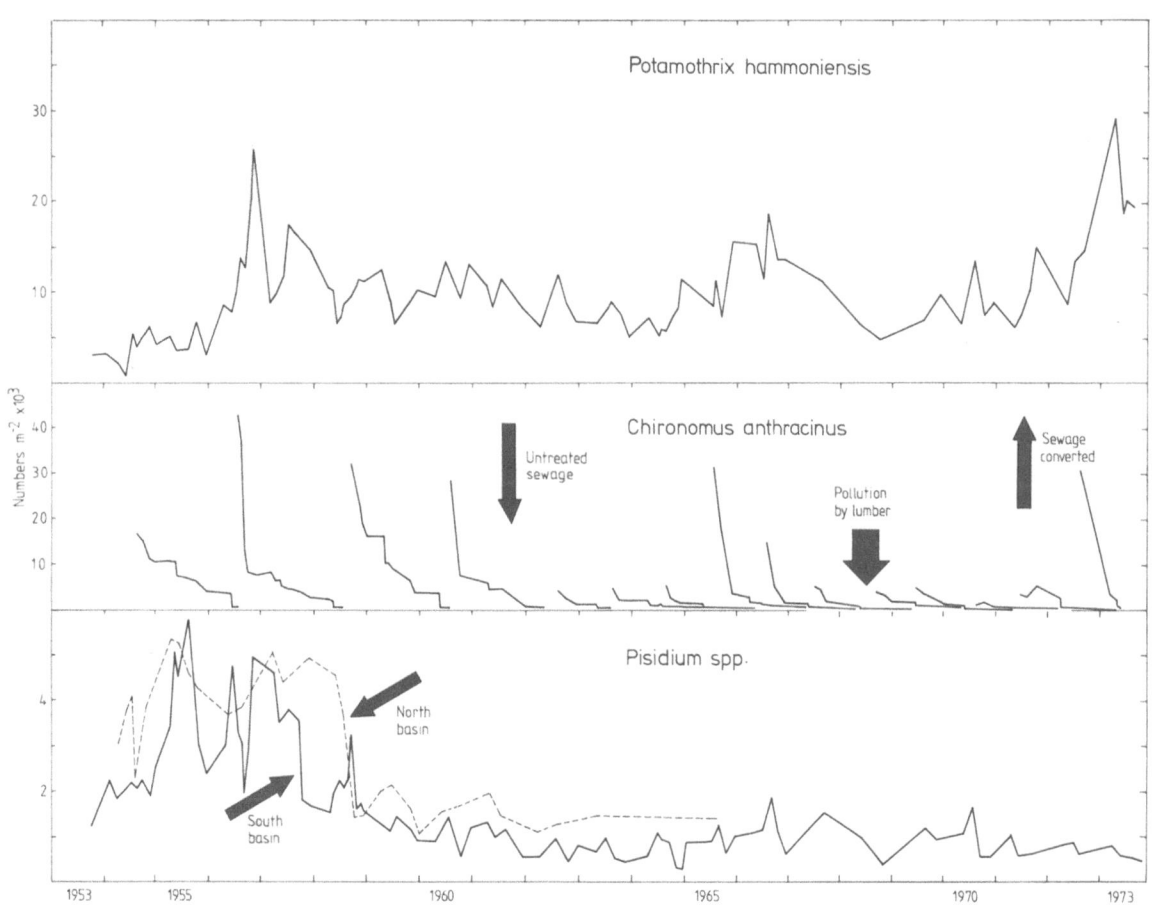

Fig. 2. Changes in abundance and population dynamics of *Potamothrix hammoniensis, Chironomus anthracinus* and 3 species of *Pisidium* (*P. casertanum, P. subtruncatum* and *P. henslowanum*) during the period 1953–1973.

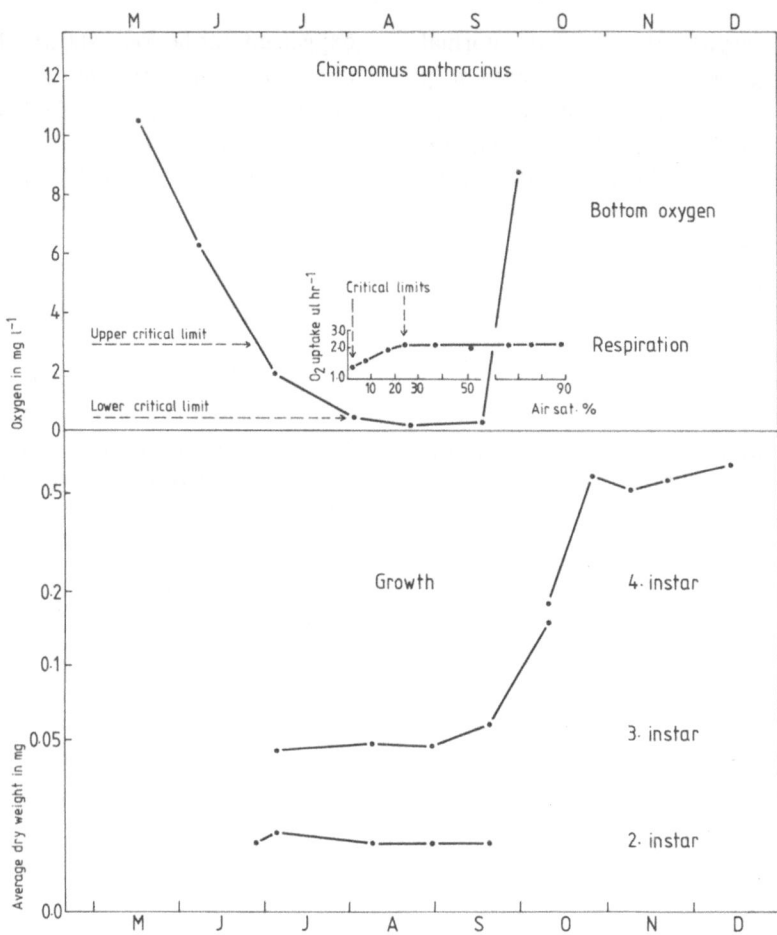

Fig. 3. The relations between declining oxygen content in the hypolimnion, decline in respiration and arrest of larval growth (*Chironomus anthracinus*).

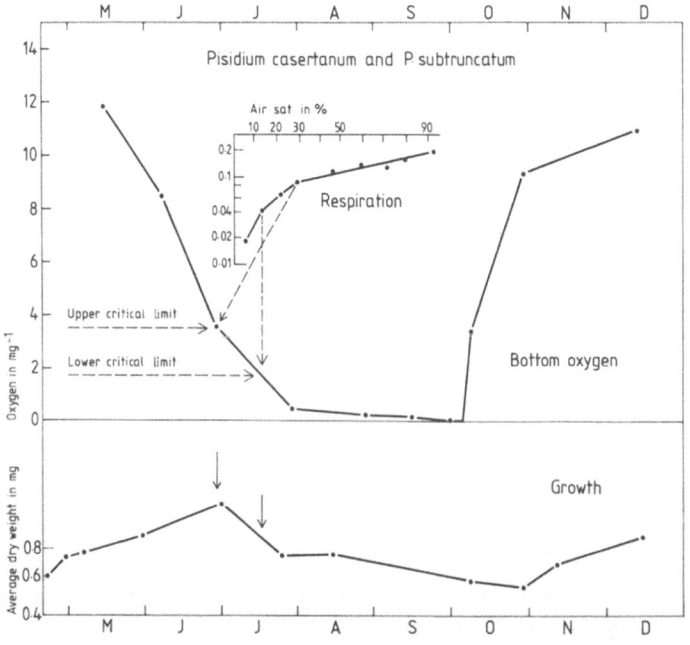

Fig. 4. The relations between declining oxygen content in the hypolimnion, decline in respiration and population growth (*Pisidium casertanum* and *P. subtruncatum*).

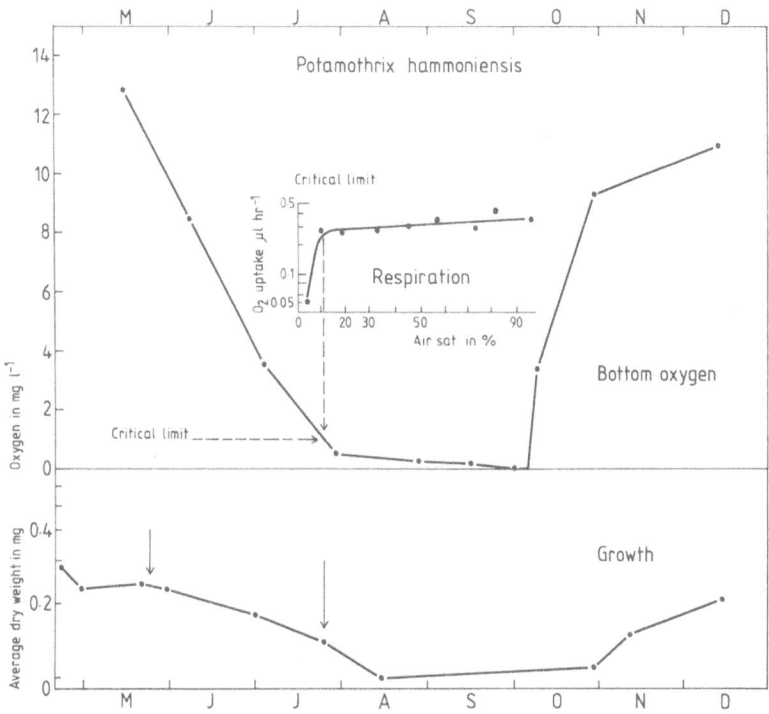

Fig. 5. The relation between declining oxygen content in the hypolimnion, decline in respiration and population growth (*Potamothrix hammoniensis*).

Lower critical limit, 2.5% air sat. equals 0.3 mg O_2 l^{-1}, 1 August.

Pisidium casertanum and *P. subtruncatum* (Fig. 4). Upper critical limit, 30% air sat. equals 3.5 mg O_2 l^{-1}, 28 June. Lower critical limit, 15% air sat. equals 1.7 mg O_2 l^{-1}, 16 July.

Potamothrix hammoniensis (Fig. 5). Critical limit, 9% air sat. equals 1.0 mg O_2 l^{-1}. 24 July.

In the case of *C. anthracinus*, growth of 2nd instar larvae stops definitively during stratification, while 3rd instar larvae increase in weight during the very last period of stratification, when tilting of the thermocline occurs.

The coincidence between seasonal decline in oxygen content of the bottom layers during summer stratification and the decline of population growth is extraordinarily clear in *Pisidium casertanum* and *P. subtruncatum* (Fig. 4) and *Potamothrix hammoniensis*. The 25% decline in oxygen uptake from 100% air saturation to 9% is also followed by a gradual decline in average weight (Fig. 5).

At the end of the stratification there is a clear difference between species. The surface feeding *C. anthracinus* utilizes immediately the increased oxygen content and increases its weight tenfold during the first 4 weeks after the autumn overturn. On the other hand, it takes the subsurface feeders *Pisidium casertanum, P. subtruncatum* and *P. hammoniensis* 3 weeks before they gain in weight after the autumn overturn.

Retardation of growth causes longer life cycles. This has been clearly demonstrated in *Chironomus anthracinus* which has a 2-year cycle in deep water and a 1-year cycle in shallower water (Jónasson, 1972). The three *Pisidium* species in Lake Esrom reached a maximum life-span of 4–5 years (Holopainen & Jónasson, 1983). The longest life-cycle of the profundal benthos was that of *P. hammoniensis* with a 6-year life-span (Jónason & Thorhauge, 1972, 1976a; Jónasson, 1978).

Discussion

This paper attempts to relate the low oxygen content of the hypolimnion during stratifiction, the

126

retardation of growth and its influence on population dynamics.

From Figs. 3–5 it is clear that growth follows oxygen decline during stratification, both summer and winter (see also Holopainen & Jónasson, 1983: Fig. 13; Jónasson & Kristiansen, 1967: Table 8). If stratification periods are long, growth and production seem negative (Jónasson & Thorhauge, 1976a; 1976b: Fig. 1). In the case of *Potamothrix hammoniensis, Pisidium casertanum* and *P. subtruncatum* growth starts 3 weeks after the autumn overturn. It follows that reproduction is delayed, e.g. in the *Pisidium* species, birth occurs 2 months after autumn overturn (Holopainen & Jónasson, 1983: Figs. 8–10). A long stratification period with low oxygen content thus has a fundamental influence on vital parameters for population dynamics, e.g. metabolism, reproduction, growth, mortality and production, which may change entirely the life-cycle and survival of different species.

However, in the present case as well as in general it is not yet clear why *Potamothrix hammoniensis* survives the oxygen free conditions better than *C. anthracinus*. The reason might well be the low metabolism of *Potamothrix hammoniensis* compared to *C. anthracinus* (cf. Jónasson, 1972: 122), and the fact that oligochaetes lower their respiration rate in mixed populations (Chapman *et al.*, 1982; Brinkhurst *et al.*, 1983). Another possible reason might be a different sulphur metabolism between oligochaetes and chironomids.

References

Berg, K., 1938. Studies on the bottom animals of Esrom Lake. K. danske Vidensk. Selsk. Skr. (Naturv. Math. Afdeling) 9(8): 1–255.

Brinkhurst, R. O., P. M. Chapman & M. A. Farrell, 1983. A comparative study of respiration rates of some aquatic oligochaetes in relation to sublethal stress. Int. Revue ges. Hydrobiol. (in press).

Chapman, P. M., M. A. Farrell & R. O. Brinkhurst, 1982. Effects of species interactions on the survival and respiration of Limnodrilus hoffmeisteri and Tubifex tubifex (Oligochaeta, Tubificidae) exposed to various pollutants and environmental factors. Wat. Res. 16: 1405–1408.

Holopainen, I. J. & P. M. Jónasson, 1983. Long-term population dynamics and production of Pisidium (Bivalvia) in the profundal of Lake Esrom, Denmark. Oikos 41: 99–117.

Jónasson, P. M., 1972. Ecology and production of the profundal benthos in relation to phytoplankton in Lake Esrom. Oikos Suppl. 14: 1–148.

Jónasson, P. M., 1978. Zoobenthos of lakes. Verh. int. Ver. Limnol. 20: 13–37.

Jónasson, P. M. & J. Kristiansen, 1967. Primary and secondary production in Lake Esrom. Growth of Chironomus anthracinus in relation to seasonal cycles of phytoplankton and dissolved oxygen. Int. Revue ges. Hydrobiol. 52: 163–217.

Jónasson, P. M. & F. Thorhauge, 1972. Life cycle of Potamothrix hammoniensis (Tubificidae) in the profundal of a eutrophic lake. Oikos 23: 151–158.

Jónasson, P. M. & F. Thorhauge, 1976a. Population dynamics of Potamothrix hammoniensis in the profundal of Lake Esrom with special reference to environmental and competitive factors. Oikos 27: 193–203.

Jónasson, P. M. & F. Thorhauge, 1976b. Production of Potamothrix hammoniensis in the profundal of eutrophic Lake Esrom. Oikos 27: 204–209.

Jónasson, P. M., E. Lastein & A. Rebsdorf, 1974. Production, insolation and nutrient budget of eutrophic Lake Esrom. Oikos 25: 255–277.

Oligochaete communities as biological descriptors of pollution in the fine sediments of rivers

Michel Lafont

Laboratoire d'Hydrobiologie du CEMAGREF; 3, Quai Chauveau, F-69009 Lyon, France

Keywords: aquatic Oligochaeta, biological quality indices, large rivers, pollution, fine sediments

Abstract

This study emphasizes the usefulness of Oligochaeta communities as descriptors of pollution in the fine sediments of large rivers. Two indices of biological quality of fine bottoms are proposed and are related to chemical parameters of the water. Tentative proposals are made for an empirical classification of polluted fine sediments.

Introduction

There are several difficulties in using Oligochaeta to assess the impact of human activities on large rivers. Each habitat category (fine sediment or stones and gravel, etc.) plays an important role in population dynamics of worms. Therefore, when evaluating the pollution tolerance of worms and describing biological pollution indicators, it is necessary to refer to the kind of bottom sediment (Lafont, 1977; Marshall & Winterbourn, 1979). The other problem is that it is almost impossible to find an unpolluted situation in large rivers. On the other hand, we can try to generalize the results obtained from a given sediment to the same sediment in other rivers considering the high degree of similarity between French oligochaete communities (Lafont, 1982).

Methods and sampling areas

The sampling and sorting protocols are to be found in two papers (Juget & Lafont, 1982; Wasson & Lafont, 1982). Samples were taken from permanent fine sediments (particles of $\leqslant 50$ μm diameter ranging from 30 to 80% in weight), at least 15 cm thick. Forty sampling sites situated in rivers of the East of France were studied (Lafont, 1982).

Results

The most damaged areas of the 40 sampling sites from a chemical and biological point of view, for example the river Saône at Pagny (Table 1), are those where fine sediments contain the lowest number of Oligochaeta species and the highest relative abundance of Tubificidae without hair setae (mature and immature worms). It is somewhat surprising to find that in large rivers such as the Rhône or Saône, Tubificidae with hair setae predominate in less polluted areas (Lafont, 1982; Lafont & Juget, 1976).

Proposals for two indices of biological quality of fine sediments

Two indices of biological quality of fine sediments are proposed:

$I_o = 10 S \cdot T^{-1}$, where S is the total number of species found in the sediment, T is the relative abundance of Tubificidae without hair setae (ma-

Hydrobiologia 115, 127–129 (1984).
© Dr W. Junk Publishers, Dordrecht.

Table 1. Saône river, seasonal values of I_o and E_o compared to minima (min.) and maxima (max.) of several chemical parameters of the water (from Wasson, 1980).

Sampling sites	Months	Number of species	Tubifi-cidae %	Tubific. without h. setae	I_o	E_o	BOD$_5$ mg l^{-1} min.	max.	NH$_4^+$ mg l^{-1} min.	max.	PO$_4^{3-}$ mg l^{-1} min.	max.	Cl mg l^{-1} min.	max.
Tillenay (slightly polluted)	May	21	73.3	27.2	7.7	E_{11}								
	Aug.	10	57	38	2.6	D_5	1.2	3.6	0.07	0.23	0.14	0.85	9	17
	Sept.	11	79	49	2.2	C_6								
Marnay (below the town of Chalon)	May	10	43.2	38.8	2.6	D_5								
	Aug.	7	86	50	1.4	C_4	1.7	3.7	0.11	0.81	0.11	0.74	25	225
	Sept.	8	97.6	72	1.1	B_4								
Pagny (below chemical outflow)	May	7	97.2	89.2	0.8	B_4								
	Aug.	9	92.7	79.3	1.1	B_5	2	3.8	0.10	0.66	0.21	0.98	44	450
	Sept.	11	87.8	75.6	1.5	B_6								

ture + immature worms) taken from the whole Oligochaeta community of the sediment.

The second index E_o is a composite one. Each letter represents the code for relative abundance of Tubificidae without hair setae. These classes are based upon observations made on the 40 sampling sites (Lafont, 1982): A = ⩾91%, B = 71 to 90%, C = 46 to 70%, D = 36 to 46%, E = 16 to 35%, F = ⩽15%. F can represent either polluted or unpolluted small streams and perhaps other situations we have not yet seen. Then Tubificidae with hair setae and *Tubifex tubifex* show mass development. This phenomenon is well known in the literature. Each number is the code for species richness: 1 = 1–2 species, 2 = 3–4 species, 3 = 5–6 species, etc. In our samples, the minimum I_o value is 0.1 and is corresponding to $E_o = A_1$ (Tubificidae without hair setae = 100%, one species found). The maximum I_o value is 7.7 and is corresponding to $E_o = E_{11}$.

Discussion and conclusion

In the river Saône (Table 1), physico-chemical measurements were made in the water column over a minimum period of at least 12 months and provide good representative data of the chemical condition of the water. As seen in Table 1, the lowest values of I_o (0.8 to 2.6) and E_o (B_4 to D_5) are related to high contents of NH$_4^+$, PO$_4^{3-}$ and Cl$^-$ in the water. Oligochaeta can be affected by toxic effects

of various substances in the sediment, and chemical effluents are well integrated by sedentary burrowing Oligochaeta. Moreover, all sediments are not contaminated in the same way and have various toxic effects on Oligochaeta (Chapman *et al.*, 1982). In our polluted fine sediments the most abundant species are: *Potamothrix hammoniensis, Limnodrilus claparedeianus, L. hoffmeisteri, L. udekemianus, Tubifex tubifex, T. ignotus, Amphichaeta leydigii, Vejdovskyella intermedia, Dero digitata, Specaria josinae, Marionina riparia.* The first six species are well known to be resistant to all kinds of pollutions in rivers (Brinkhurst, 1966, 1980; Aston, 1973; Howmiller & Scott, 1977; Ladle, 1971; Marshall & Winterbourn, 1979).

In contrast, the other five species are not known to be pollution tolerant in large rivers. It is necessary to consider that the number of resistant Oligochaeta is probably much greater than generally believed. All these observations emphasize results obtained by Chapman *et al.* (1982).

Tentative proposals are made in Table 2 for an empirical classification of polluted fine sediments. With indices I_o and E_o, we try to define pollution impact (I_o) and a biological parameter (E_o) of sediment. In my opinion, definition by means of the Oligochaeta communities of biological parameters of the muds (Lafont & Juget, 1981) is a way of assessing the pollution impact and can give an approach to a typology of the fine sediments of rivers.

Table 2. Biological classification of fine sediments of large rivers according to the degree of pollution (from Lafont, 1982).

E_o	I_o	Comments – Diagnosis
E_{11} E_{12}	5.1–8	slightly polluted sediments
D_{14}		*21–28 species* – Tub. without hair setae = *15–45%*
E_7 D_8	3.1–5	medium level of pollution – Sediments begin to become toxic but species richness can be still high.
C_{11} B_{12}		*14–24 species* – Tub. without hair setae = *15–90%*
E_5 D_6 D_9	2 –3	high level of pollution – Toxicity increases; species richness decreases.
C_7 C_6		*10–20 species* – Tub. without hair setae = *15–90%*
B_{10} B_9 B_8		
C_6 C_5	1.5–1.9	– idem – very high level of pollution.
B_7 B_6		*10–14 species* – Tub. without hair setae = *46–90%*
C_4 B_5 B_4	1 –1.4	reduced biocenosis – sediments are very toxic
A_5		*8–10 species* – Tub. without hair setae = *46–100%*
B_4 B_3	0.1–0.9	ultimate stage of pollution before azoic sediments
A_4 A_3 A_1		*1–8 species* – Tub. without hair setae = *71–100%*

Acknowledgements

I am indebted to Dr. J. Juget, Dr. J. G. Wasson, Dr. P. M. Chapman and Dr. M. Ladle for stimulating discussions and for criticism of my manuscript; to Ms. C. Duc and Dr. M. Ladle for revision of the English text; and to Ms. M. Taillole for typing the manuscript.

References

Aston, R. J., 1973. Tubificids and water quality: a review. Envir. Pollut. 5: 1–10.

Brinkhurst, R. O., 1966. The Tubificidae (Oligochaeta) of polluted waters. Verh. int. Ver. Limnol. 16: 854–859.

Brinkhurst, R. O., 1980. Pollution biology – the North American experience. In R. O. Brinkhurst & D. G. Cook (eds.), Aquatic Oligochaete Biology. Plenum Press, N.Y.; Lond.: 471–475.

Chapman, P. M., M. A. Farrel & R. O. Brinkhurst, 1982. Relative tolerances of selected aquatic oligochaetes to individual pollutant and environmental factors. Aquat. Toxicol. 2: 47–67.

Howmiller, R. P. & M. A. Scott, 1977. An environmental index based on relative abundance of oligochaete species. J. Wat. Pollut. Cont. Fed. 49: 809–815.

Juget, J. & M. Lafont, 1982. L'échantillonnage de la faune benthique: revue des techniques de prélèvements, d'extraction et de tri; application aux oligochètes. Sci. Eau 1: 243–254.

Ladle, M., 1971. The biology of Oligochaeta of Dorset chalk streams. Freshwat. Biol. 1: 83–97.

Lafont, M., 1977. Les oligochètes et la détection des pollutions dans les cours d'eau. Eaux Ind. 17: 84–85.

Lafont, M., 1982. Etat des connaissances acquises sur les peuplements d'Oligochètes. In Etude des méthodes biologiques d'appréciation quantitative de la qualité des eaux. Rapp. CEMAGREF, Lyon, 152–196.

Lafont, M. & J. Juget, 1976. Les Oligochètes du Rhône. I. Relevés faunistiques généraux. Ann. Limnol. 12: 253–268.

Lafont, M. & J. Juget, 1981. Les Oligochètes de quelques lacs jurassiens et leur utilisation pour apprécier l'état biologique des sédiments profonds. 1981. Ann. scient. Univ. Besançon 4: 47–57.

Marshall, J. W. & M. J. Winterbourn, 1979. An ecological study of a small New Zealand stream with particular reference to the Oligochaeta. Hydrobiologia 65: 199–208.

Wasson, J. G. & M. Lafont, 1982. Les Oligochètes et les Chironomidae: techniques d'échantillonnage. In Etude des méthodes biologiques d'appréciation quantitative de la qualité des eaux. Rapp. CEMAGREF, Lyon, 134–151.

Wasson, J. G., 1980. Etude écologique de la Saône préalable à l'implantation d'une centrale électronucléaire sur le site de Sennecey-Boyer. Rapp. CTGREF, 14 pp. + annexes.

Eutrophication of Lakes Léman and Neuchâtel (Switzerland) indicated by oligochaete communities

Claude Lang
Conservation de la Faune, Ch. du Marquisat 1, CH-1025 St-Sulpice, Switzerland

Keywords: aquatic Oligochaeta, benthos, eutrophication, indicator communities, lake, tubificid

Abstract

In 1978–80, oligochaete communities of meso-eutrophic Lake Léman (Lake of Geneva) were compared to those of mesotrophic Lake Neuchâtel. Worm species were classified into three groups corresponding to their increasing tolerance to eutrophication: (1) oligotrophic species, mostly *Peloscolex velutinus, Stylodrilus heringianus;* (2) mesotrophic species, mostly *Potamothrix vejdovskyi, P. bedoti;* (3) eutrophic species, mostly *Potamothrix hammoniensis, P. heuscheri, Tubifex tubifex.* In both lakes, eutrophic species constituted the bulk of the communities in terms of absolute abundance. However, relative abundance of mesotrophic and eutrophic species was higher in Lake Léman; oligotrophic species were more important in Lake Neuchâtel. These data confirmed the trophic classification of lakes based on chemical parameters. The number of zero values, which perturbated statistical analysis, was reduced by using species groupings instead of isolated species. Thus, making the lakes more comparable even if different species were present in each one. Relative density values based on all samples were distributed among 4 density classes for the 3 species groupings. The 12 resulting frequencies described the community structure expressed in terms of eutrophication. Furthermore, these frequencies may be used for comparison of eutrophication levels in several lakes.

Introduction

In the present study, oligochaete communities of meso-eutrophic Lake Léman (Lake of Geneva) were compared to those of mesotrophic Lake Neuchâtel. In both lakes, oligochaete communities changed strongly according to depth and to area (Lang & Lang-Dobler, 1980a; Lang & Cuvit, 1981). Therefore, a whole set of oligochaete communities, reflecting a large range of environmental conditions, was available for comparison. This comparison resulted in: (1) definition of the specific data necessary to compare the lakes; (2) the evaluation of oligochaetes as an indicator of eutrophication.

Different approaches were possible to show the relationship between benthic communities and the trophic level of lakes. Community structure may be defined either directly by component species or indirectly by regrouping species with similar sensitivity to eutrophication. The first approach – using directly component species – was feasible only if the species were restricted to narrow trophic ranges; as were the species constituting the 15 characteristic chironomid communities described by Saether (1979).

The second approach – regrouping species – appeared more practical with species being present in a broad range of trophic conditions and whose relative density increased in some point of this range. This kind of method has been used for diatoms in rivers, for polychaetes in the sea, for oligochaetes in lakes (Lange-Berthalot & Lorbach, 1979; Bellan, 1980; Lang & Lang-Dobler, 1980b), and it has been applied in the present study.

Hydrobiologia 115, 131–138 (1984).

132

Stations and methods

Characteristics of the lakes (Table 1)

Phosphorus concentrations were increasing with time in both lakes. Oxygen concentrations were decreasing in the deepest area of Lake Léman whereas they remained high in the profundal of Lake Neuchâtel (Fahrni, 1982). The persistence of high oxygen concentrations in Lake Neuchâtel was attributed to its orientation in the axis of the two prevailing winds, favoring a strong mixing.

Sampling stations and methods

In Lake Léman, 294 stations were distributed regularly in the whole lake (1 station per 1.96 km^2). In Lake Neuchâtel, 77 stations were located on 20 equidistant transects perpendicular to the coast, distributed in the whole lake. On every transect, stations were situated at a depth of 40, 60, 90 and 120 m. The mean depth of the sampling stations corresponded to the mean depth of each lake (Table 1). Stations were visited during the spring 1978 in Lake Léman, some during the summer 1979 and most (68.8%) during the summer 1980 in Lake Neuchâtel.

Stations were sampled with a Shipek grab in Lake Léman, with a corer (type Kajak-Brinkhurst) in Lake Neuchâtel. Furthermore in 1979, 15 stations (10 to 35 m deep) were sampled by a diver using hand-pushed cores in Lake Neuchâtel. Sedi-

ment samples were constituted by 16 cm^2 cores, 10 to 20 cm long, in both lakes. One 16 cm^2 core was collected from the sediment, inside the Shipek grab, at every station of Lake Léman. In Lake Neuchâtel, 10 cores were taken by the corer per station in 1979, 4 per station in 1980.

Worms were present in 197 of the 294 samples taken in Lake Léman. Cores without oligochaetes, attributed to low sampling efficiency of the Shipek grab (Brinkhurst, 1974), were not included in subsequent analysis. Numbers of worms collected by the diver and by the corer were comparable. The core, taken from inside the Shipek grab, tended to underestimate the number of worms, but less than if the whole Shipek was sampled.

Identification of oligochaetes

For both lakes, tubificid and lumbriculid worms, stained in Bengal Rosa, were separated from the sediment with a sieve (0.2 mm mesh size aperture). All worms which were present in the cores from Lake Léman were mounted in Hydramount of Gurr. A random subsample of 50 worms was used if more than 50 individuals were present. Worms were counted in every core from Lake Neuchâtel and their mean density per core was calculated. A random subsample of 30 to 60 worms was taken from all the cores pooled together and mounted in Water Mounting Medium of Gurr. The relative density of every species in the subsample was related to the total mean worm density per core to calculate their

Table 1. Characteristics of Lake Léman and of Lake Neuchâtel.

Variable	Unit	Léman	Neuchâtel
Lake surface	km^2	581.5	214.6
Mean depth	m	153	64
Maximum depth	m	310	153
Lake volume	km^3	89.0	13.8
Theoretical water residence time	year	11.9	8.2
Area of drainage basin	km^2	7 975	2 672
Average altitude of the drainage basin	m	1 670	780
Altitude of lake surface	m	372	429
Number of inhabitants in the drainage basin	10^3	760	245
Average concentration of total phosphorus at 1979 spring overturn	mg m^{-3}	90	56
Lowest oxygen concentration recorded during 1979 in the profundal	mg l^{-1}	1.7	8.0
More than 4 mg O$_2$ l^{-1} up to a depth of	m	260	153
Lowest average concentration recorded in the whole lake	mg l^{-1}	7.4	9.0
Trophic level		meso-eutrophic	mesotrophic

Sources: Fahrni (1982), Lachavanne (1980), Sollberger (1974), B. Pokorni (pers. commun.).

absolute density per m². Immature worms were identified according to the shape of setae and other characters.

Previous studies

The studies of Monard (1919) in Lake Neuchâtel and of Juget (1967) in Lake Léman were used as references. Monard's study consisted of 71 stations sampled in 1917 and 1918 with a modified Steinmann dredge. Stations were located in the northern part of the lake; their mean depth was 67 m. Juget's study consisted of 58 samples taken with an Ekman grab in 10 stations visited between 1958 and 1967; the mean depth of the stations was 95 m.

Data analysis methods

Species groupings

Some worm species, frequent and abundant in one lake, were absent in the other. Also, most species were absent in too many samples (too many zeros) to perform statistical analysis of their density (Green, 1979). To overcome these problems, oligochaete species, with similar tolerance to eutrophication, were regrouped. Species 1 to 3 of Table 2 – pooled together – represented the oligotrophic species, i.e. the species typical of oligotrophic lakes. Species 4 to 11 represented the mesotrophic species; species 12 to 18, the eutrophic species. The used classification of species in terms of eutrophication was discussed elsewhere (see 'Oligochaetes as indicators' in the discussion). Using species groupings instead of the isolated species reduced the number of zero values. Furthermore, the lakes were more comparable even if different species were present in each one.

Types of data

Two types of data were analysed in this study, the frequency of occurrence and the density. The frequency of occurrence represented the number of samples where a given worm species or a species grouping were present, irrespective of their abundance (Table 2). The absolute density was based on the number of individuals of a given species grouping per sample. The relative density was based on the number of individuals of a given species grouping expressed as a percentage of the total density per sample. The relative density represented the community structure, i.e. the relationship between species groupings.

Median density

The value of absolute (or relative) density of a species grouping may be summarized by the arithmetic mean or the median. In this study, the median was used because this statistic was more robust than the mean, especially if extreme values were present (Reckhow, 1980). For example, the mean value of the absolute density was higher in Lake Léman than in Lake Neuchâtel; the reverse was true for the median. Furthermore, the median corresponded to the assumption of the Mann-Whitney test which was used to compare the density of species groupings in both lakes (Conover, 1971).

Classes of density

To analyse the data differently, density classes were defined and the frequency of samples was calculated for each class. One problem, however, was being objective in defining the limits for the density classes. Results reviewed by Wiederholm (1980) helped to solve this problem for the absolute density: they indicated that densities exceeding 10 000 worms m⁻² were characteristic or organic pollution. This value was selected for the lower limit of the second density class, and this value was doubled for the lower limit of the third density class.

The density classes for the relative density were defined from results of previous factorial correspondence analysis (Lang & Lang-Dobler, 1980b). In this analysis, samples were described by the relative density of the three species groupings. Relationship between samples and the three species groupings were graphically displayed on a factorial plot. Samples close to a species grouping were characterized by high relative density of this species grouping; samples far from this species grouping were characterized by low relative density of this species grouping. Distribution of samples on the factorial plot according to their distance from species groupings permitted us to select limits for density. This selection was more objective than for the absolute density.

134

Typical communities

According to these data, the minimal value of the relative density necessary to consider one worm community as typical of a given species grouping was equal or superior to 50% for typical oligotrophic and mesotrophic communities, and 100% for typical eutrophic communities. This difference was due to the resistance of eutrophic species to a broad range of conditions. In fact, the so-called eutrophic species make up a significant portion of the community in oligotrophic lakes. For example, *Potamothrix hammoniensis,* a typical eutrophic species, was very frequent in the oligotrophic Lake Neuchâtel in 1918 (Table 2).

Exclusion of zero values

In Table 3, worm densities were analysed in two ways. First, the samples with zero values were included to describe the whole lake. Second, these samples were excluded to describe exclusively the communities where a given species grouping was present (selected samples of Table 3). Densities of oligotrophic and of mesotrophic species groupings were higher in selected samples than in the whole lake because these groupings were absent in many samples.

Calculations and statistical tests were performed with the SPSS package (Nie, *et al.,* 1975; Hull & Nie, 1981).

Results

Frequency of occurrence (Table 2)

Some species, frequent in one lake, were absent in the other. *Bythonomus lemani* (in 1958–67), *Potamothrix vejdovskyi* and *P. heuscheri* were frequent in Lake Léman, absent in Lake Neuchâtel. *P. moldaviensis* and *P. bedoti* were frequent in Lake Neuchâtel, but absent in Lake Léman (Table 2). At

Table 2. Percentage of samples where a given worm species is present and mean relative density (%) of worm species in Lake Léman and in Lake Neuchâtel.

No	Worm species	Percentage of presence in samples				Significance of Chi^2	Mean relative density (%)	
		Neuchâtel 1918[a]	Léman 1958–67[b]	Neuchâtel 1979–80	Léman 1978		Léman 1978	Neuchâtel 1979–80
1	*Bythonomus lemani* Grube	0	63.8	0	1.5	NS	0.5	0
2	*Stylodrilus heringianus* Claparède	33.8	62.1	50.6	9.6	*	2.9	9.8
3	*Peloscolex velutinus* Grube	77.5	62.1	31.2	5.1	*	1.1	3.5
4	*Potamothrix vejdovskyi* Hrabě	0	46.6	0	36.5	*	16.5	0
5	*Peloscolex ferox* Eisen	14.0	27.6	15.6	10.7	NS	2.4	0.9
6	*Psammoryctides barbatus* Grube	11.2	13.8	7.8	1.5	*	0.2	0.6
7	*Aulodrilus limnobius* Bretscher	0	0	0	0.5	NS	0.1	0
8	*Potamothrix moldaviensis* Vejdovsky & Mrázek	0	0	16.9	0	*	0	4.6
9	*Potamothrix bedoti* Piguet	0	(+)0	32.5	(+)0	*	0	6.7
10	*Ilyodrilus templetoni* Southern	0	6.9	3.9	0.5	NS	0.3	0.1
11	*Aulodrilus pluriseta* Piguet	1.4	34.5	0	4.6	NS	0.5	0
12	*Limnodrilus profundicola* Verrill	1.4	10.3	5.2	0.5	*	0.1	0.1
13	*Limnodrilus claparedianus* Ratzel	2.8	–	3.9	0	*	0	0.1
14	*Limnodrilus hoffmeisteri* Claparède	5.6	–	23.4	4.6	*	0.2	1.0
15	*Limnodrilus* species 12 to 14	–	74.1	74.0	35.0	*	6.9	8.4
16	*Potamothrix heuscheri* Bretscher	0	93.1	0	44.7	*	19.4	0
17	*Potamothrix hammoniensis* Michaelsen	63.5	58.6	89.6	34.0	*	13.9	27.0
18	*Tubifex tubifex* Müller	18.3	77.6	94.8	62.4	*	35.3	38.2
19	Oligotrophic species 1 to 3	90.1	74.1	59.7	13.7	*	4.6	13.3
20	Mesotrophic species 4 to 11	26.6	60.3	57.1	40.1	*	20.0	13.0
21	Eutrophic species 12 to 18	76.1	98.3	100.1	94.4	NS	75.4	73.4
	Number of samples	71	58	77	197		197	77

The Chi^2 test is calculated from data Léman, 1978 and Neuchâtel, 1979–80. NS not significant, * p $\leqslant$ 0.05, – missing data; [a] Monard (1919), [b] Juget (1967), (+) species present in the lake, not in the above samples.

the beginning of this century, *Bythonomus lemani* was very frequent in Lake Léman, though absent in Lake Neuchâtel (Monard 1919). The other above species were more or less recent immigrants.

In 1978–80, *Stylodrilus heringianus* became more frequent than *Peloscolex velutinus* or *Bythonomus lemani* in both lakes. In 1918, *Peloscolex velutinus* was absent from the area directly influenced by the River Areuse (Monard, 1919). This species and *Bythonomus lemani* seemed to be the ultra-oligotrophic species which were very frequent in most oligotrophic Swiss lakes at the beginning of the century (Piguet & Bretscher, 1913). Their replacement by *Stylodrilus heringianus* indicated a subtle shift from very sensitive towards more tolerant oligotrophic species.

The frequency of worm species in lakes was in some measure related to their mean relative density (Table 2). The most frequent species were, in general, the most abundant in the community. Four species constituted the bulk of the communities in Lake Léman, three species in Lake Neuchâtel. In both lakes, *Tubifex tubifex* was the most abundant

species, followed by *Potamothrix hammoniensis* in Lake Neuchâtel, by *P. heuscheri* in Lake Léman.

In 1978–80, oligotrophic and mesotrophic species were more frequent in Lake Neuchâtel than in Léman; the eutrophic species were present in almost all samples. The frequency of oligotrophic species decreased strongly from Lake Neuchâtel 1918 to Lake Léman 1978 (Table 2). During the same period, frequency of mesotrophic species showed first an increase and then a decrease. Eutrophic species constituted rapidly the bulk of the community. The frequency of these species groupings was a robust criterion to compare eutrophication of several lakes. However, many small samples were necessary to assess precisely species frequency in a whole lake.

Absolute density of species groupings (Table 3, No 1–6)

Overall, the total absolute density of three species groupings and the absolute density of oligotrophic and eutrophic species were higher in Lake Neuchâ-

Table 3. Median absolute and relative density of 3 groupings of worm species in Lakes Léman (L) and Neuchâtel (N) and percentage of samples in four density classes.

No.	Species groupings	Lake	Median	Significance Mann-Whitney	Upper limits of density classes				Units $n\,m^2$ %	Significance of Chi^2
					0 / 0	9 999 / 9	19 999 / 49	500 000 / 100		
1.	Total	L	5919.6	*	–	62.4	16.2	21.3		*
	(All samples)	N	13750.0		–	40.3	33.8	26.0		
2.	Oligotrophic	L	53.0	*	86.3	12.7	0.5	0.5		*
	(All samples)	N	712.0		40.3	57.1	2.6	0		
3.	Oligotrophic	L	933.8	NS	–	92.6	3.7	3.7		NS
	(If present)	N	1206.5		–	95.7	4.3	0		
4.	Mesotrophic	L	233.3	NS	59.9	31.0	6.6	2.5		*
	(All samples)	N	351.0		42.9	55.8	1.3	0		
5.	Mesotrophic	L	3068.2	*	–	77.2	16.5	6.3		*
	(If present)	N	2121.0		–	97.7	2.3	0		
6.	Eutrophic	L	3631.4	*	5.1	68.5	11.7	14.7		*
	(All samples)	N	10000.0		0	49.4	35.1	15.6		
7.	Oligotrophic	L	0.004	*	86.3	4.1	5.6	4.1		*
	(All samples)	N	4.649		40.3	28.6	20.8	10.1		
8.	Oligotrophic	L	21.700	NS	–	29.6	40.7	29.6		NS
	(If present)	N	10.700		–	47.8	34.8	17.4		
9.	Mesotrophic	L	0.055	NS	59.9	3.6	14.7	21.8		*
	(All samples)	N	4.167		42.9	14.3	35.1	7.8		
10.	Mesotrophic	L	52.200	*	–	8.9	36.7	54.4		*
	(If present)	N	15.500		–	25.0	61.4	13.6		
11.	Eutrophic	L	99.943	*	5.1	2.0	16.8	76.1		NS
	(All samples)	N	83.333		0	0	22.1	77.9		

No. 1 to 6 absolute density, 7 to 11 relative density.
* p ⩽ 0.05; NS not significant, – not included.

tel than in Lake Léman according to the values of the median, whereas mesotrophic species had the same density in both lakes. However, in selected samples, absolute densities of oligotrophic species was the same in both lakes, whereas mesotrophic species predominated in Lake Léman. These differences were clearly due to the higher number of zero values recorded in Lake Léman (Table 2). In this case, estimation of the absolute density was modified by different frequencies of occurrence.

Most absolute densities of oligotrophic and mesotrophic species were included in the low density class, while, on the other hand, eutrophic species were more abundant in the high density classes. This difference demonstrated the pratical significance of these classes to indicate eutrophication.

Values of total densities were distributed differently in both lakes: proportionately there were more low and high values present in Lake Léman than in Lake Neuchâtel.

These data were difficult to interpret because eutrophication contributed to either the increase or the decrease of the absolute density of the worms. The absolute density of the worms increased with organic sedimentation only if the oxygen was not limited (Lang & Hutter, 1981); if this element was deficient, absolute density decreased as was shown in the deepest area of Lake Léman. Total densities were higher than 50 000 worms m^{-2} (up to a maximum of 430 000 m^{-2}) in 7.6% of the samples from Lake Léman. These high density stations were exposed to the inputs of the Rhône River. Such high densities were not encountered in Lake Neuchâtel.

Relative density of species groupings (Table 3, No 7–11)

Typical oligotrophic communities were scarce and their numbers were the same in both lakes, whereas typical mesotrophic and eutrophic communities were better represented in Lake Léman than in Lake Neuchâtel. Percentages of typical eutrophic communities were, respectively, 53.3% and 14.3% (p < 0.001). In Lake Léman, typical communities (oligotrophic, mesotrophic or eutrophic) were present in 79.2% of the samples versus 32.2% in Lake Neuchâtel. Intermediate communities were represented in the other samples.

The frequencies of samples in 4 classes of relative (or absolute) density for the three species groupings

(i.e. 12 samples frequencies) described fairly well the structure of oligochaete communities expressed in terms of eutrophication (Table 3).

Factorial correspondence analysis, based on these 12 frequencies, may be used for comparison of several lakes. These data clearly demonstrated how a single value, such as the mean or the median, was not enough to define the eutrophication level of a lake by the oligochaete communites.

Discussion

Oligochaetes as indicators

Data presented in this paper were based on the fact that oligochaete species may be classified according to their tolerance to eutrophication. Classification of species in terms of tolerance presented in Table 2 (No. 19–21) was very similar to that of Milbrink (1980) with two exceptions. Firstly, *Peloscolex ferox* was considered in my study as mesotrophic rather than oligotrophic, and secondly, *Limnodrilus profundicola* became eutrophic instead of oligotrophic.

Two important modifications of the previous classification of worm species in terms of eutrophication (Lang & Lang-Dobler, 1980b) derived from a study made in two 35 m deep stations in Lake Léman (Lang & Hutter, 1981). In this study, the annual organic sedimentation was equal to 157 g C m^{-2} in station 1, to 214 g in station 2. The absolute and the relative densities of oligotrophic species (Table 2, No 1–3) decreased strongly in station 2 compared to station 1. On the contrary, the absolute and the relative densities of *Potamothrix hammoniensis* and *P. heuscheri* increased in station 2. The relative density of *Potamothrix vejdovskyi* decreased in station 2 whereas its absolute density increased. The intermediate position of *P. vejdovskyi* between oligotrophic and eutrophic species suggested for this species a sensitivity corresponding to mesotrophic conditions. *Potamothrix heuscheri* displayed the same reaction as *P. hammoniensis* to increasing organic inputs. Therefore, this species was classified as eutrophic.

In a second study, in Lake Morat, a typical eutrophic lake (Davaud, 1976), *Limnodrilus profundicola* constituted 16.2% of worm communities, the other two species being *Potamothrix hammoniensis* and *Tubifex tubifex* (Lang & Cuvit, 1981).

Table 4. Compared eutrophication of Lakes Léman and Neuchâtel according to oligochaete communities.

Type of data	Species groupings	Observed trend Léman	Neuchâtel	More eutrophic lake according to the trend	Trend increased by sampling design
Occurrence frequency	Oligotrophic	-	+	Léman	- in Léman
	Mesotrophic	-	+	Léman	
	Eutrophic	=		=	
Absolute density	All	-	+	Neuchâtel	- in Léman
	Oligotrophic	=		=	
	Mesotrophic	+	-	Léman	
	Eutrophic	-	+	Neuchâtel	
Relative density	Oligotrophic	=		=	+ in Léman
	Mesotrophic	+	-	Léman	
	Eutrophic	+	-	Léman	
Frequency of typical communities	Oligotrophic	=		=	+ in Léman
	Mesotrophic	+	-	Léman	
	Eutrophic	+	-	Léman	

Trend: + high; - low; = same in both lakes.
Absolute and relative density was considered exclusively for samples where a given species grouping was present (Table 3).

Peloscolex ferox, classified as a typical oligotrophic species in Swedish lakes (Milbrink, 1980), was dominant in Vierwaldstättersee at the beginning of the century (Obermayer, 1922). *Peloscolex velutinus,* which was absent from this lake, was dominant in the other Swiss lakes. This observation suggests a kind of competitive interaction between these two species.

Therefore, classification of a given worm species according to eutrophication depended also on the other species present. For example, a species may be absent from a lake for mere zoogeographical reasons. This absence permitted another equivalent species to fill up its ecological niche and to take its role as an indicator of eutrophication. Consequently, the attribution of a worm species to a trophic category must be reevaluated for each lake.

Differences in sampling design (Table 4)

According to several criteria, oligochaete communities indicated that Lake Léman is more eutrophic than Lake Neuchâtel. However, some of these results might have been biased by differences in sampling design. For example, Shipek grab tended to underestimate absolute density in Lake Léman. In Lake Neuchâtel on the other hand, results of several cores pooled together tended to: (1) increase the frequency of the presence of scarce species, (2) reduce the frequency of extremely high and low values of the absolute density and (3) reduce the frequency of communities dominated by one species grouping.

Advantages of species groupings

Regrouping worm species according to their tolerance to eutrophication presented several advantages: (1) the number of zero values was reduced, (2) the absolute and the relative density of worm categories were quantifiable in both lakes, therefore amenable to statistical tests, (3) the use of species groupings instead of the isolated species bypassed the identification problem of immature individuals for some species, (4) the seasonal variations of species groupings were also less than those of the component species (Lang & Hutter, 1981) and (5) the use of species groupings overcame problems due to different zoogeographical distributions of worm species.

Acknowledgments

Mrs B. Lang identified oligochaetes of Lake Léman, and Mrs L. Faravel provided technical assistance. Mr J. M. Helbling assisted with statistical methods. Mrs S. Hurni corrected the English, and Miss J. Sonnay and V. Nicole typed the manuscript. The comments of two anonymous reviewers helped to improve the paper. Mr R. Ducret helped to sample Lake Neuchâtel. Samples were taken in Lake Léman by the 'Laboratoire de sédimentologie de l'Université de Genève' (Prof. J. P. Vernet). Studies were supported in both lakes by the 'Office fédéral de la protection de l'environnement'.

References

Ballon, G., 1980. Relationship of pollution to rocky substratum polychaetes on the French Mediterranean. Coast. mar. Pollut. Bull. 11: 318–321.

Brinkhurst, R. O., 1974. The benthos of lakes. Macmillan Press, Ltd, Lond., Basingstoke, 190 pp.

Conover, W. J., 1971. Practical nonparametric statistics. J. Wiley & Sons, N.Y., 462 pp.

Davaud, E., 1976. Contribution à l'étude géochimique et sédimentologique de dépôts lacustres récents (Lac de Morat, Suisse). Thèse Univ. Genève, 129 pp.

Fahrni, H. P., 1982. Die Belastbarkeit von Seen. Gas Wass. Abwass. 62: 122–134.

Green, R. H., 1979. Sampling design and statistical methods for environmental biologists. J. Wiley & Sons. N.Y. 257 pp.

Hull, C. H. & N. H. Nie, 1981. SPSS update 7–9. New procedures and facilities for releases 7–9. McGraw-Hill Publishing Co Ltd, N.Y., 402 pp.

Juget, J., 1967. La faune benthique du Léman: modalités et déterminisme écologique du peuplement. Thèse doct. Univ. Lyon, 360 pp.

Lachavanne, J. B., 1980. Les manifestations de l'eutrophisation des eaux dans un grand lac profond: le Léman (Suisse). Schweiz. Z. Hydrol. 42: 127–154.

Lang, C. & B. Lang-Dobler, 1980a. Niveau d'eutrophisation du lac Léman en 1978 évalué à partir des communautés de vers (tubificidés et lumbriculidés). Rapport présenté à l'office fédéral de la protection de l'environnement (disponible sur demande).

Lang, C. & B. Lang-Dobler, 1980b. Structure of tubificid and lumbriculid worm communities, and three indices of trophy based upon these communities, as descriptors of eutrophication level of Lake Geneva (Switzerland). In R. O. Brinkhurst & D. G. Cook (eds.), Aquatic oligochaete biology. Plenum Press, N.Y. Lond.: 457–470.

Lang, C. & L. Cuvit, 1981. Degré d'eutrophisation du lac de Neuchâtel en 1979-80 et du lac de Morat en 1980 évalué à partir de la faune des sédiments. Rapport présenté à l'office fédéral de la protection de l'environnement (disponible sur demande).

Lang, C. & P. Hutter, 1981. Structure, diversity and stability of two oligochaete communities according to sedimentary inputs in Lake Geneva (Switzerland). Schweiz. Z. Hydrol. 43: 265–276.

Lange-Bertalot, H. & K. D. Lorbach, 1979. Die Diatomeenbesiedlung des Rheins in Abhängigkeit von Abwasserbelastung. Arch. Hydrobiol. 87: 347–363.

Milbrink, G., 1980. Oligochaete communities in pollution biology: the european situation with special reference to lakes in Scandinavia. In R. O. Brinkhurst & D. G. Cook (eds.), Aquatic oligochaete biology. Plenum Press, N.Y., Lond.: 433–455.

Monard, A., 1919. La faune profonde du lac de Neuchâtel. Thèse Univ. Neuchâtel. Imprimerie centrale, Neuchâtel, 176 pp.

Nie, N. H., C. H. Hull, J. G. Jenkins, K. Steinbrenner & P. H. Bent, 1975. SPSS, Statistical package for the social sciences. McGraw-Hill Publishing Ltd, Co. N.Y., 675 pp.

Obermayer, H., 1922. Beitrage zur Kenntnis der Litoralfauna des Vierwaldstättersees. Schweiz. Z. Hydrol. 2: 1–105.

Piguet, E. & K. Bretscher, 1913. Oligochètes. Catalogue des invertébrés de la Suisse. Georg & Cie, Genève, 215 pp.

Reckhow, K. H., 1980. Techniques for exploring and presenting data applied to lake phosphorus concentration. Can. J. Fish. aquat. Sci. 37: 290–294.

Saether, O. A.,1979. Chironomid communities as water quality indicators. Holarct. Ecol. 2: 65–74.

Sollberger, H., 1974. Le lac de Neuchâtel (Suisse). Ses eaux, ses sédiments, ses courants sous-lacustres. Thèse Univ. Neuchâtel, 415 pp.

Wiederholm, T., 1980. Use of benthos in lake monitoring. J. Wat. Pollut. Cont. Fed. 52: 537–547.

Lethal and sublethal tolerances of aquatic oligochaetes with reference to their use as a biotic index of pollution

Peter M. Chapman[1] & Ralph O. Brinkhurst[2]

[1]E.V.S. Consultants Ltd., 195 Pemberton Avenue, North Vancouver, B.C., Canada V7P 2R4

[2] Ocean Ecology Laboratory, Institute of Ocean Sciences, P.O. Box 6000, 9860 West Saanich Road, Sidney, B.C., Canada V8L 4B2

Keywords: aquatic Oligochaeta, pollution, indicators, toxicity

Abstract

A series of recent studies have been completed by the authors involving: 1) determining the lethal tolerances of 12 oligochaete species classified (from ecological studies) as tolerant, moderately tolerant and intolerant to selected chemical toxicants and environmental factors under defined bioassay conditions with and without sediment; 2) determining lethal tolerances of candidate species to toxicants in combination with a range of abiotic factors; 3) measuring respiratory stress imposed by exposure to individual and combined sublethal concentrations of toxicants and environmental factors; and, 4) determining differences in lethal tolerance and respiratory stress between individual and mixed species. Surprisingly few previous studies have been done in this area considering the importance of oligochaetes as field pollution indicators. The results of the above major studies coupled with histopathological work are reviewed. Data from these studies substantiate the present use of oligochaete species assemblages as indicators of organic pollution and suggest their use in the laboratory for toxicant screening tests. The range of responses of different oligochaete species to individual and combined stress is complex, particularly in mixed species, which provides useful indications of specific stress factors. The application of these experimental laboratory studies to field situations is described.

Introduction

At the First Aquatic Oligochaete Symposium three years ago, we each noted areas of oligochaete biology requiring future work. One of us (Chapman et al., 1980) emphasized the need for toxicity studies. The other (Brinkhurst, 1980a, b) noted a lack of traditional tolerance tests and recommended that mixed species studies be undertaken.

Having followed our own advice, we can now present detailed information on toxicological tests with oligochaetes and the application of this information to field studies. The following account is thus an overview and synthesis of a number of studies representing state-of-the-art information.

Lethal tolerance tests

The relative tolerances of indicator species to a series of individual chemical toxicants and environmental factors were determined by Chapman et al. (1982a). Methodology involved 96 h acute lethal bioassays with 24 h solution replacement. Experiments were run both with and without sediment. Mortalities were compared using a basic t-test.

A rank order of tolerance was determined for all combinations tested (Fig. 1), which indicated that relative tolerances were toxicant specific. The use of present oligochaete species assemblages (tolerant, moderately tolerant, intolerant) to indicate degree of trophy was substantiated. However, the same rank order did not hold for chemical toxicants. Sediments were observed to modify toxicity, in many cases by an order of magnitude.

Hydrobiologia 115, 139–144 (1984).

140

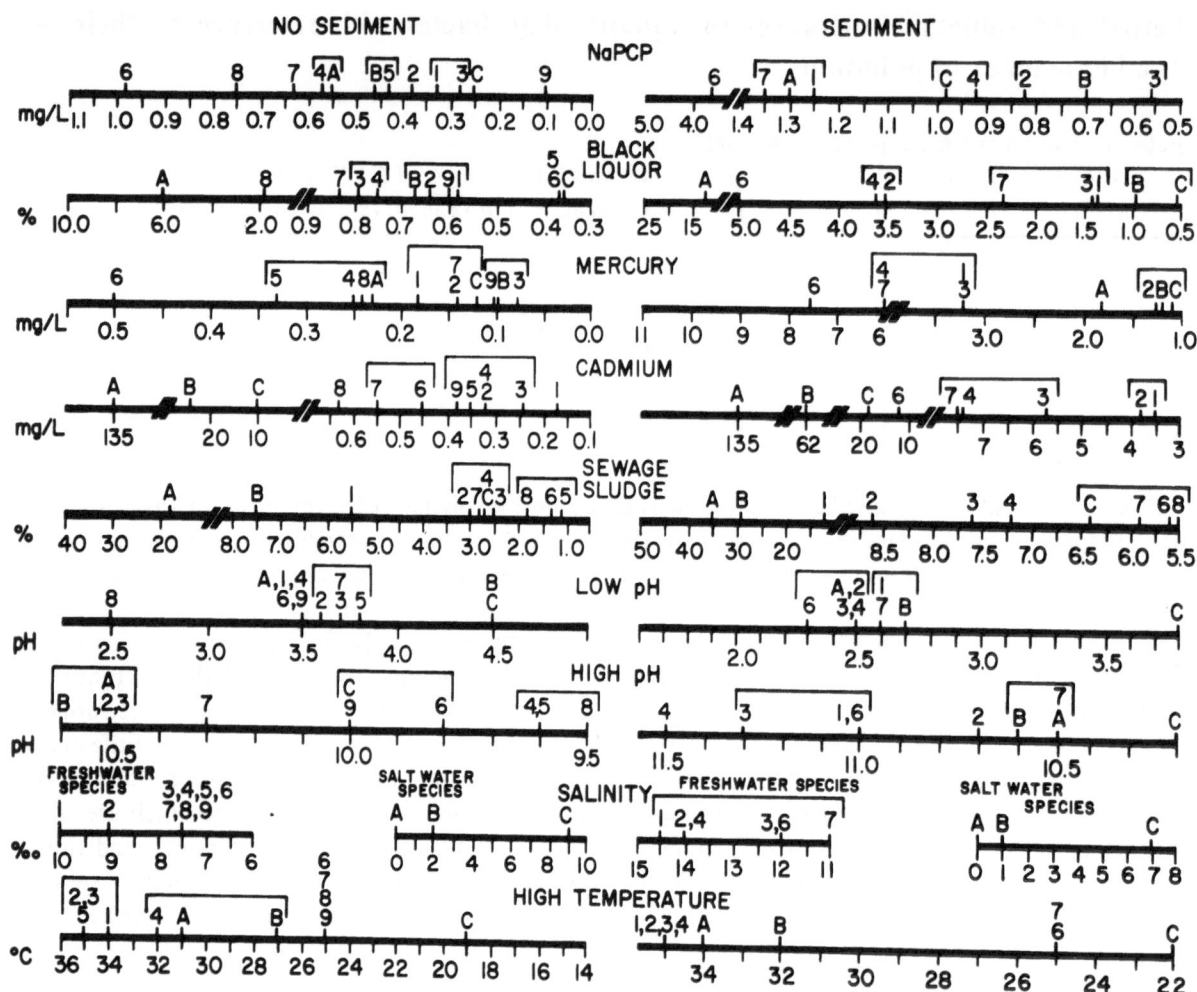

Fig. 1. Relative tolerances and 96 h LC50 values (non-significant differences are bracketed). (Freshwater species: 1 = *Limnodrilus hoffmeisteri*; 2 = *Tubifex tubifex*; 3 = *Branchiura sowerbyi*; 4 = *Quistadrilus multisetosus*; 5 = *Spirosperma ferox*; 6 = *Spirosperma nikolskyi*; 7 = *Stylodrilus heringianus*; 8 = *Rhyacodrilus montana*; 9 = *Varichaeta pacifica*. Saltwater species: A = *Monopylephorus cuticulatus*; B = *Tubificoides gabriellae*; C = *Limnodriloides verrucosus*.)*

Taking this study a step further, Chapman *et al.* (1982b) combined toxicants and environmental factors in multivariate bioassay tests. Although a great deal of variation was noted in toxicity under different pH, salinity and temperature regimes (Table 1), the study verified the relative tolerances determined previously for Cd, NaPCP and black liquor (a toxic component of pulp mill effluent). However, the rank order of species tolerance to mercury was highly variable, indicating the unpredictability of this toxicant's effect in the aquatic environment.

One significant result was the effect of salinity on

*Taxonomic revisions now refer to *Varichaeta* as *Varichaetadrilus, T. gabriellae* as *T.* sp. and *L. verrucosus* as *Tectidrilus diversus.*

Cd toxicity. The previous study had shown that salt water oligochaetes were 1–2 orders of magnitude more tolerant to Cd than freshwater species. Exposing freshwater species to Cd at 5 ppt salinity increased their apparent tolerance by an order of magnitude mainly due to Cd complexation. It is not unreasonable to suspect that other metals behave similarly, suggesting that discharges of metals to estuarine areas have far less effect on the aquatic ecosystem than similar discharges to freshwater.

In a further study, Chapman *et al.* (1982c) compared the tolerances of individual and mixed *Limnodrilus hoffmeisteri* and *Tubifex tubifex*. Mixed species were significantly more tolerant to toxicants under baseline conditions (Table 2), suggesting that survival under stress is greater in mixed compared

Table 1. Percent change in 96 h LC50 values following toxicant exposure under varying environmental conditions. Standard conditions (10 °C, pH 7, 0 or 20‰ salinity) are varied with changes in temperature (1 °C, 20 °C), pH (6, 8) and salinity (5 or 10‰).

Toxicants	Freshwater species			Saltwater species	
	Limnodrilus hoffmeisteri	*Tubifex tubifex*	*Stylodrilus heringianus*	*Monopylephorus cuticulatus*	*Limnodriloides verrucosus*
NaPCP	*0/+88	−3/+74	+19/+186	−30/+218	−69/+208
Black liquor	−8/+116	+12/+88	−10/+117	−76/+33	−72/−6
Cd	−35/+4 841	−34/+4 275	−56/+3 900	−51/0	+130/+190
Hg	−39/+178	−29/+100	+22/+157	−39/+83	+8/+25

* Maximum (and minimum) LC50 values under varied environmental conditions minus and divided by LC50 under standard conditions.

Table 2. Percent increase in 96 h LC50 values for individual versus mixed species (at 10 °C, pH 7, 0‰ salinity).

	Toxicant		
	NaPCP	Hg	Cd
Limnodrilus hoffmeisteri vs mixture	76	22	241
Tubifex tubifex vs mixture	53	64	81

to pure populations. Multivariate comparative tests remain to be conducted.

Since current theory holds that certain oligochaetes survive gross organically polluted conditions due to their ability to survive low oxygen conditions, the relative tolerances of individual and

Table 3. Relative tolerances to anoxia.

Species	Mean survival (d)
Freshwater	
Rhyacodrilus montana/Varichaeta pacifica*	26
Limnodrilus hoffmeisteri	23
Tubifex tubifex	16
Branchiura sowerbyi	6
Limnodrilus hoffmeisteri/Tubifex tubifex*	5
Stylodrilus heringianus	4
Saltwater	
Monopylephorus cuticulatus	42
Limnodriloides victoriensis	14
Tubificoides gabriellae	6
Limnodriloides verrucosus	0

* Mixed species.

mixed species to anoxia were determined by Chapman *et al.* (1982a, c) (Table 3). Individual species tolerances followed and supported the present rank order of tolerance to organic pollution. However, the two species mixtures tested showed unexpected differences. Combinations of the oligotrophic species *Rhyacodrilus montana* and *Varichaeta pacifica* were surprisingly tolerant to anoxia while combinations of the eutrophic species *L. hoffmeisteri* and *T. tubifex* were surprisingly intolerant. We presently have no explanation for these differences.

Sublethal tolerance tests

The bioassays described above provided new and necessary data on relative tolerances. However, lethality tests are not sensitive to subtle toxicant effects and low toxicant concentrations. Accordingly, more dynamic studies were conducted into the physiological effects of toxicants on selected oligochaetes. Experiments were conducted using respiration to indicate stress as this is a physiological process that is responsive to sublethal effects while being relatively easy to monitor in live animals.

Respiration rates were determined by enclosing oligochaetes in syringes and allowing them to reduce the ambient oxygen. Periodic (1–2 h) injections of triplicate aliquots into a blood-gas analyser (Radiometer model PHM73) were made with minimum disturbance of the worms which were kept in the dark under constant environment conditions between measurements. Method checks included retesting and the use of controls.

The first step in this research was to determine baseline respiration rates for various species. The

142

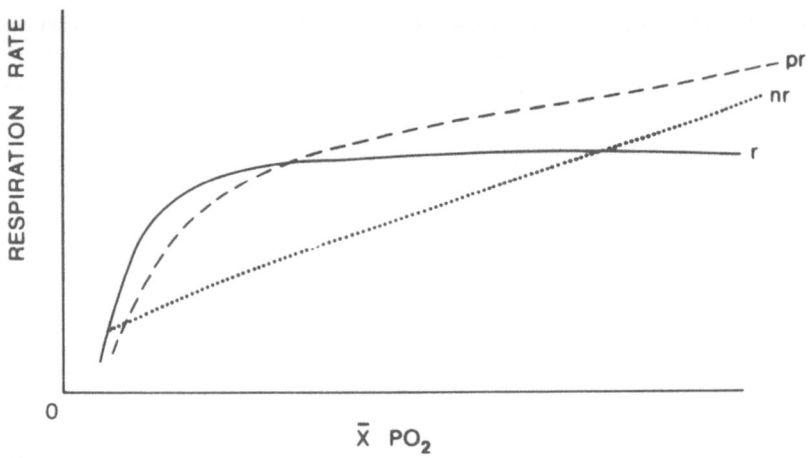

Fig. 2. Types of respiratory patterns observed in aquatic oligochaetes. (r = regulators, e.g. *L. hoffmeisteri, T. tubifex, R. montana/V. pacifica* mixtures; pr = partial regualtors, e.g. *B. sowerbyi, S. heringianus, M. cuticulatus, S. nikolskyi;* nr = non regulators, e.g. *L. hoffmeisteri/T. tubifex* mixtures.)

second step was to examine changes occurring as a result of exposure to toxicants and environmental factors in univariate and multivariate tests (Brinkhurst *et al,* 1983). Rates were also determined for mixed species in baseline and univariate tests (Chapman *et al.,* 1982c).

Three basic respiration patterns were observed (Fig. 2): regulation, partial regulation and nonregulation. Under sublethal stress, all theoretically possible changes were observed, with the anticipated changes in critical point (external oxygen level at which regulation ceases) being the least common variant (Table 4).

We found that the ability to regulate respiration was not correlated with tolerance. For example, some stresses unexpectedly improved regulation while mixed species, which were more tolerant to toxicants than individual species, were nonregulators. Moreover, increases or decreases in 96 h LC50

Table 4. Changes in respiratory pattern observed under sublethal stress.

Type of change	Regulators	Partial regulators	Non-regulators
No change	×	×	×
Rate increase	×	×	×
Rate decrease	×	×	×
Change in critical point	×	×	–
Loss of regulation	×	×	–
Gain of regulation	–	×	×

values were not correlated with changes in respiration rate under the same test conditions. Thus sublethal testing showed a far wider range of response to toxicant and environmental stress than did lethality testing.

An example of the relatively low information content of lethal tolerance tests is illustrated by the results of a histopathological study undertaken in part by the present authors (Thompson *et al.,* 1982). The saltwater species *Monopylephorus cuticulatus* was exposed to high and low cadmium concentrations for 4 or 8 days followed by preservation, sectioning, staining and microscopic examination. Gel chromatography was used to determine protein types associated with metal levels.

Compared to controls, worms exposed to sublethal cadmium levels exhibited rupturing of the body wall and degeneration of both digestive epithelium and phagocytic amoebocytes. Measurements by Brinkhurst *et al.* (1983) indicated a severe depression of respiration rate at these cadmium levels also providing indications of distress not apparent from LC50 testing. Though more time consuming and expensive than respiration measurements, histopathological techniques provided a detailed assessment of sublethal effects through inspection of the entire organism. In addition, a metallothionein-like protein was isolated from the cadmium-exposed worms. Metallothionein detoxifies metals and longer exposure to cadmium resulted in greater synthesis of this protein. This preliminary study raises conjecture regarding the possible relationship between relative tolerances of different species to metals and their relative abilities to produce metallothionein-like proteins. This area of research merits further work.

Application to field situations

The results of our laboratory experiments substantiate the continued use of presently-defined oligochaete species assemblages to delimit areas of organic pollution. Mixed species work to date indicates an enhanced survival to toxicants (but not to anoxia), which is generally consistent with previous work. Although the definition of usable species assemblages responsive to specific toxicants seems less promising following the above studies, further studies particularly with non-tubificid oligochaetes, is merited. In recent work, Bailey & Liu (1980) tested lethality of a variety of metals and organic toxicants to the lumbriculid *Lumbriculus variegatus* and obtained a 96 h LC50 value much lower $(0.074 \text{ mg l}^{-1})$ than values obtained with a variety of other species by Chapman *et al.* (1982a) (range = $0.17\text{--}0.63 \text{ mg l}^{-1}$).

Our respiration studies indicate a wide range of response to different toxicants, and it may be possible to use these varied responses to define specific toxicant effects. We can use a progression of laboratory tests ranging from lethal to sensitive sublethal to determine the possible effects of present or proposed effluent discharges and new chemicals. Moreover, the environment itself can be tested by ranking sediments and waters according to their level of toxic effect, as has recently been done for Puget Sound, Washington, using tests including oligochaete respiration (Chapman *et al.*, 1982d).

The Puget Sound study represents a successful application of laboratory methodology to field situations and points the way, we hope, to more dynamic approaches to pollution assessment. Future developmental studies should consider not only physiological reactions such as respiration but also histopathological studies. Research on morphological changes induced by toxicants is also merited. In the previous Symposium, Milbrink (1980) told us how worms exposed to high sediment mercury levels exhibited setal abnormalities. Similar abnormalities have recently been noted in

worms collected from an area of copper and zinc contamination (R. D. Kathman, pers. commun.).

We feel that a great deal more research is needed in all aspects of oligochaete biology to not only answer academic questions, but also to take full advantage of the utility of this group as pollution indicators. The melding of traditional field distribution studies and dynamic experimental approaches holds great promise in this regard and we look forward to future practical applications of these and other research efforts.

Acknowledgements

We thank Dr. Gary Vigers and the staff of E.V.S. Consultants Ltd. for their assistance in all stages of these studies. Experimental studies were supported under the Canadian Unsolicited Proposal program. Partial support for PMC was provided through an NSERC Industrial Post Doctoral Fellowship and subsequently through an Industrial Research Fellowship.

References

Bailey, H. C. & D. H. W. Liu, 1980. *Lumbriculus variegatus,* a benthic oligochaete, as a bioassay organism. In J. C. Eaton, P. R. Parrish & A. C. Hendricks (eds.), Aquatic Toxicology. ASTM STP707: 205–215.

Brinkhurst, R. O., 1980a. Production biology of the Tubificidae (Oligochaeta). In R. O. Brinkhurst & D. G. Cook (eds.), Aquatic Oligochaete Biology. Plenum Press, N.Y.: 205–209.

Brinkhurst, R. O., 1980b. Pollution biology – the North American experience. In R. O. Brinkhurst & D. G. Cook (eds.), Aquatic Oligochaete Biology. Plenum Press, N.Y.: 471–475.

Brinkhurst, R. O., P. M. Chapman & M. A. Farrell, 1983. A comparative study of respiration rates of some aquatic oligochaetes in relation to sublethal stress. Int. Rev. ges. Hydrobiol. 68: 683–699.

Chapman, P. M., L. M. Churchland, P. A. Thomson & E. Michnowsky, 1980. Heavy metal studies with oligochaetes. In R. O. Brinkhurst & D. G. Cook (eds.), Aquatic Oligochaete Biology. Plenum Press, N.Y.: 477–502.

Chapman, P. M., M. A. Farrell & R. O. Brinkhurst, 1982a. Relative tolerances of selected aquatic oligochaetes to individual pollutants and environmental factors. Aquat. Toxicol. 2: 47–67.

Chapman, P. M., M. A. Farrell & R. O. Brinkhurst, 1982b. Relative tolerances of selected aquatic oligochaetes to combinations of pollutants and environmental factors. Aquat. Toxicol. 2: 69–78.

Chapman, P. M., M. A. Farrell & R. O. Brinkhurst, 1982c. Effects of species interactions on the survival and repiration of *Limnodrilus hoffmeisteri* and *Tubifex tubifex* (Oligochaeta, Tubificidae) exposed to various pollutants and environmental factors. Wat. Res. 16: 1405–1408.

Chapman, P. M., G. A. Vigers, M. A. Farrell, R. N. Dexter, E. A. Quinlan, R. M. Kocan & M. Landolt, 1982d. Survey of biological effects of toxicants upon Puget Sound biota I. Broad-scale toxicity survey. U.S. Dept. Commerce, NOAA tech. Memo. OMPA-25, 97 pp.

Milbrink, G., 1980. Oligochaete communities in pollution biology: the European situation with special reference to lakes in Scandinavia. In R. O. Brinkhurst & D. G. Cook (eds.), Aquatic Oligochaete Biology. Plenum Press, N.Y.: 433–456.

Thompson, K. A., D. A. Brown, P. M. Chapman & R. O. Brinkhurst, 1982. Histopathological effects and cadmium-binding protein synthesis in the marine oligochaete *Monopylephorus cuticulatus* following cadmium exposure. Trans. am. micros. Soc. 101: 10–26.

Oligochaeta of the middle Po River (Italy): principal component analysis of the benthic data

Andreina Paoletti[1] & Beatrice Sambugar[2]
[1]*Dipartimento di Biología, Università degli Studi di Milano, Via Celoria, 26, I-20133 Milano, Italy*
[2]*Museo Civico di Storia Naturale, Lungadige Porta Vittoria, 9, I-37129 Verona, Italy*

Keywords: aquatic Oligochaeta, benthos, principal component analysis

Abstract

The application of principal component analysis to two types of habitat (the benthos of macrophytes and of central river bed) enabled us to single out some of the factors that affect the dynamics and the structure of the oligochaete population and its various reactions to environmental conditions. As regards macrophytes, the distribution of the variables on the basis of the first component is correlated, to a certain extent, with a seasonal factor without any significant differences among sites. The largest population is most closely correlated with the summer months. In fact, we found that the Naididae and Tubificidae species generally develop in larger numbers at higher temperatures. For the Tubificidae, we could detect a precise seasonal cycle. In the central river bed habitat, the first component was correlated with the river discharge, which determines the granulometric characteristics of the sediment; we noticed a correlation among the sites that have the same characteristics, regardless of sampling site or date. The species which correlate most closely among themselves are the Tubificidae *Limnodrilus hoffmeisteri, Tubifex tubifex, L. udekemianus* and *L. profundicola*, which are very characteristic of environments that contain abundant organic matter. The second component is correlated with temperature, and hence with the availability of oxygen, which determines the presence and the abundance of more sensitive species.

Introduction

The oligochaete community of a river depends on the interaction of numerous factors: the most important are the rate of flow and the seasonal cycle of the different species (Hynes, 1970; Wachs, 1967; Schwank, 1981a–b, 1982a–b).

Moreover, the presence in a river of aquatic plants determines the settlement of different communities on the bank, where macrophytes grow, and in the central river bed.

A section of the course of the River Po between Piacenza and Cremona, near the little town of Caorso, was investigated: here the river moves with a slowness typical of alluvial rivers – a slowness which is increased by the presence of the Isola Serafini dam, where there is a hydroelectric power sta-tion. This work formed part of the River Po hydro-biological research program, promoted by ENEL-DCO Piacenza (Italy).

The discharge ranges from 300 m^3 s^{-1} to about 8000 m^3 s^{-1}; freshets occur, in general, in spring and autumn; the temperature of the water ranges from a minimum in winter of 5 °C to a maximum in summer of 25 °C. Deposits are characterized by sand and mud-sand substrata, continuously changing according to hydrological events (Bertonati & Joannilli, 1981).

The course of the river is characterized by broad bends and islets, whose banks consist of shoals where macrophytes, in particular *Typha latifolia* and *Phragmites communis*, grow.

Hydrobiologia 115, 145–152 (1984).

Stations and sampling plan

Our study covered 17 stations, 3 of which were taken from banks of macrophytes and 14 from along the section of central river bed (see Fig. 1).

Among the macrophyte stations, numbers 72 and 73 were on the left bank of the Po, while number 71 was in the area least affected by the current of the islet 'Isola de Pinedo'. The samples, taken seasonally, were collected from the summer of 1974 up to the winter of 1977: 4 samples for each of the two groups (*Typha latifolia* and *Phragmites communis*) were taken from each station.

The sediment of the central river bed was sampled every two months from July 1974 to May 1976 and later on every three months until February 1977: 4 samples were taken from each station.

Material and methods

The macrophyte area was sampled within a zone of 907 cm² using a sharp-rimmed steel cylinder. The inside of this cylinder was completely emptied using a water ejector with a wide diameter. The central river bed samples were collected with a 530 cm² Ponar grab. Sediment was washed through a sieve (300 μm mesh). The organisms were later sorted from the preserved debris and fixed in FAA (formalinacetonalcohol). If worm numbers exceeded 100 individuals, a random subsample was examined. The specimens were identified and counted, after mounting in polyvinyl lactophenol.

Results and discussion

The Oligochaeta which were found belong to the following families: Aeolosomatidae, Lumbriculidae, Enchytraeidae, Naididae, Tubificidae and Lumbricidae. Table 1 lists the taxa found and their subdivision into the two habitats. Very few Aeolosomatidae, Lumbriculidae and Enchytraeidae were found and then only in the macrophyte habitat. Sixteen species of Naididae were found among the

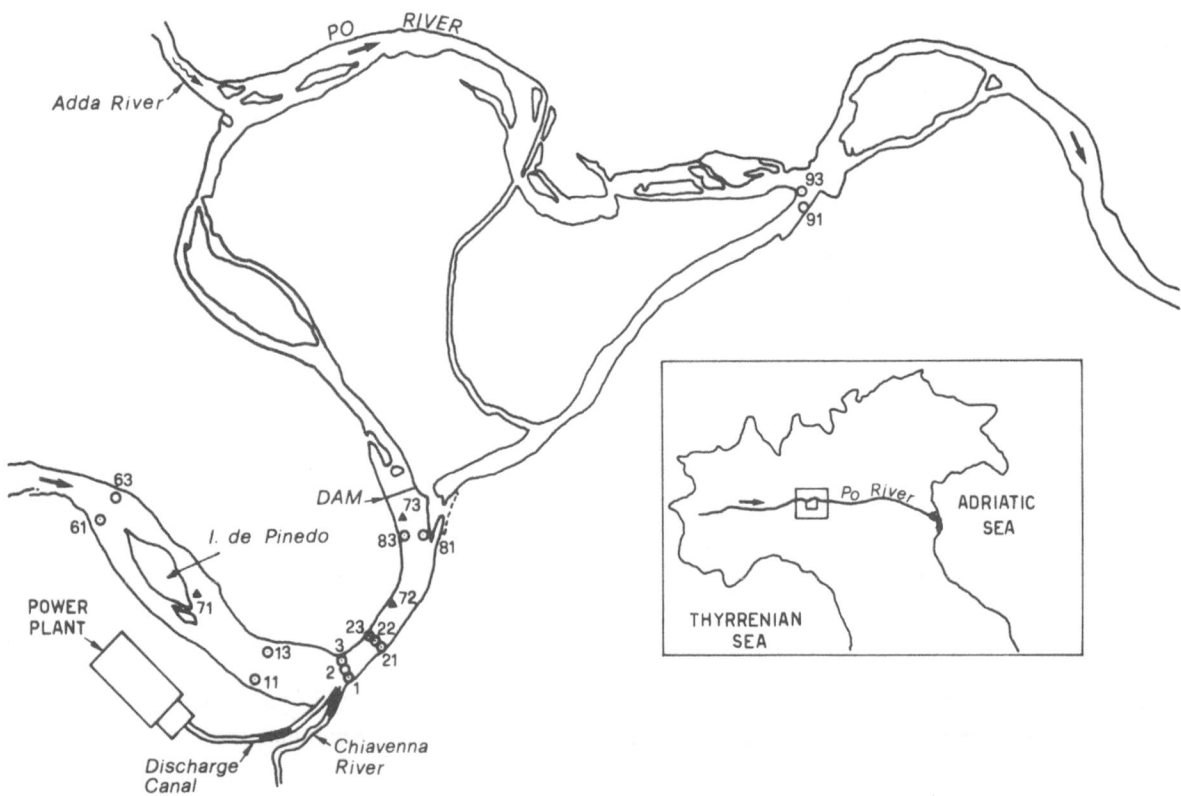

Fig. 1. Study area and sampling stations.

Table 1. List of the taxa.
m = macrophyte, b = central river bed.

Species	m	b
AEOLOSOMATIDAE		
Aeolosoma sp.	+	
LUMBRICULIDAE		
Lumbriculus variegatus	+	
Stylodrilus sp.	+	
ENCHYTRAEIDAE		
Enchydraeidae gen.sp.	+	
NAIDIDAE		
Chaetogaster diastrophus	+	
C. diaphanus	+	
C. cristallinus	+	
Paranais frici	+	+
Uncinais uncinata	+	++
Ophidonais serpentina	++	+
Nais communis	+	+
N. variabilis	+	+
N. simplex	+	+
N. bretscheri	+	+
N. pardalis	+	
N. elinguis	+	+
Slavina appendiculata	+	
Stylaria lacustris	++	+
Dero digitata	++	+
Pristina foreli	+	
TUBIFICIDAE		
Tubifex tubifex	++	++
T. ignotus	+	+
Limnodrilus hoffmeisteri	++	++
L. udekemianus	++	++
L. claparedeianus	+	+
L. profundicola	+	++
Psammoryctides barbatus	+	+
P. albicola	+	
Ilyodrilus templetoni		+
Aulodrilus pluriseta	+	
A. pigueti	+	
Branchiura sowerbyi	+	+
LUMBRICIDAE		
Eiseniella tetraedra	+	

macrophytes, of which *Nais variabilis, N. bretscheri, N. pardalis, Chaetogaster cristallinus* and *C. diastrophus* were very sporadic, while *Stylaria lacustris, Ophidonais serpentina* and *Dero digitata* were very abundant. In the central river bed there were fewer species of Naididae which were generally scarce in any case, with the exception of *Uncinais uncinata*, which was fairly abundant.

The Tubificidae constituted the most prevalent part of the fauna in both macrophyte and central river bed habitats; the predominating species were *Limnodrilus hoffmeisteri, L. udekemianus* and *Tubifex tubifex*. In the central river bed *L. profundicola* was also abundant.

The presence of a greater variety of fauna in the macrophyte habitat is due to the characteristics of the habitat itself, in that it is very complex and rich in microhabitats, consisting as it does of a thinly-particled bottom, which is the domain of Tubificidae, and of the banks of water plants which are colonized above all by Naididae and some species of Lumbriculidae and Enchytraeidae. Moreover, the macrophytes provide a less perturbed environment, which is affected less by variations in water flow, whose granulometric texture is constant and whose sediment is enriched by organic substance deriving from vegetal components (Sambugar, 1981).

The area of central river bed is, on the other hand, less rich in microhabitats and is subject to disturbance due to the speed of current which changes the structure of the sediment and quantity of organic matter (Paoletti Di Chiara, 1981).

A principal component analysis was carried out on a Honeywell 66-60 computer, with multivariate data analysis (Cooley & Lhones, 1971) for the populations living in the two different habitats, using all the data collected over the three year period.

The analysis showed that the total variance for the first four principal components of macrophyte population was 39.23% and in the central river bed population, 45.29%. A projection of the observation points (sampling sites and species abundance) on the plane of the first and second components produces two scatter diagrams.

As regards the first of these (Fig. 2), which relates to the macrophyte habitat, we can draw the following conclusion: the values for the sampling sites during the three year study follow a line which is inclined with respect to both axes and we can see a seasonal pattern that separates the sampling stations in winter from those in the summer. The pattern shifted for the summer of 1975, as shown in the lower right quadrant. This would indicate that, to some extent, the first component is correlated with seasonal change. We then investigated the importance of temperature in this connection, plotting the values for the first component of the three different sites, 71, 72 and 73, as well as the temperatures of

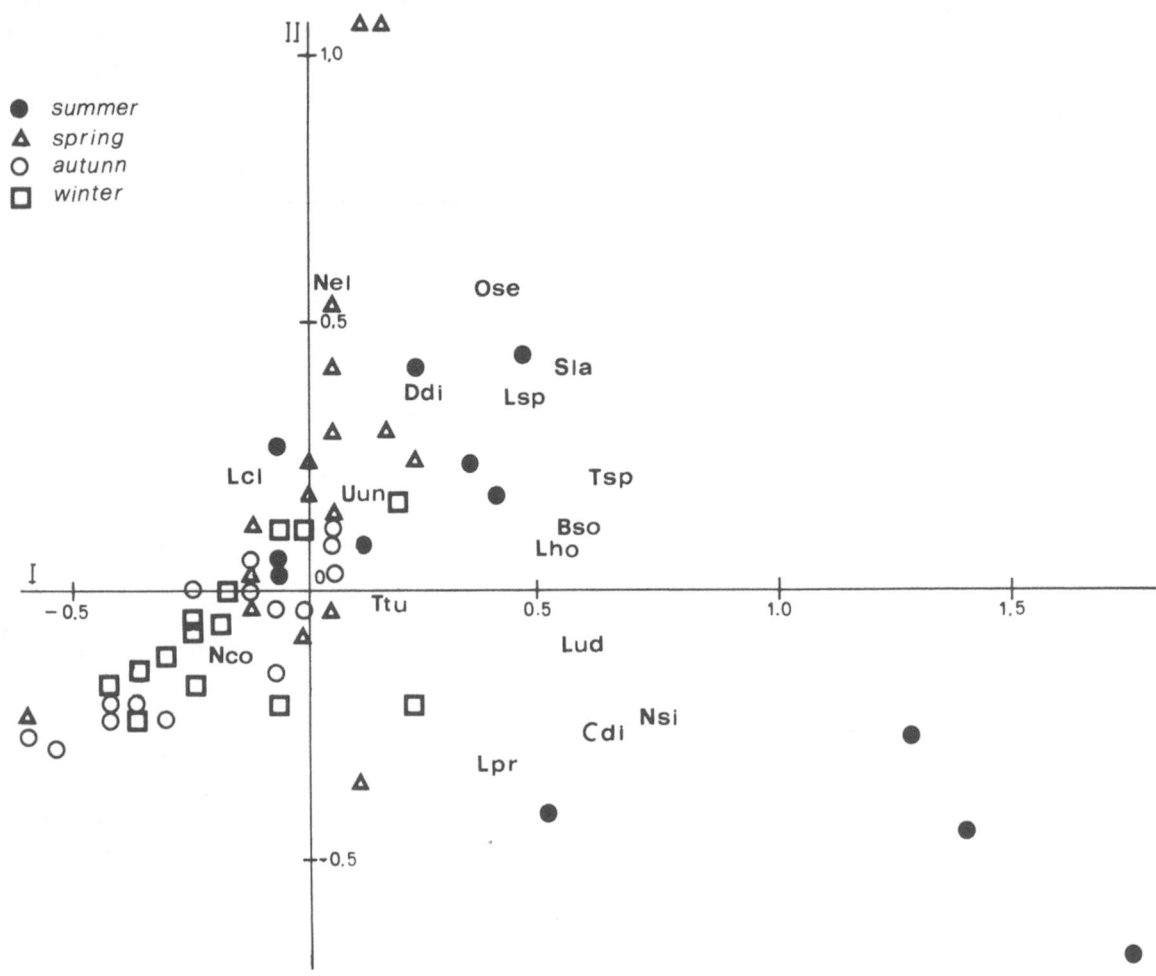

Fig. 2. Plot of factors I and II of the macrophyte habitats. Graphic symbols indicate seasonal samples and the three-letter species abbreviations are as follows:

Cdi = *C. diaphanus*, Ddi = *D. digitata*, Nsi = *N. simplex*, Nco = *N. communis*, Nel = *N. elinguis*, Ose = *O. serpentina*, Sla = *S. lacustris*, Uun = *U. uncinata*, Ttu = *T. tubifex*, Tsp = *Tubifex* imm., Lho = *L. hoffmeisteri*, Lud = *L. udekemianus*, Lcl = *L. claparedeianus*, Lpr = *L. profundicola*, Lsp = *Limnodrilus* imm., Bso = *B. sowerbyi*.

the water in the river (Fig. 3). The patterns are similar, although there is some divergence in the first period (June–September 1974) and in the last (December 1976–April 1977). In fact, during the last period the water was exceptionally deep, which undoubtedly brought about changes in the fauna in the macrophyte banks.

For this reason the correlation coefficients between the temperature and the first components have been computed on 7 points excluding the values of the first and the last time periods. The results are: 0.68, 0.89, 0.80, for the three stations (71, 72, 73).

In the pattern so far discussed, the most abundant

species occur mostly during the summer months. In fact *Limnodrilus hoffmeisteri, Dero digitata, Ophidonais serpentina, Stylaria lacustris, L. udekemianus, Limnodrilus* spp. and immature *Tubifex tubifex* are found near these months in the diagram. It is well known that in the Tubificidae the maturation and deposit of cocoons accelerate as the temperature increases. For both *Limnodrilus* and *Tubifex*, we noticed the repetition of precise biological cycles during three years: the greatest number of immature forms follow the greatest number of mature specimens. We found the greatest number of individuals of the *Limnodrilus* genus (mature and im-

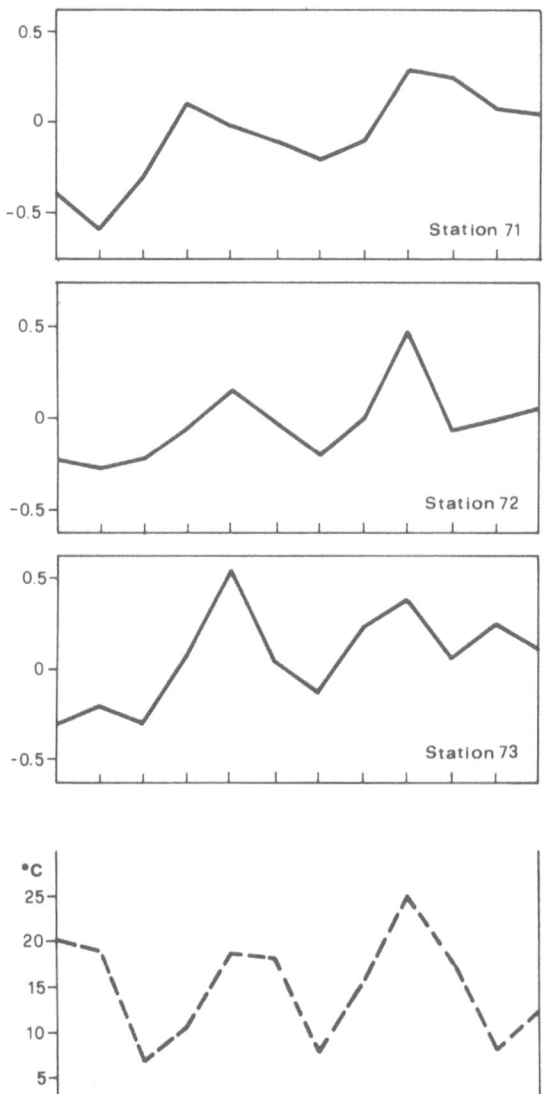

Fig. 3. Seasonal variations of first component of three macrophyte stations (71, 72 and 73) and water temperature during a three-year period.

mature) in the summer months, while *T. tubifex* was abundant in mature forms from December to April, which explains its more central position in the diagram.

For the Naididae also, the correlation with summer months is mostly related to their biological cycle, since their asexual reproduction is usually accelerated by increases in temperature. However, some species have been reported by Learner *et al.* (1978) to have annual cycles that have winter or springtime maxima. *N. communis* and *N. elinguis,* two of these, were indeed found more frequently in samples taken in the winter or spring.

For central river bed fauna, the scatter diagram of variables for the principal components (Fig. 4) shows that distribution of most of the samples depends on a gradient based on the granulometric structure of the substrata. They vary from sites, on the left, in which there are none of the finer particles (mud and clay <0.074 mm) to sites with constantly high percentages of these fine fractions. Under these conditions, one cannot detect any correlations between samples taken at the same time or samples taken from the same site at different times.

Since the granulometric characteristic of the sediment depends on changes in water flow, the first component is correlated with hydrological changes in the river. The lower rate of flow of the current causes mud to be deposited during periods of low water, which does not happen during periods of full flow. In addition, because of the contour of the river along this tract, different sampling sites which are exposed to the water flow to a greater or lesser degree have different granulometric characteristics at the same sampling time.

Since the oligochaete population is largely dependent on the presence of sediment rich in organic matter, the worms will be abundant only at those sites and at those times in which these conditions exist. In fact, if we compare the first component with the percentage of mud at the sites, we find a positive correlation. For example, if we plot the values for the first component, during the three year period, against the percentage of mud and the discharge of the Po river for stations 21 and 83, we can see a similar pattern (Fig. 5; r = 0.70 and r = 0.67 for stations 21 and 83).

The species most strictly correlated with the mud percentage are the Tubificidae *T. tubifex, L. hoffmeisteri, L. udekemianus* and *L. profundicola.* These species, especially the first three, are common above all in environments rich in organic matter (Brinkhurst, 1966; Timm, 1970). The development of a population is also characterized by simultaneous increases in all the age classes and, in fact, the numbers of immature forms of *Limnodrilus* spp. and *T. tubifex* are closely correlated with the numbers of adults. The changes in water flow that occur

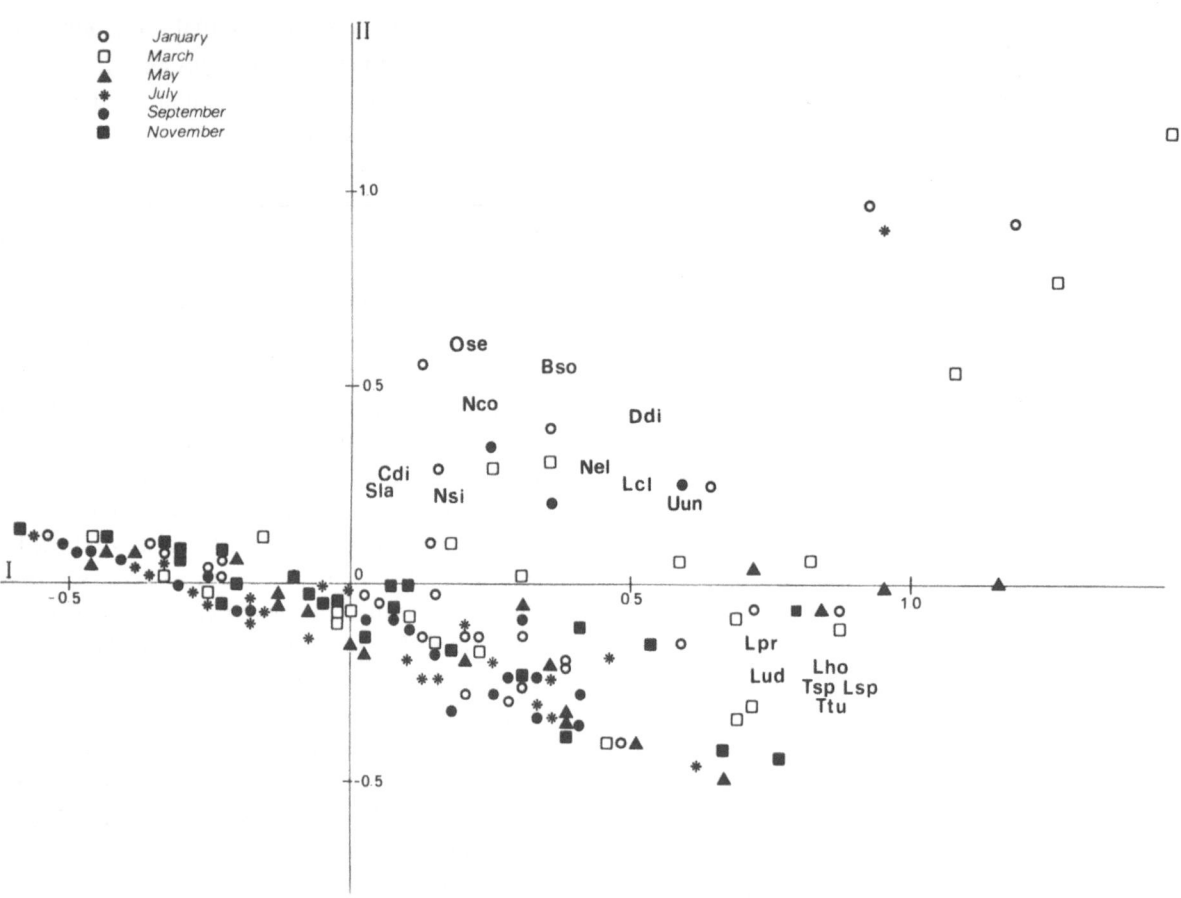

Fig. 4. Plot of factors I and II of central river bed habitats. Graphic symbols indicate monthly samples; species abbreviations as for Fig. 2.

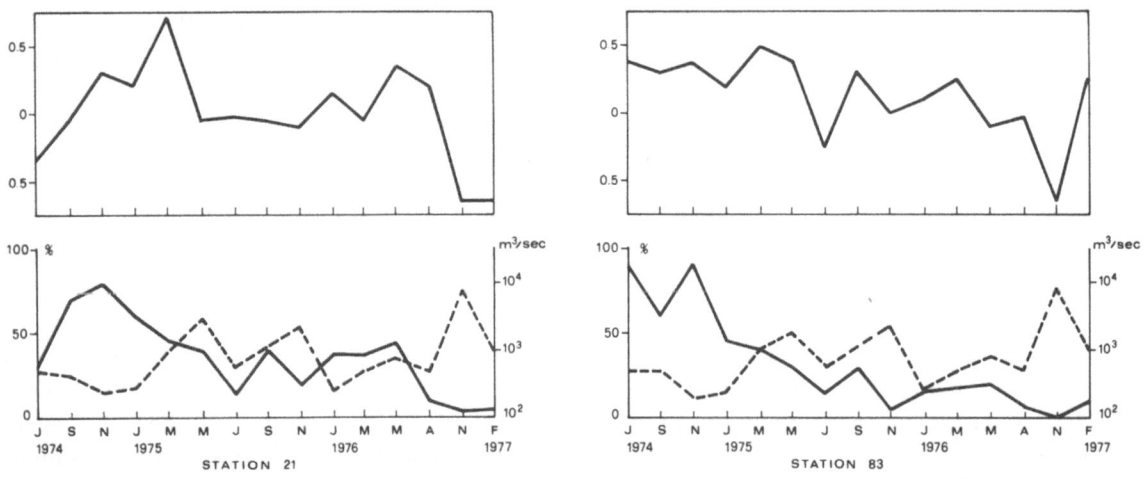

Fig. 5. Seasonal variations of the first component of central river stations (21 and 83), the mud fractions connected with them and the Po discharge (broken lines) during a three-year period.

over a year prevent the establishment of a seasonal population cycle.

From the variables which are clustered according to the granulometric gradient, there occurs a separation of a group of several species of Naididae (*Chaetogaster diaphanus, Stylaria lacustris, Ophidonais serpentina, Nais communis, N. simplex, N. elinguis, Dero digitata* and *Uncinais uncinata*) and the tubificids *Branchiura sowerbyi* and *L. claparedeianus.* The species, isolated in this way, can be separated according to the correlation factor of the first component, into two groups: one with the values lower than 0.4 and the other with values higher than 0.4.

The first group contains species which are sporadic in the central river bed and are, generally, found in the macrophyte banks, while the second consists of species usually dwelling in the soft bottom.

A certain number of samples relative to the months of January and March are also correlated with those species. The separation of these variables is due to the secondary component. Since the samples are correlated with the winter months, the conclusion to be drawn is that the second component is related to temperature.

Neither group includes species with winter peaks, but all the species have precise needs regarding oxygen and certainly the lower temperatures during the winter months make possible a greater availability of oxygen in the sediment also.

The first group consists of species which live in the macrophyte banks where the necessary oxygen is present. Their presence in the central river bed is accidental, and they may be seen only in those months when, owing to good oxygenation conditions in the sediment, these species can survive for a certain length of time.

The second group also consists of species requiring oxygen. In fact, the naidids *U. uncinata* and *N. elinguis* can survive only in well oxygenated sediment. *N. elinguis* can be found, where there is organic pollution, only if there is no scarcity of oxygen (Lafont, 1977). *L. claparedeianus* also requires high oxygen concentration (Kennedy, 1965) and is, in fact, more common in lothic than in lentic environments (Milbrink, 1973), where it can find oxygen more easily.

Therefore, the deviation of variables, owing to the second component, is based on the greater need for oxygen of these species; a more abundant supply of oxygen can be found in winter months, when the lower temperature makes possible better sediment oxygenation.

Acknowledgements

We are grateful to Dr. R. Cironi of the ENEL DCO of Piacenza, for his statistical analysis and data interpretation. We are indebted also to Dr. B. Rossaro for his constructive suggestions.

References

Bertonati, M. & E. Joannilli, 1981. Considerazioni sulle caratteristiche dell'acqua e dei sedimenti nel triennio 1974–1977. Riv. Idrobiol. 20 (1): 123–138.

Brinkhurst, R. O., 1966. The Tubificidae (Oligochaeta) of polluted waters. Verh. int. Ver. Limnol. 16: 854–859.

Cooley, W. W. & P. R. Lohnes, 1971. Multivariate data analysis. John Wiley & Sons, New York, 364 pp.

Hynes, H. B. N., 1970. The ecology of running waters. Liverpool Univ. Press, 555 pp.

Kennedy, C. R., 1965. The distribution and habitat of Limnodrilus Claparède (Oligochaeta: Tubificidae). Oikos 17: 158–168.

Lafont, M., 1977. Les Oligochètes d'un cours d'eau montagnard pollué: le Bief Rouge. Ann. Limnol. 13: 157–167.

Learner, M. A., G. Lochhead & B. D. Hughes, 1978. A review of the biology of British Naididae (Oligochaeta) with emphasis on the lotic environment. Freshwat. Biol. 8: 357–375.

Milbrink, G., 1973. On the use of indicator communities of Tubificidae and some Lumbriculidae in the assessment of water pollution in Swedish lakes. Zoon 1: 125–139.

Paoletti Di Chiara, A., 1981. Gli Oligocheti del benthos del medio Po presso Caorso (Piacenza). Riv. Idrobiol. 20: 173–178.

Sumbugar, B., 1981. Gli Oligocheti raccolti tra i banchi di macrofite nel medio Po a Caorso (Piacenza). Riv. Idrobiol. 20: 179–186.

Schwank, P., 1981a. Turbellarien, Oligochaeten und Archianneliden des Breitenbachs und anderer oberhessischer Mittelgebirgsbäche, 1. Lokalgeographische Verbreitung und die Verteilung der Arten in den einzelnen Gewässern in Abhängigkeit vom Substrat. Arch. Hydrobiol., Suppl. 62: 1–85.

Schwank, P., 1981b. Turbellarien, Oligochaeten und Archianneliden des Breitenbachs und anderer oberhessischer Mittelgebirgsbäche, 2. Die Systematik und Autökologie der einzelnen Arten. Arch. Hydrobiol., Suppl. 62: 86–147.

Schwank, P., 1982a. Turbellarien, Oligochaeten und Archianneliden des Breitenbachs und anderer oberhessischer Mittelgebirgsbäche, 3. Die Taxozönosen der Turbellarien und

Oligochaeten in Fliessgewässern – eine synökologische Gliederung. Arch. Hydrobiol., Suppl. 62: 191–253.

Schwank, P., 1982b. Turbellarien, Oligochaeten und Archianneliden des Breitenbachs und anderer oberhessischer Mittelgebirgsbäche, 4. Allgemeine Grundlagen der Verbreitung von Turbellarien und Oligochaeten in Fliessgewässern. Arch. Hydrobiol., Supp. 62: 254–290.

Timm, T., 1970. On the fauna of the Estonian Oligochaeta. Pedobiologia 10: 52–78.

Wachs, B., 1967. Die Oligochaeten Fauna der Fliessgewässer under besonderer Berücksichtigung der Beziehungen zwischen der Tubificiden-besiedlung und dem Substrat. Arch. Hydrobiol. 63: 310–386.

Short term dynamics of the dominant annelids in a polyhaline temperate estuary*

Robert J. Diaz

Department of Estuarine and Coastal Ecology, Virginia Institute of Marine Science and School of Marine Science, The College of William and Mary, Gloucester Point, VA 23062, U.S.A.

Keywords: aquatic Oligochaeta, long term changes, population dynamics, polychaetes, recruitment

Abstract

Weekly sampling over a two year period from a muddy sand bottom in the polyhaline York River, Virginia, U.S.A., clearly identified the pattern of recruitment and survival of the dominant annelid species. Three intermingled recruitment strategies and two survival patterns were observed, ranging from the classic opportunistic life style of mass recruitment over short time periods followed by mass mortality to prolonged recruitment with lower mortality. Qualitatively the annelid assemblage was very similar from year to year with most of the changes being quantitative. Oligochaetes, *Tubificoides* spp., were the most stable and characteristic members of the annelid assemblage.

Introduction

While it is well known that estuarine and marine benthic communities vary seasonally (Boesch, 1973; Eagle, 1975; Nichols, 1975) and over longer periods of time (Buchanan *et al.*, 1974; Eagle, 1975; Boesch *et al.*, 1976) the exact nature and timing of events responsible for much of the variation are less well known. Much of the seasonal and long term dynamics of benthic communities revolve around recruitment and subsequent survival of these recruits. The sampling interval chosen to characterize populations through time may sometimes misrepresent recruitment of survival if the interval is large in relation to the scale of events in the benthos. This is particularly true in temperate estuarine areas where the life span of many of the species is typically a year or less.

Without detailed information on natural population dynamics, fluctuations within the benthos could be mistaken for the effects of a pollutant or other disturbance (Boesch *et al.*, 1976). The results

of any field experimental work will also be influenced by imprecise information on what is causing observed population variation.

This paper documents the short term (weekly) variation of the dominant annelid species in a polyhaline soft-bottom muddy sand benthic community. Yearly variations are also examined for 1980 and 1981. Factors responsible for the community variations are discussed.

Methods

Starting in September 1979 one 0.017 m² grab sample was collected weekly from off a pier at the Virginia Institute of Marine Science. The sample was sieved (250 μm) and preserved in 10% buffered formalin for further analysis. The Institute is located near the mouth of the York River estuary, a major tributary of the Chesapeake Bay, USA. The site characteristics are:

Location: 37°14′45″ N, 76°30′00″ W
Depth: 2 m at MLW
Salinity: 18 to 22‰

---

*Contribution No. 1118 of the Virginia Institute of Marine Science

Hydrobiologia 115, 153–158 (1984).
© Dr W. Junk Publishers, Dordrecht.

Temperature: 2 to 28 °C
Sediment: muddy sand
RPD depth: 0.5 to 2.0 cm

The weekly sampling represents an attempt to document the natural population dynamics of the benthos with short interval sampling over a long period of time. While one grab per week does not identify spatial variation for the week, the closeness of samples in the time series and length of the series provide an excellent means of factoring out the spatial component of the total variation. In this paper data on the dominant annelids for 117 weeks, from September 7, 1979 to January 5, 1982, are considered. During this period 4 weeks were missed. A complete listing of the data will be provided upon request.

Reciprocal averaging ordination was conducted using ORDIFLEX of the Cornell Ecological Program series (Gauch, 1977). Cluster analysis was done using COMPAH developed at the Virginia Institute of Marine Science. Bray-Curtis similarity coefficient and flexible sorting strategy (Beta = -0.25) were used (Boesch, 1977). Data were square root transformed for both the ordination and clustering to subdue very large recruitment pulses. The run test, a nonparametric test that identifies nonrandom trends, was conducted using the SPSS package (Hull & Nie, 1981). Plots of the species were smoothed using the median smoothing technique described by Mosteller & Tukey (1977).

Results

A cluster of the entire data set (150 taxa and 117 weeks) isolated a single group of seven dominant annelids (Table 1). Six were polychaetes and one was an oligochaete. The oligochaetes were mainly *Tubificoides brownae*, but since many of them were immature and could not be positively identified, they were all referred to as *Tubificoides* spp.

While *Tubificoides* spp. population peaks were not the largest, oligochaetes were the most consistent members of the community, occurring in all 117 samples. Their standard deviation, an indication of variability through time, was less than their mean abundance. For all polychaetes the standard deviation was greater than the mean. *Nereis succinea*, while low in mean abundance, was the most consistently occurring polychaete, being present 90% of the time. For approximately 40% of the time both *Tubificoides* spp. and *N. succinea* were present in densities greater than or equal to their mean. The other species exceeded their mean abundance only 13 to 30% of the time (Table 1). The less time a species population abundance was above its mean abundance the more pulsed its recruitment was. The run test indicated that all the polychaetes had highly significant nonrandom trends in the number of consecutive weeks that their population levels were above or below mean values. *Tubificoides* spp. showed no significant trends (Table 1).

From examining the population changes over the entire period (Fig. 1) it is apparent that short duration, large recruitment pulses interspersed between extended periods of low population densities are typical for *Streblospio benedicti*, *Heteromastus filiformis*, *Glycinde solitaria*. Recruitment pulses for *N. succinea* and *Polydora ligni* were more frequent and populations did not decline as much as the other polychaetes. *Mediomastus ambiseta* had a completely different pattern. Starting in September 1980 *M. ambiseta* populations increased and con-

Table 1. Descriptive statistics for the dominant annelids.

Species	mean abundance (ind. 0.017 m⁻²)	SD	Occurrences out of 117	Occurrences greater than the mean	Maximum abundance (ind. 0.017 m⁻²)	Median abundance	Run test Z score	Run test prob.
Tubificoides spp.	32.7	31.1	117	46	155	20	-1.39	0.165
S. benedicti	26.9	89.5	76	15	726	1	-8.50	0.000
M. ambiseta	17.0	29.1	79	35	176	4	-4.94	0.000
H. filiformis	13.6	26.0	93	26	133	3	-6.36	0.000
P. ligni	11.7	18.9	92	33	102	3	-2.42	0.015
N. succinea	6.8	7.6	105	43	42	4	-2.53	0.011
G. solitaria	3.3	5.5	77	33	33	1	-3.81	0.000

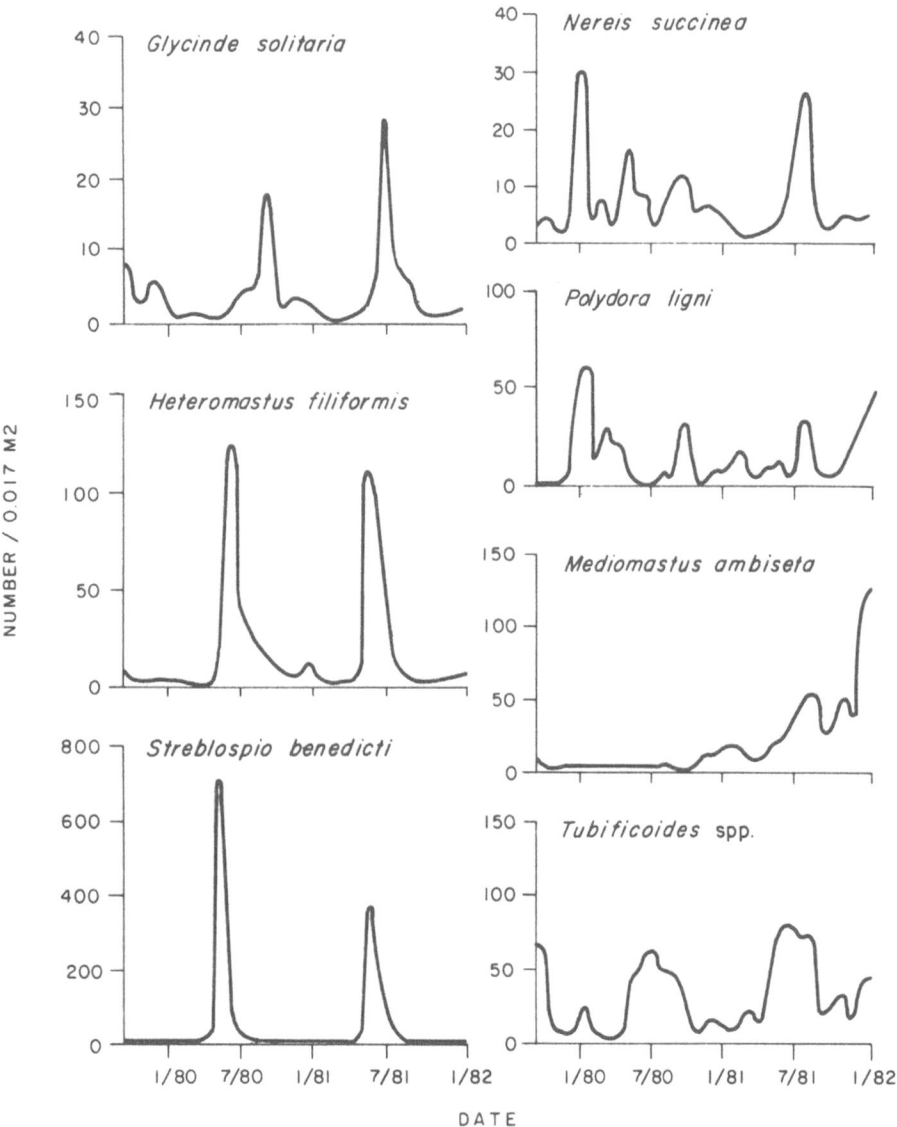

Fig. 1. Population dynamics of the dominant annelids from Sept. 1979 to Jan. 1982.

tinued to increase to January 1982. *Tubificoides* spp. did have protracted recruitment peaks, but these were punctuated by periods of declining densities. However, oligochaete populations between recruitments did not decline to as low a level as the polychaetes.

To better identify the relationships between the annelid populations through time, reciprocal averaging ordination was used. The ordination of the seven taxa and 117 weeks was very efficient with three axes accounting for a total of 80% of the

variation expressed in the data (axis I accounted for 37, axis II 27, and axis III 16% of the variance). When the species and seasons were plotted in the first three ordination axes a new set of interpretive axes were superimposed that explain much of the population dynamics (Fig. 2). Interpretive axis A graded from eruptive mass settling of very short duration and subsequent population crash to protracted recruitment at lower rates with increased survival. The species at the extremes of axis A were *S. bendicti* and *M. ambiseta*. Interpretive axis B

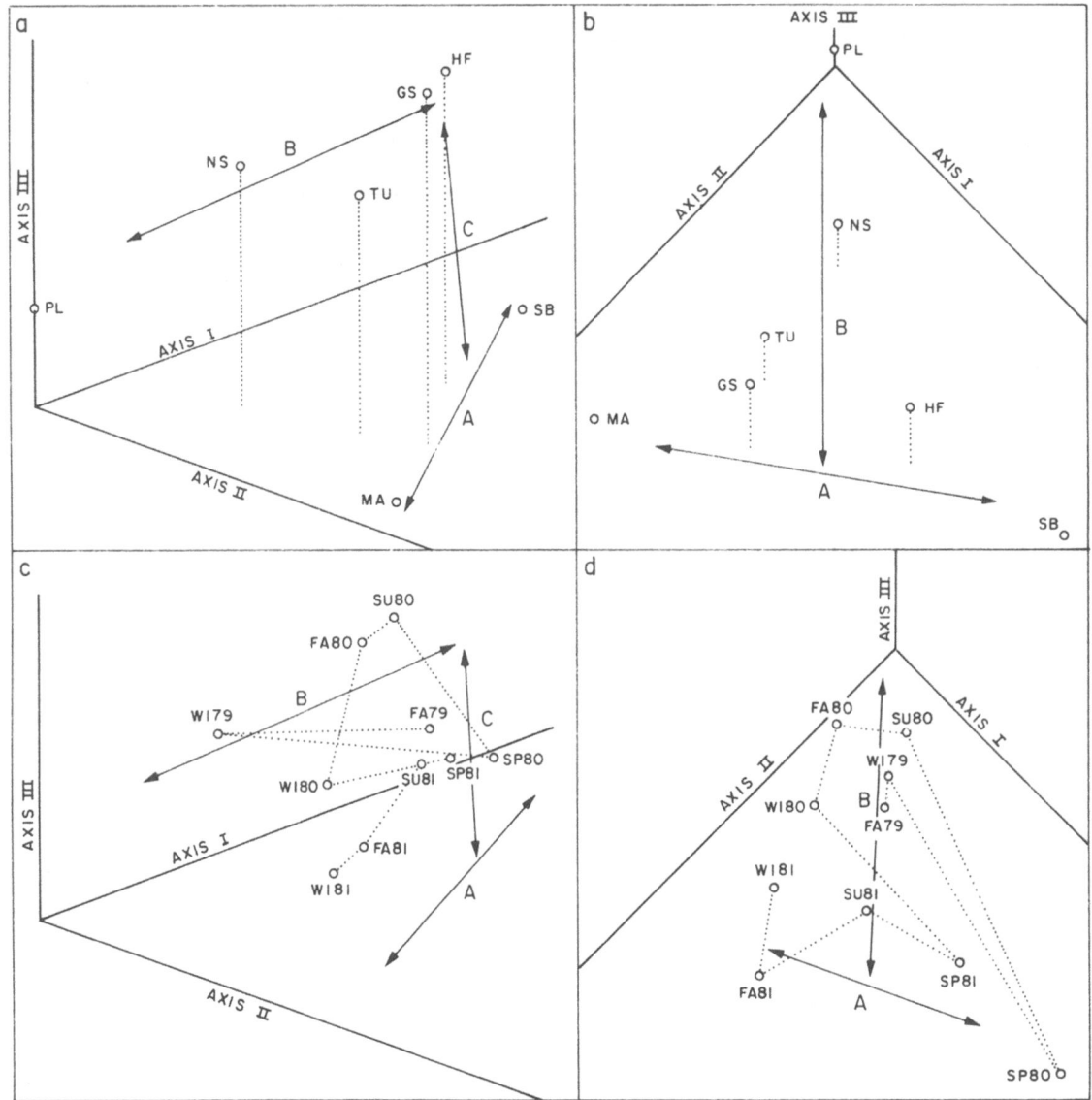

Fig. 2. Perspective views of the species and seasons placed in the first three axes of the ordination space. View a and c are form behind and above axis II. View b and d are from above axes I and II away from axis III. PL = *P. ligni*, NS = *N. succinea*, TU = *Tubificoides* spp., GS = *G. solitaria*, HF = *H. filiformis*, MA = *M. ambiseta*, SB = *S. benedicti*, FA = Fall (Sept. to Nov.), WI = Winter (Dec. to Apr.), SP = Spring (May), SU = Summer (June to Aug.).

graded from primarily fall-winter recruitment with several peaks per year, represented by *Polydora ligni*, to predominantly summer recruitment, represented by *Tubificoides* spp. Interpretive axes A and B were not orthagonal nor did they connect in three dimensions. A third axis, C, relates mainly to season connected. At one end is summer and at the other fall, winter, and spring. Thus axis C appears

to be a third dimensional distortion of A and B. At the low end of axis C were *S. benedicti* and *M. ambiseta*, and at the high end *G. solitaria*.

When the centroid of each season from fall 1979 to winter 1981 is plotted in the same ordination space as the taxa (this is the main advantage of reciprocal averaging, that it gives coordinates for both taxa and stations in the same ordination space), it is evident that the dominant annelid community spends most of its time near the ends of interpretive axes B and C, near the middle of axis A (Fig. 2c, d). This emphasizes the importance of the oligochaetes as the most typical and consistent members of the community since the centroid of the entire 117 weeks lies near the position of *Tubificoides* spp. maintained the highest population densities throughout the two years (Table 1). This pattern is in sharp contrast to most of the polychaetes which had recruitment and subsequent mortality within a single season. The only exception was *M. ambiseta* which increased almost monotonically during 1981.

Discussion

Life history patterns

Among these annelids there were three intermingled recruitment strategies and two survival patterns. The range was from the classic opportunistic life style (Grassle & Grassle, 1974) of large recruitment over a very short period of time followed by mass mortality, as shown by *S. benedicti*, to prolonged recruitment with lower mortality, as shown by *M. ambiseta*. In between these two styles is a third, of prolonged recruitment punctuated by periods of high mortality, as shown by *Tubificoides* spp.

Although both *H. filiformis* and *S. benedicti* appear to be classic opportunists, there are differences. The slope of the population curve for *H. filiformis* indicates declining mortality after initial recruiments settlement. This is most pronounced in 1980. For both years *S. benedicti* showed little change in mortality after settlement.

Both *S. benedicti* and *P. ligni* are shallow infaunal tube dwellers that are vulnerable to predation (Virnstein, 1979). They tended to recruit before predators became active. In the case of *P. ligni*, recruitment was after predators left the area. Through most of the year these two species, particularly *S. benedicti*, were present in very low densities. *Tubificoides* spp., which were found by Virnstein (1977, 1979) to be unaffected by predation, showed the opposite trend of large prolonged recruitment through the summer when predators were most active. The major predators in the system are blue crabs (*Callinectes sapidus*) and bottom feeding fishes (such as juvenile spot and croakers).

Two year trends

Over the two year period the seven annelid taxa fell into two groups based on periodicity of recruitment. *S. benedicti, H. filiformis, G. solitaria, Tubificoides* spp., and with some variation *P. ligni*, all recruited during the same times of the year each year. *S. benedicti* and *H. filiformis* had a very distinct recruitment period each spring. *Tubificoides* spp. and *P. ligni* had both summer and winter recruitments. The winter *Tubificoides* spp. recruitment was much smaller and temporally shorter than in summer. The reverse was true for *P. ligni. G. solitaria* recruited once a year in the summer with small but very distinct peaks. The other two species, *M. ambiseta* and *N. succinea* did not have repeatable recruitment patterns. *N. succinea* seemed to recruit aperiodically during any season. *M. ambiseta*, which was virtually absent for the first year, showed a trend of increasing population densities through the second year (Fig. 1).

While from year to year there was periodicity in five of the seven annelids the magnitude of settlement was different. *S. benedicti, H. filiformis*, and *P. ligni* were more abundant in 1980. *Tubificoides* spp. and *G. solitaria* were more abundant in 1981. This quantitative yearly variation accounted for most of the community changes among the annelids. Boesch *et al.* (1976) also found the macrobenthic community at a 9 m mud site about 500 m from the pier site to be more quantitatively than qualitatively variable through time.

Overall, summing all seven species, there were about the same density of individuals in both years. Buchanan *et al.* (1974) found that over a four-year period at a mud site off the Northumberland coast, the population densities more than doubled, but there was no essentially change in the estimated productivity of the community. They found that production was differently partitioned between the species each year with the total yearly production being the same. It seem likely that the total productivity of the seven annelids was approximately the same in 1980 and 1981, based on the assumptions of Buchanan *et al.* (1974) and that the average size of a worm was the same both years (Zaika, 1971).

The factors that make one year better then

another for a particular species are difficult to identify and would require a longer time-series of data. The complexity of species interactions from predation (Virnstein, 1979) to trophic group amensalism (Rhoads & Young, 1970) combined with climatic (Buchanan et al., 1978), or salinity, or substrate changes (Boesch et al., 1976) may overshadow any progressive tendencies in the benthos. To date no long-term study has demonstrated a cyclic (other than seasonal) nature to the trends in the benthos.

Acknowledgments

This work was supported by the Commonwealth of Virginia as part of its long-term benthic monitoring program. I am grateful to B. Meehan, D. Penry, M. Kravitz, T. Fredette, E. Koepfler and E. Wilkins (in order of participation on the project) for collection and processing of the samples, and P. Chapman and L. Schaffner for their review of this manuscript.

References

Boesch, D. F., 1973. Classification and community structure of macrobenthos in the Hampton Roads area, Virginia. Mar. Biol. 21: 226–244.

Boesch, D. F., 1977. Application of numerical classification in ecological investigations of water pollution. Ecol. Rec. Ser. EPA-600/3-77-033, 115 pp.

Boesch, D. F., M. L. Wass & R. W. Virnstein, 1976. The dynamics of estuarine benthic communities. In: Estuarine Processes, 1. Academic Press, N.Y.: 177–196.

Buchanan, J. B., P. F. Kingston & M. Sheader, 1974. Long term population trends of the benthic macrofauna in the offshore mud of the Northumberland Coast. J. mar. biol. Ass. U.K. 54: 785–795.

Buchanan, J. B., M. Sheader & P. F. Kingston, 1978. Sources of variability in the benthic macrofauna off the south Northumberland Coast, 1971–1976. J. mar. biol. Ass. U.K. 58: 191–209.

Eagle, R. A., 1975. Natural fluctuations in a soft bottom benthic community. J. mar. biol. Ass. U.K. 44: 864–878.

Gaugh, H. G. Jr., 1977. Ordiflex. Ecology and Systematics. Cornell University, Ithaca, N.Y. 185 pp.

Grassle, J. F. & J. P. Grassle, 1974. Opportunistic life histories and genetic systems in marine benthic populations. J. mar. Res. 32: 253–284.

Hull, C. H. & N. H. Nie, 1981. SPSS update 7.9. McGraw-Hill Publishing Co. Ltd, N. Y., 402 pp.

Mosteller, F. & J. W. Tukey, 1977. Data analysis and regression. Eddison-Wesley Publishing Co., Reading, Mass., 588 pp.

Nichols, F. H., 1975. Dynamics and energetics of three deposit-feeding benthic invertebrate populations in Puget and Washington. Ecol. Monogr. 45: 57–82.

Rhoads, D. C. & D. K. Young, 1970. The influence of deposit-feeding organisms on sediment stability and community trophic structure. J. mar. Res. 28: 150–178.

Virnstein, R. W., 1977. The importance of predation by crabs and fishes on benthic infauna in Chesapeake Bay. Ecology 58: 1199–1217.

Virnstein, R. W., 1979. Predation on estuarine infauna: response patterns of component species. Estuaries 2: 69–86.

Zaika, V. E., 1973. Specific production of aquatic invertebrates. J. Wiley & Sons, N.Y., 154 pp.

The occurrence of species of semi-aquatic Enchytraeidae (Oligochaeta) in Ireland

Brenda Healy & Thomas Bolger
Department of Zoology, University College, Belfield, Stillorgan Road, Dublin 4, Ireland

Keywords: aquatic Oligochaeta, Enchytraeidae, environmental parameters, ordination techniques

Abstract

The Enchytraeidae are essentially terrestrial oligochaetes but many species have marked aquatic tendencies. Over two thirds of recorded Irish species were found in soils which were submerged or frequently flooded and 35% showed a distinct preference for these conditions. Relatively few species were living in soils subject to drought. Red blood was present in 28 species, all but one from soils with more than 55% water. *Cognettia sphagnetorum* and *C. glandulosa* developed red blood in very wet conditions. In a survey of Irish wetlands, samples were taken from bog, heath, marsh, fen, margins of lakes and rivers, and saltmarsh. The influence of various environmental parameters was determined using ordination techniques. Magnesium and pH were found to be the most important factors. A high level of magnesium distinguished coastal sites and pH 5.2 separated two clusters representing acid peat and marsh-fen-aquatic sites. Groups of indicator species characterized each of the three clusters. The ecological distribution of the indicator species is described, and their usefulness in classifying enchytraeid communities is discussed.

Introduction

For the purposes of this study, a semi-aquatic habitat is defined as one in which the ground is waterlogged for at least part of the year but where there is no permanent cover of water. We include lands subject to flooding, areas of impeded drainage and a variety of peatlands. Ireland is particularly rich in such habitats. The frequency of rain throughout the year – 180 days in the south-east and 250 in the west – coupled with the high average humidity, leave little opportunity for the soils to dry out. In addition, there is poor drainage throughout much of the country, particularly in the central plain. At the present time, 16% of the land surface is occupied by peatlands and of the 50% under pasture a large proportion is in need of drainage.

In these waterlogged soils, a terrestrial fauna comprising groups such as Lumbricidae, Pulmonata, Collembola, Acarina and various other ar-thropods coexists with aquatic groups such as Turbellaria, Rotifera, Copepoda, Ostracoda, Trichoptera, Chironomidae, Ceratopogonidae and other Diptera, and aquatic molluscs such as *Lymnaea* and *Pisidium*. The relative importance of terrestrial and aquatic components depends on both the amount of water present and its permanence. During the present survey, Copepoda and chironomid larvae were frequently taken in woodland leaf litter and Ostracoda, Cladocera, trichopteran larvae and *Pisidium* in rushy pastures. Discussions with colleagues from other countries indicate that such occurrences are by no means confined to Ireland.

Within the faunal sequence aquatic – terrestrial, the Enchytraeidae occupy a median position. While the family as a whole is essentially terrestrial and relatively few species are habitually found in sub-aquatic substrates, many species are characteristic of wet soils and both numbers and diversity in such environments can be high.

Hydrobiologia 115, 159–170 (1984).

This paper analyses some records of enchytraeids in a wide range of Irish wetlands, relating species distribution to a number of environmental factors.

Methods

The main body of samples analysed in this account formed part of a survey designed to determine the ecological limits and preferences of enchytraeids in all types of terrestrial and aquatic habitats in Ireland. Details of the sampling programme and descriptions of sampling sites may be found in Healy (1976, 1979). Only the sites with greater than 45% water were used for the analysis. These sites, 107 in all, were of the following types:

Bog (ombrotrophic mire) 16
Flush (rheotrophic mire) 9
Wet heath 12
Fen 7
Marsh 10
Woodland 12
Margins of lakes and rivers 15
Salt marshes 16

Not all of these were waterlogged at the time of sampling but all had a high water content or could be subjectively described as 'wet'.

Two sample units measuring 32 cm^2 × 5 cm, together with material for measurement of environmental parameters, were taken from an extensive range of sites. Enchytraeids were extracted using the wet funnel method of O'Connor (1955) and identified live. Other oligochaetes are extracted by this method, though less efficiently. Where they occurred their presence was noted but they were not identified further than family level.

The environmental factors measured were pH, water content (% of wet weight), organic matter (loss on ignition), organic carbon, total nitrogen and exchangeable (i.e. soluble) calcium, potassium, phosphorus, magnesium and sodium. The estimation of water content by weight was the simplest method applicable to the range of soils sampled. Unfortunately, values can be misleading when used to compare different types of wet soil owing to the differences in the specific gravity of the solids. For example, waterlogged *Sphagnum* peat may contain up to 96% water whereas a waterlogged sand or clay may have less than 30%. Consequently water content expressed as a percentage of the wet weight does not always reflect the degree of waterlogging. A volumetric measurement of the fraction of available pore space occupied by water might have been more appropriate.

Results

Occurrence of enchytraeids in relation to soil water content

The occurrence of enchytraeids in relation to water content of the full range of habitats sampled, including dry soils and submerged sites, is summarized in Table 1. Substrates are classified according to a subjective estimation of the average conditions as this appeared to be more meaningful than the actual water content at the time of sampling, e.g. 'flooded' soils are those liable to frequent flooding, 'saturated' soils are those likely to be waterlogged for a considerable part of the year and 'dry' soils are those subject to summer drought. Over two thirds of all recorded species were found in habitats which were submerged or frequently flooded and twenty nine species, or 35%, showed a distinct preference for these conditions. Among the species found in non-coastal sites 85% were taken from soils which were frequently waterlogged. The number of spe-

Table 1. Occurrence of enchytraeids in sites with varying amounts of water.

	Inland sites					Coastal sites				Total
	Submerged	Flooded	Saturated	Moist	Dry	Dry	Moist/wet	Intertidal	Subtidal	
Number of sites	9	32	39	43	26	17	20	25	18	229
Number of species	14	39	47	44	32	23	27	23	3	82
Mean enchytraeids:										
32 cm^2	62.7	99.3	177.5	129.5	19	169.2	37.1	37.1	2.2	113.7
Mean spp.: site	5.3	8.9	9.1	8.3	8.5	5.3	5.7	4.4	0.78	7.4

cies taken from non-coastal sites subject to drought was only thirty-two compared with thirty-nine from soils subject to flooding. The number of species per site (i.e. in 64 cm^2) was also greater in flooded than in dry soils.

The terrestrial sites in coastal areas (mainly sand dunes and machair) are considered separately because they contain a high proportion of species which only occur on the coast. The sand dunes were among the driest habitats investigated and this is reflected in the lower enchytraeid density and diversity. Many of the species which are usually considered to be characteristic of intertidal sands such as *Marionina spicula*, *M. southerni*, *M. subterranea*, *M. achaeta* and *Enchytraeus albidus*, do, in fact, extend supralittorally into the dune face where moisture levels are frequently below 1%. These species, together with others characteristic of sand dunes, e.g. *Fridericia callosa*, *Enchytraeus buchholzi* and *Buchholzia fallax* var. *arenaria* (Healy, 1976), must therefore have a high resistance to desiccation, or be able to avoid its effects.

Very few species can be described as characteristic of aquatic habitats. Only two species of *Grania* were entirely sublittoral and no freshwater species was confined to limnic conditions. Most of the commonly occurring species had a wide range of distribution in relation to soil water content. Some had a narrower range, however, and were found more frequently in wet, medium or dry soils as follows:

Preferring wet soils:
 Mesenchytraeus armatus
 M. sanguineus
 Cernosvitoviella atrata
 C. goodhui
 C. sphaerotheca
 C. palustris
 Achaeta aberrans
 Cognettia sphagnetorum
 C. glandulosa
 C. hibernica
 Henlea perpusilla
 Fridericia perrieri
 F. polychaeta
 Marionina argentea
 M. riparia
 M. filiformis
Preferring moist soils:
 Achaeta affinis

 Fridericia bisetosa
 Marionina clavata
Preferring dry soils:
 Achaeta bohemica
 A. eiseni
 Fredericia connata
 F. paroniana
 F. aurita
 F. sylvatica
 F. discifera
 Enchytronia parva
 Marionina communis

In coastal habitats, *Lumbricillus* spp., *Cernosvitoviella immota* and *Marionina appendiculata* preferred wet conditions while the remainder preferred moist or dry habitats. A few species, notably *Marionina argentea* and *Enchytraeus buchholzi*, occured in both freshwater and salt marshes, but most were restricted to one or the other. Among typical marine littoral forms, only *Marionina southerni* and an undetermined *Lumbricillus* are known to penetrate into freshwater in Ireland.

Enchytraeid abundance was, on average, highest in the saturated soils followed by the moist ones. In both flooded and dry soils the number of enchytraeids was reduced. The figures for moist and dry soils may be unrealistic, however, owing to greater depth penetration (see Discussion). Species diversity, expressed as the mean number of species in 64 cm^2, was highest in the wet and flooded soils in inland sites but higher in terrestrial than intertidal sites in coastal areas (Table 1).

Occurrence of enchytraeids with red blood

The majority of enchytraeids have blood which appears colourless although low concentrations of pigment may be present. In some species, however, the blood is red and as this is likely to be an adaptation to low oxygen concentrations, the distribution of red-blooded species is of interest. Red blood was noted in twenty-eight of the recorded species, representing eight genera. All were found in aquatic habitats or soils with more than 55% water, and most occurred in soils with more than 80% water (Fig. 1). This distribution coincided with the distribution of the aquatic oligochaete families, Tubificidae, Naididae and Lumbriculidae, which also have red blood (Fig. 1). Among the species com-

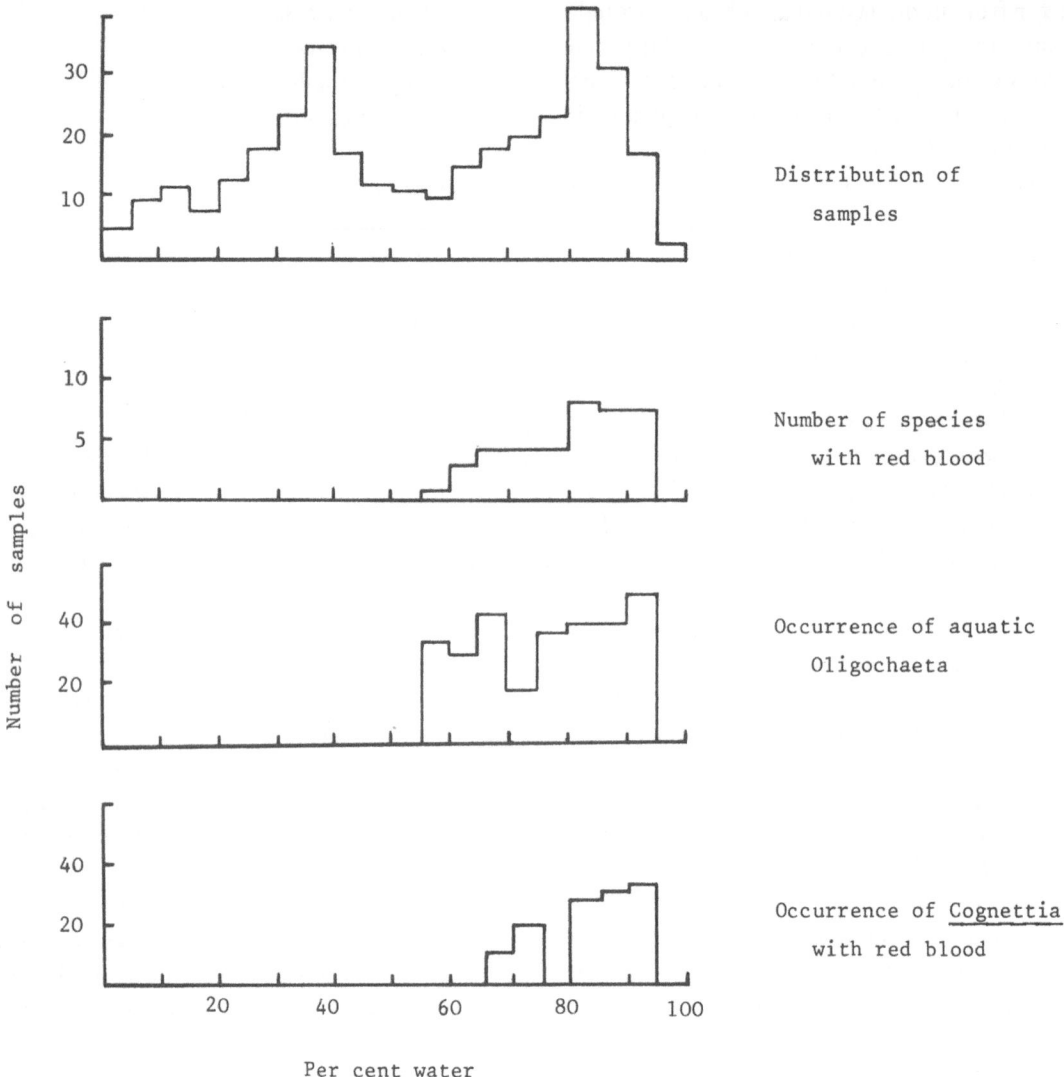

Fig. 1. Distribution of enchytraeids with red blood and aquatic Oligochaeta (Tubificidae, Naididae and Lumbriculidae) in relation to soil water content.

monly found in aquatic habitats, only *Marionina argentea* in freshwater and *M. appendiculata* in the marine littoral do not have red blood.

In a few species there was a marked ecophenotypic variation in the amount of haemoglobin present. *Cernosvitoviella* spp. were often only faintly coloured when living in damp soils while the red colour tended to be more pronounced in individuals from submerged sites. *Cognettia sphagnetorum* and *C. glandulosa*, which normally had colourless blood, developed a red coloration in waterlogged conditions, although not, apparently, in very acid peat. Typically, red-blooded individuals occurred in flushed areas of bog (rheotrophic mire) where pH and nutrient levels are higher than in the surrounding ombrotrophic bog. These worms tend to be bigger than their counterparts from drier soils, they sometimes have supplementary septal glands and there is a higher proportion of sexually mature individuals. Red blood in *C. sphagnetorum* occurred in substrates with more than 70% water and at pH > 3 (Fig. 2). This distribution coincided with that of aquatic oligochaetes which were also absent from very acid bogs (Fig. 3).

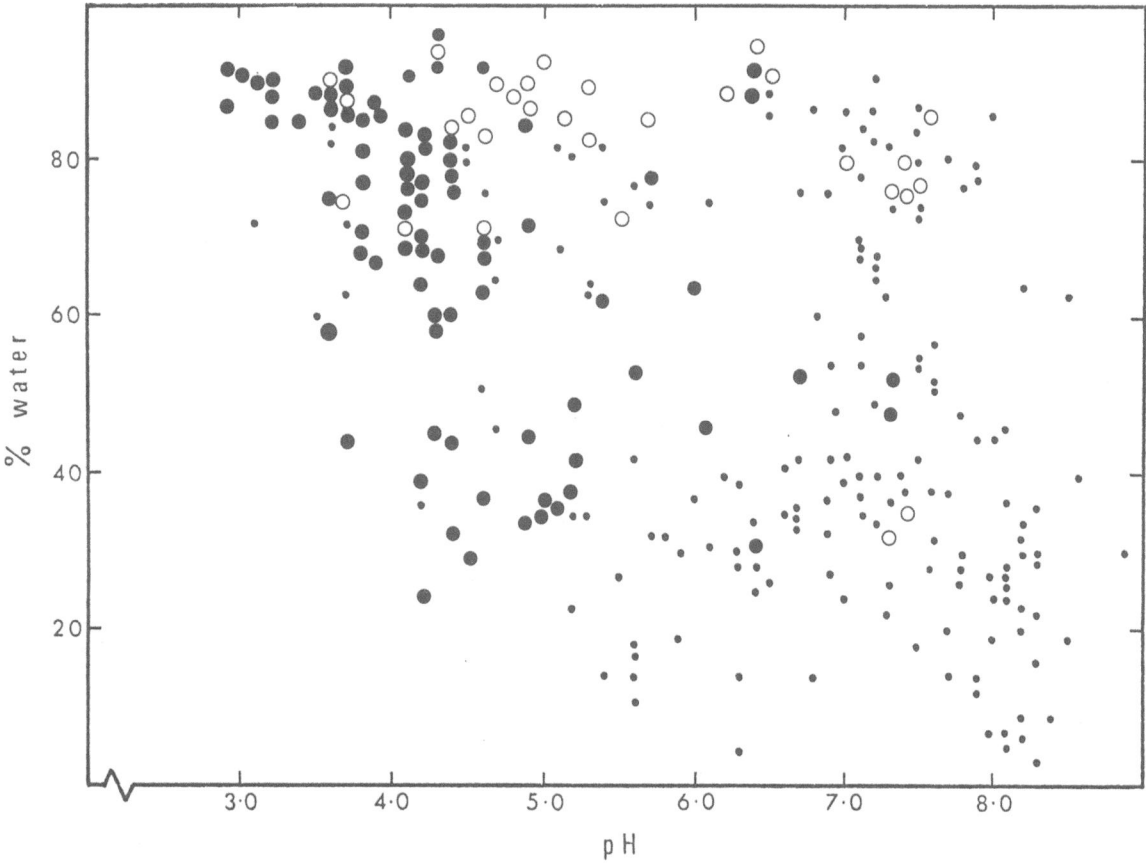

Fig. 2. Occurrence of *Cognettia sphagnetorum* (large dots) and individuals with red blood (circles) in relation to soil water content and pH. Small dots indicate samples without *Cognettia*.

Factors affecting the occurrence of enchytraeids in semi-aquatic sites

When biological and physical variables are measured in a series of samples, statistical analysis usually proceeds in two stages: (1) a reduction of the biological data to fewer variables which are efficient carriers of information, and (2) relating these biological variables to the physical variables in an explanatory manner (Green, 1979). Several methods of carrying out such analyses have been described, however, those most commonly used are ordination followed by correlation analysis (e.g. Sprules, 1977) and classification followed by discriminant analysis (e.g. Green & Vascatto, 1978). We chose the latter. The method of classification used was two-way indicator species analysis (Hill, 1979b). Data are first ordinated using reciprocal

averaging. Species are selected which characterize the extremes of the ordination axis and these are weighted in order to polarize the samples. The samples are then classified into two groups using the species with maximum value to indicate the poles of the ordination axis. This divisive procedure is repeated until a level determined by the investigator is reached.

Figure 4 shows an ordination diagram produced by detrended correspondence analysis (Hill, 1979a) upon which the classification has been superimposed. It shows three clusters which represent a separation of samples into those from acid peat sites, marsh-fen-aquatic sites and coastal sites. The acid peat sites are characterized by the presence of *Marionina clavata, Achaeta affinis, Cognettia sphagnetorum, C. cognettii, C. hibernica* and *Mesenchytraeus sanguineus. Henlea perpusilla, Mesenchy-*

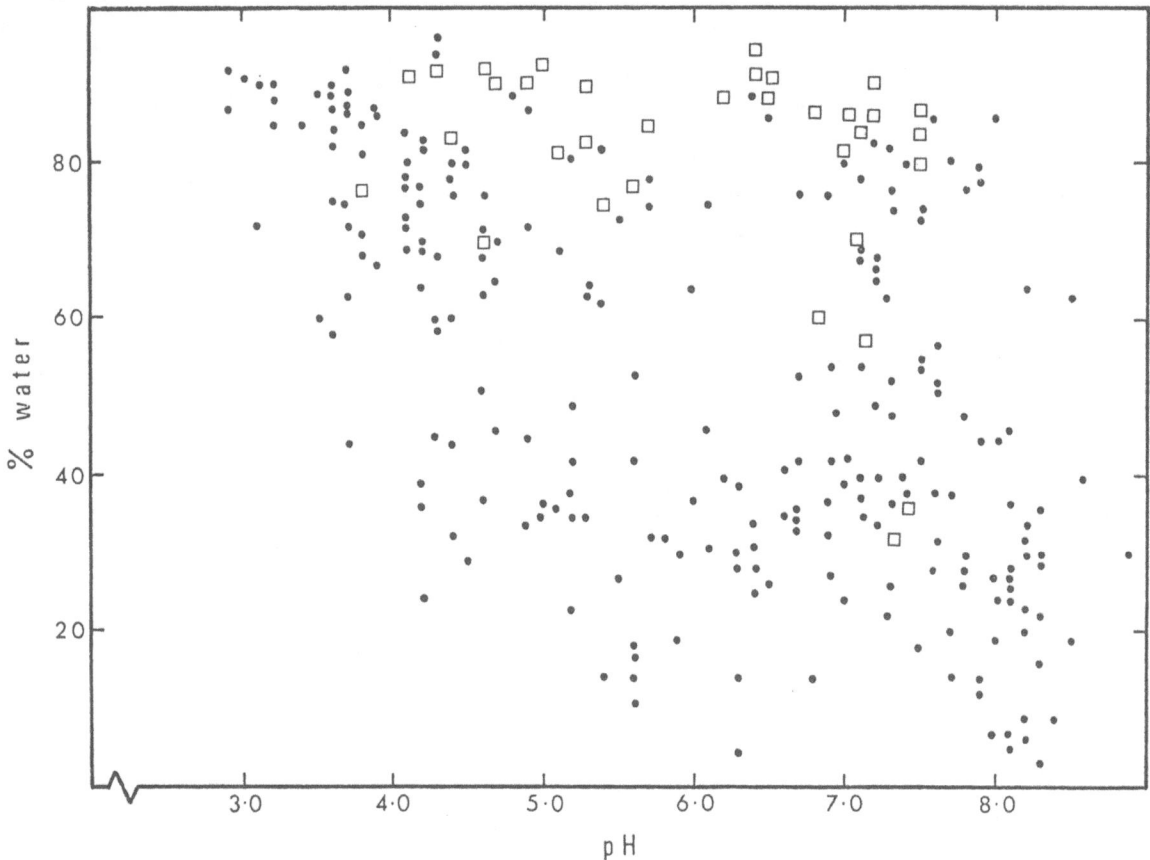

Fig. 3. Occurrence of Tubificidae, Naididae and Lumbriculidae (squares) in relation to soil water content and pH. Small dots indicate samples without aquatic Oligochaeta.

traeus armatus, Fridericia bulbosa, F. galba, F. perrieri, Buchholzia fallax, Achaeta bohemica, Marionina argentea, Enchytraeus buchholzi and *Cognettia glandulosa* characterize the marsh-fen-aquatic cluster and *Lumbricillus kaloensis, Marionina subterranea, M. appendiculata, M. spicula, M. southerni* and *Fridericia callosa* show preference for the coastal sites. The frequency of occurrence of the more important species in the three groups is shown in Table 2.

Standardized discriminant function coefficients indicated that magnesium levels and pH were the most important factors in the first and second functions respectively (Table 3). When these factors are plotted against the coordinates of the samples on the first ordination axis it becomes apparent that Mg level can be used to identify the coastal sites and that pH can be used to distinguish between the remaining clusters (Figs. 5–6). A simple dichotomous key based on these two factors is 96.96% successful in placing samples in the correct clusters.

1. Mg levels > 780 ppm .... Coastal site
   Mg levels < 780 ppm 2
2. pH < 5.2 .... Acid peat site
   pH > 5.2 .... Marsh-fen-aquatic site

Notes on the distribution of species within the three groups of sites

In acid peat, i.e. pH < 5.2, *Cognettia sphagnetorum* is a constant and usually abundant species with *Marionina clavata* and *Achaeta affinis* as frequent companions in the drier sites and *Mesenchytraeus sanguineus, Cernosvitoviella atrata* and *Cognettia hibernica* in the wetter ones. In peat with pH < 4.0 these may be the only species present. Above pH 5.0

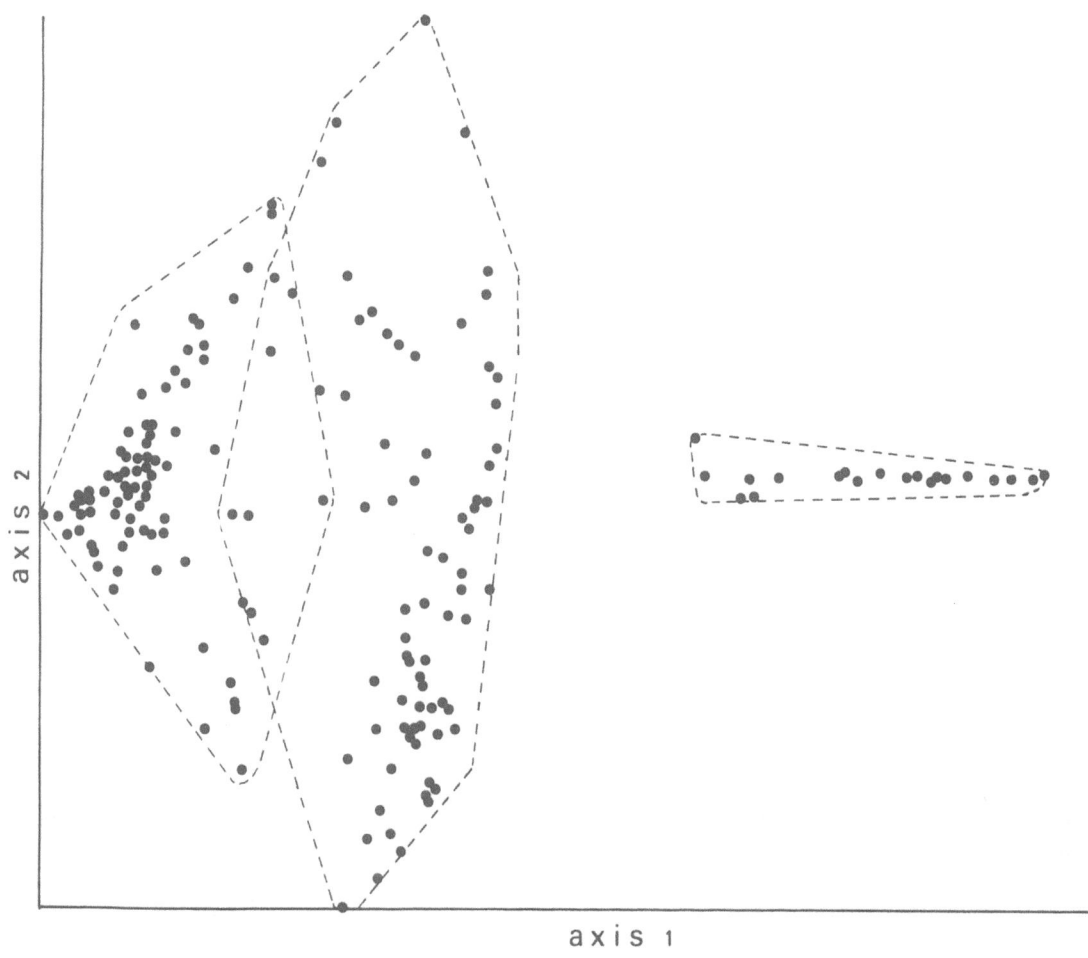

Fig. 4. Classification based on two-way indicator species analysis superimposed on an ordination produced using detrended correspondence analysis.

species characteristic of marsh and fen, such as *Enchytraeus buchholzi, Cognettia glandulosa, Fridericia perrieri* and *Marionina argentea*, accompany the typical species of acid peat. Rheotrophic sites in bog and heath are characterised by high diversity and the presence of aquatic oligochaetes.

In marsh, fen and aquatic sites, with pH > 5.2, the dominant species are usually *Marionina argentea, Enchytraeus buchholzi, Henlea perpusilla, Fridericia perrieri, Buchholzia fallax* and various *Fridericia* species. In the drier areas, grassland species such as *Achaeta bohemica, Fridericia galba* and *Fridericia bulbosa* may occur. These drier marshes have a very high diversity and up to eighteen species have been recorded in two samples. With frequent flooding the diversity falls and *Cer-*

nosvitoviella spp., *Cognettia glandulosa, Marionina argentea* and *M. riparia* become the most frequent and abundant species. In sub-aquatic sites only *M. argentea, M. riparia* and *C. atrata* are frequent.

Marionina spicula and *Enchytraeus albidus* are characteristic of coastal habitats. These are accompanied by *Fridericia callosa* in the drier areas, *Marionina southerni* and *M. subterranea* around HWS and *M. preclitellochaeta* and *Lumbricillus viridis* around HWN. Enchytraeids are rare below MTL on sandy beaches and are poorly represented in salt marshes. *Lumbricillus kaloensis, Marionina appendiculata* and *Cernosvitoviella atrata* occur in salt pans and creeks and on mud flats, while on rocky shores *M. appendiculata* and *Lumbricillus*

Table 2. Frequency of occurrence (in %) of some species under various environmental conditions.

Species	Water content		
	45–80%	> 80%	Flooded
Acid tolerant	(n = 14)	(n = 33)	(n = 44)
Cognettia sphagnetorum (Vejdovsky)	100	100	100
Cognettia hibernica Healy	71	24	36
Mesenchytraeus sanguineus N. & C.	57	33	9
Marionina clavata N. & C.	7	66	82
Achaeta affinis N. & C.	0	55	73
Cognettia cognettii (Issel)	0	18	39
Acid intolerant	(n = 20)	(n = 24)	(n = 20)
Marionina riparia (Bretscher)	65	17	0
Marionina argentea (Michaelsen)	90	79	95
Enchytraeus buchholzi Vejdovsky	45	67	85
Cognettia glandulosa (Michaelsen)	25	79	45
Henlea perpusilla Friend	25	92	75
Fridericia perrieri (Vejdovsky)	20	88	70
Buchholzia fallax Michaelsen	5	63	75
Mesenchytraeus armatus (Levinsen)	15	33	35
Fridericia galba Hoffmeister	10	79	45
Fridericia bulbosa (Rosa)	0	42	50
Achaeta bohemica (Vejdovsky)	0	21	40
Wide-ranging in inland sites	(n = 34)	(n = 57)	(n = 64)
Cernosvitoviella atrata (Bretscher)	34	56	76
Salt tolerant	(n = 16)	(n = 5)	(n = 9)
Marionina appendiculata N. & C.	88	60	11
Lumbricillus kaloensis N. & C.	81	20	11
Marionina spicula (Leuckart)	56	100	56
Enchytraeus albidus Henle	56	20	67
Marionina southerni (Černosvitov)	0.6	40	44
Fridericia callosa (Eisen)	0	60	65
Cernosvitoviella immota (Knöllner)	0	60	0

semifuscus may be important in crevices and dense algal turf with *Grania macrochaeta* in *Corallina* turf.

Table 3. Variables and standardized discriminant function coefficients.

	Function	
	I	II
pH	.66	.86
Water content	.17	.37
Organic carbon	.02	−.43
N	−.09	.29
C/N	.02	.02
Ca	−.48	.08
P	−.33	.25
K	−.23	−.21
Mg	.99	−.47
Na	.13	.04

Discussion

The Enchytraeidae have a wider ecological distribution than other oligochaete families being found in freshwater and marine sublittoral sediments as well as in all kinds of terrestrial habitats including those subject to severe drought like young sand dunes. A number of species occur across most of the water content range and it must be concluded that either individuals have a very wide range of tolerance or there is much ecophenotypic variation, for there are chemical as well as physical differences between aerated and saturated soils. For example, flooding causes anoxia, pH shifts, denitrification and changes in the stability of important salts (Ponnamperuna, 1972).

A characteristic feature of many species inhabiting wet environments is the presence of red blood.

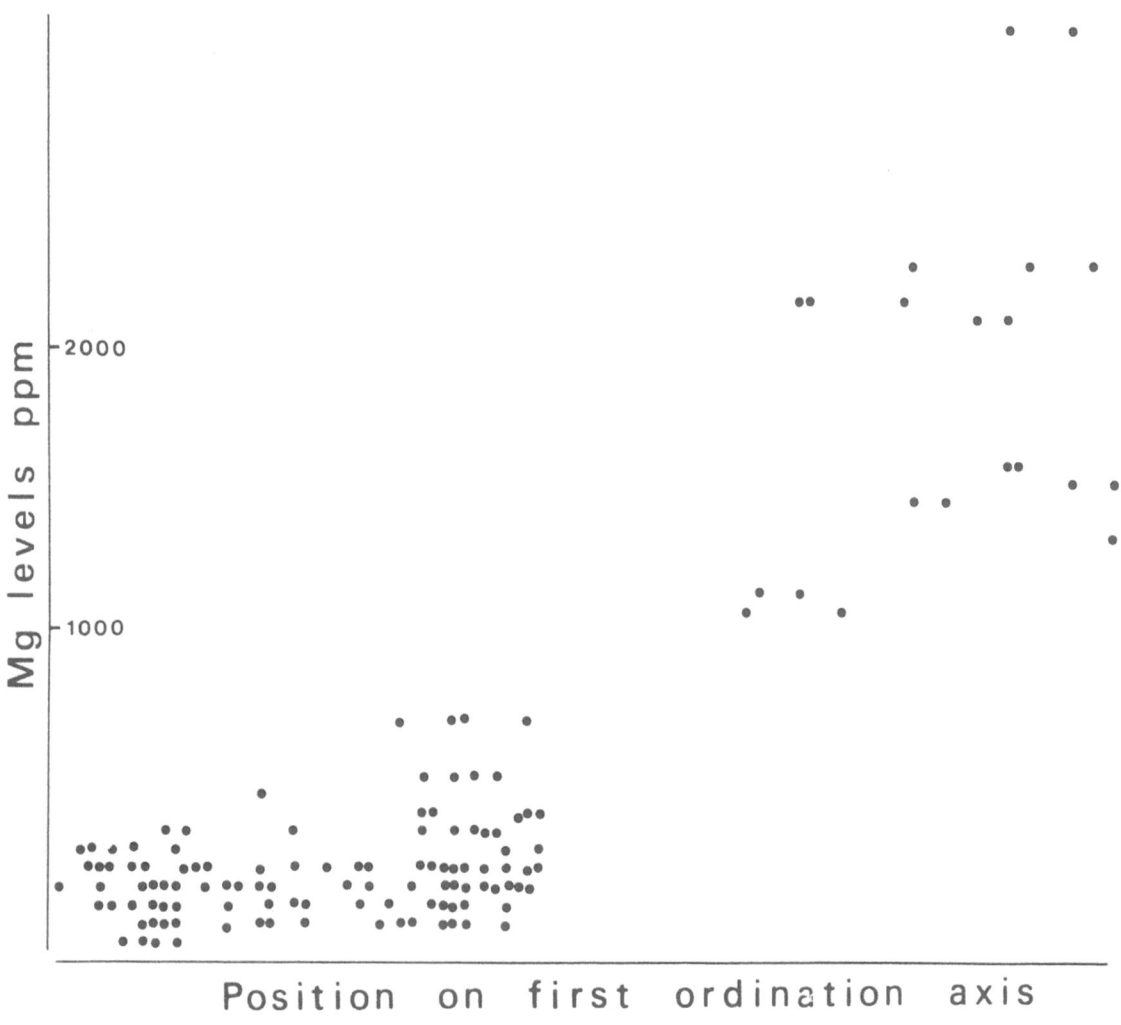

Fig. 5. Relationships between position of samples on the first ordination axis and magnesium levels.

This is undoubtedly an adaptation to low oxygen concentrations. Almost all the species living habitually in water have red blood, but none of those living in drier soils do. The presence of increased haemoglobin in specimens from wet soils suggests that there may be an individual response to oxygen deficiency. Earthworms living in submerged sites also develop more haemoglobin than do those from aerated soils (Omodeo, 1984). Some other peculiarities of enchytraeid species inhabiting very wet soils may be adaptations to the environment, for example, the smaller amount of collagen in the cuticle of *Mesenchytraeus* and *Lumbricillus* in comparison with more terrestrial species (Richards, 1977) and the highly contractile, rapid movement of *Cernosvitoviella* spp. which enables them to migrate verti-

cally over greater distances than their companion species and so avoid surface desiccation (Springett, *et al.,* 1970).

The depth penetration of enchytraeids, as for other groups, is less in waterlogged soils than in aerated ones. In wet peat, enchytraeids are mainly found in the top 6 cm (Peachey, 1963; Nurminen, 1967; Healy, 1976), but in soils with free drainage worms penetrate to a greater depth (Nielsen, 1955; O'Connor, 1957). Springett, *et al.* (1970), in a study of moorland enchytraeids, demonstrated vertical migration related to changes in humidity with different species showing different degrees of response. Some species, e.g. *Cernosvitoviella briganta,* traversed several centimetres in a few hours and *Cognettia sphagnetorum* sometimes migrated into the

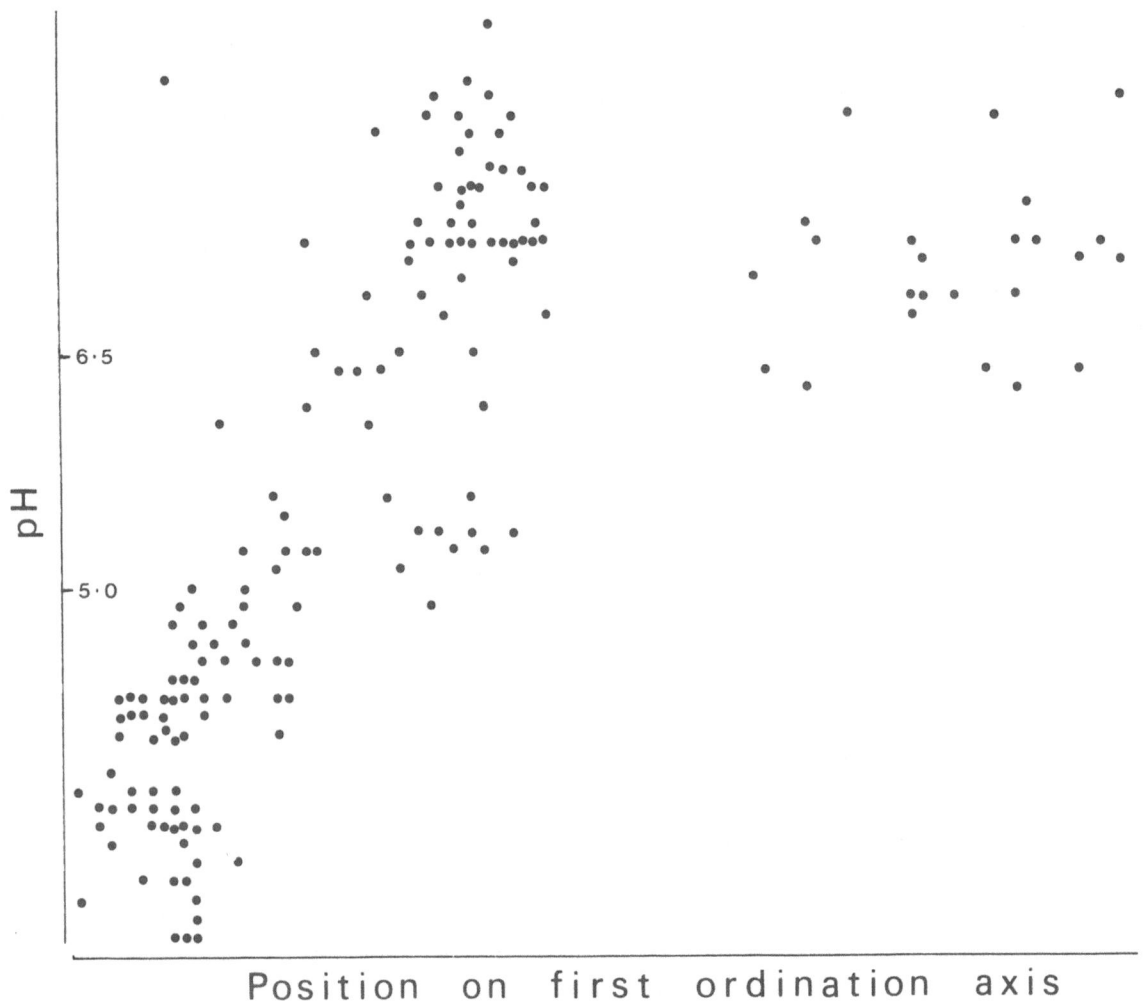

pH

6·5

5·0

Position on first ordination axis

Fig. 6. Relationships between position of samples on the first ordination axis and pH.

above ground vegetation. Seasonal segregation by depth, with *Cernosvitoviella atrata* migrating to deeper layers in summer, occurs in Alaskan tundra (MacLean, *et al.* 1977). Such responses allow individuals to avoid anoxia and, perhaps more importantly, drought.

Variations in the water content are obviously important in determining the distribution of enchytraeid species and the preference of some species for wet or dry soils has been demonstrated. Two other factors appear to have an influence in semi-aquatic sites, namely ions derived from seawater and pH. Coastal habitats are highly specialised environments and both the littoral and immediate supralittoral are dominated by a specially adapted fauna.

Some inland species can occur in the intertidal zone but they are never plentiful. Only *Marionina argentea* and *Enchytraeus buchholzi* were found during the survey, but *Marionina communis, Fridericia paroniana, Henlea ventriculosa, H. perpusilla* and *Enchytraeus lacteus* have been taken from marine and brackish-water sites in other regions of western Europe (Lasserre, 1971; Tynen, 1972; Rasmussen, 1973).

The importance of pH as a limiting factor, demonstrated here for wet soils, appears to be true for all inland habitats and has been noted in several studies (Standen & Latter, 1977; Healy, 1979; Hågvar & Abrahamsen, 1980; Standen, 1980, 1982), although it is still not clear whether it is pH itself or some related factor or factors which are responsible for apparent pH preferences. In general, *Cognettia sphagnetorum* dominates the enchytraeid fauna in acid soils but becomes less abundant above pH 5.0 when diversity increases and *Fridericia* spp. usually become prominent.

In this study, three principal types of semi-aquatic habitat have been distinguished on the basis of their enchytraeid fauna: (1) acid peats with a pH less than 5.2 (bog, flush, wet heath), (2) neutral or base-rich soils and peats (fen, marsh and aquatic sites), and (3) coastal habitats (salt marshes, beaches and brackish wetlands). Each type is characterized by a group of species which are exclusive to that type or at least rare in other types. The use of enchytraeids as indicators of broad ecological conditions appears to be justified by the high level of correlation in this study. Their use in distinguishing more precise conditions may be limited, however, by the wide tolerance exhibited by most species.

Enchytraeids appear to have a preference for wet soils, but this may be due more to the absence of drought in these situations than to any positive advantage gained by living in saturated surroundings. The vulnerability to drought of most enchytraeid species is demonstrated by reports of summer population decline (Nielsen, 1955; O'Connor, 1957). Dry soils usually have lower numbers than wet ones (Healy, 1979) and while a greater depth penetration in dry soils may be responsible for some of the observed differences, the pattern of population growth which has been observed, for example in Norwegian dry meadow (Solhøy, 1975), suggests that mortality due to desiccation and reproductive failure are the chief causes. Abrahamsen (1971) found that maximum population growth for *Cognettia sphagnetorum* occurred in peat saturated to 50–95% of the water holding capacity and that populations were not reduced by prolonged saturation. Dózsa-Farkas (1977) also gives 60–95% WHC as the optimum for survival of *Fridericia galba* but found that for this species mortality increased with saturation. At twelve of the marsh sites included in the present study, adjacent grassland with free drainage and a lower water content were also sampled. The mean number of enchytraeids per sample was 76 for the dry sites and 133 for the wet. At ten sites diversity was also greater in the marshy areas (Healy, 1976).

While enchytraeids appear to find optimal condi-

tions in waterlogged or temporarily flooded soils, they are not so successful in aquatic habitats. In lakes and rivers numbers are generally low by comparison with those of Tubificidae and Naididae. Many, if not most, of the species recorded in small numbers in the present survey could be accidental, i.e. derived from populations inhabiting the banks. Only *Marionina argentea, M. riparia* and *Cernosvitoviella atrata* were sufficiently frequent to be considered as belonging to the subaquatic community. In some situations, however, enchytraeids may become more important, in dense vegetation in the littoral zone, for example, or in small streams and ditches. In one river system in the Pyrénées, Naididae and Tubificidae dominated the oligochaete fauna below 1600 m especially where there was a certain amount of silt, but in the torrents at higher altitude enchytraeids constituted one third of the oligochaete population (Giani & Lavandier, 1977).

The Enchytraeidae is then essentially a terrestrial family with aquatic tendencies. Their success in semi-aquatic habitats seems to be due to a combination of characteristics. Firstly, they are eurytopic, relatively undemanding and to some extent opportunistic (Giere, 1980; Giani & Lavandier, 1977) and nearly all can survive periods of immersion. Secondly, no species, as far as is known, has evolved a physiological system for resisting drought such as the summer diapause in earthworms, hence populations in dry soils are subject to periodic decline. Thirdly, most species seem to need some kind of open framework in which to live, such as soil with a crumb structure, living vegetation or leaf litter in various stages of decomposition. The presence of escape routes provided by emergent vegetation may be important in allowing individuals to avoid toxic or anoxic conditions. Finally, most of the semi-aquatic habitats investigated are to some extent temporary and do dry out for a time, at least at the surface. Enchytraeids may be able to take advantage of these periods to grow and multiply. An investigation of the level of activity of submerged individuals and, in particular their ability to reproduce would clarify this point.

References

Abrahamsen, G., 1971. The influence of temperature and soil moisture on the population density of Cognettia sphagnetorum (Oligochaeta: Enchytraeidae) in cultures with homogenised raw humus. Pedobiologia 11: 417-424.

170

Dózsa-Farkas, K., 1977. Beobachtungen über die Trockenheits-toleranz von Fridericia galba (Oligochaeta, Enchytraeidae). Opusc. zool., Bpest. 14: 77–83.

Giani, N. & P. Lavandier, 1977. Les oligochètes du torrent d'Estaragne (Pyrénées Centrales). Bull. Soc. Hist. nat. Toulouse, 113: 234–243.

Giere, O., 1980. Tolerance and preference reactions of marine Oligochaeta in relation to their distribution. In R. O. Brinkhurst & D. G. Cook (eds.), Aquatic Oligochaete Biology. Plenum Press, New York: 385–409.

Green, R. H., 1979. Sampling methods and experimental design for environmental biologists. Wiley, New York, xi + 257 pp.

Green, R. H. & G. L. Vascatto, 1978. Analysis of environmental factors controlling spatial patterns in species composition. Water Res. 12: 583–590.

Hågvar, S. & G. Abrahamsen, 1980. Colonisation by Enchytraeidae, Collembola and Acari in sterile soil samples with adjusted pH levels. Oikos 34: 245–258.

Healy, B., 1976. Taxonomy and ecology of Enchytraeidae (Oligochaeta) in Ireland. Ph.D. Thesis, Natn. Univ. Ireland, 284 pp.

Healy, B., 1979. Distribution of terrestrial Enchytraeidae in Ireland. Pedobiologia 20: 159–175.

Hill, M. O., 1979a. DECORANA. A FORTRAN program for detrended correspondence analysis and reciprocal averaging. Cornell Univ., Ithaca, New York, iii + 52 pp.

Hill, M. O., 1979b. TWINSPAN. A FORTRAN program for arranging multivariate data in an ordered two-way table by classification of the individuals and attributes. Cornell Univ., Ithaca, New York, iv + 90 pp.

Lasserre, P., 1971. The marine Enchytraeidae (Annelida, Oligochaeta) of the eastern coast of North America with notes on their geographic distribution and habitat. Biol. Bull. mar. biol. Lab., Woods Hole 14: 440–460.

MacLean, S. F., G. K. Douce, E. A. Morgan & M. A. Steel, 1977. Community organisation in the soil invertebrates of Alaskan Arctic. In U. Lohm & T. Persson (eds.). Soil organisms as components of ecosystems. Ecol. Bull. (Stockholm) 25: 90–101.

Nielsen, C. O., 1955. Studies on the Enchytraeidae. 2. Field Studies. Natura jutl. 4: 5–58.

Nurminen, M., 1967. Ecology of enchytraeids (Oligochaeta) in Finnish coniferous forest soil. Ann. zool. fenn. 4: 147–157.

O'Connor, F. B., 1955. Extraction of enchytraeid worms from a coniferous forest soil. Nature, Lond. 175: 815–816.

O'Connor, F. B., 1957. An ecological study of the enchytraeid worm population of a coniferous forest soil. Oikos 8: 161–199.

Omodeo, P., 1984. On aquatic Oligochaeta Lumbricomorpha in Europe. In G. Bonomi & C. Erséus (eds.), Aquatic Oligochaeta. Proceedings of the Second International Symposium on Aquatic Oligochaeta Biology. Dev. Hydrobiol. (this volume).

Peachey, J. E., 1963. Studies on the Enchytraeidae (Oligochaeta) of moorland soils. Pedobiologia 2: 81–95.

Ponnamperuma, F. N., 1972. The chemistry of submerged soils. Adv. Agron. 24: 24–95.

Rasmussen, E., 1973. Systematics and ecology of the Isefjord marine fauna. Oligochaeta. Ophelia 11: 125–131.

Richards, K. S., 1977. Structure and function in the oligochaete epidermis (Annelida). Symp. zool. Soc. Lond. 39: 171–193.

Solhøy, T., 1975. Dynamics of Enchytraeidae populations on Hardangarvidda. In F. Wielgolaski (ed.), Fennoscandian Tundra Ecosystems, 2. Animals and Systems Analysis. Springer-Verlag, Berlin, New York: 60–65.

Springett, J. A., J. E. Brittain & B. P. Springett, 1970. Vertical movements of Enchytraeidae (Oligochaeta) in moorland soils. Oikos 21: 16–21.

Sprules, W. G., 1977. Crustacean zooplankton communities as indicators of limnological conditions: An approach using principal component analysis. J. Fish. Res. Bd Can. 34: 962–975.

Standen, V., 1980. Factors affecting the distribution of Enchytraeidae (Oligochaeta) in associations of peat and mineral sites in northern England. Bull. Ecol. 11: 599–608.

Standen, V., 1982. Associations of Enchytraeidae (Oligochaeta) in experimentally fertilized grasslands. J. anim. Ecol. 51: 501–522.

Standen, V. & P. M. Latter, 1977. Distribution of a population of Cognettia sphagnetorum (Enchytraeidae) in relation to microhabitats in a blanket bog. J. anim. Ecol. 46: 213–229.

Tynen, M. J., 1972. The littoral Enchytraeidae (Oligochaeta) of Anglesey and the Menai Strait with notes on habitats. J. nat. Hist. 6: 21–29.

The oligochaetes (Annelida, Oligochaeta) in a lake and a canal in the agricultural landscape of Poland

K. Kasprzak

Provincial Administration, Department of Environment Protection, Stalingradzka 16/18, PL-60-967 Poznań, Poland

Keywords: aquatic Oligochaeta, diversity, respiration, production

Abstract

Data are presented concerning an investigation into the diversity, abundance, production and respiration of the oligochaete fauna in the eutrophic, polymictic Lake Zbechy and in a melioration canal in the Wielkopolska Region, an area of intensive agriculture. It was found that in the canal the average biomass of oligochaetes was about four times higher than in the lake. Oligochaetes expend 1.1–4 times more energy in respiration than in tissue production. Species diversity and species number are positively correlated, while the correlation between diversity and abundance is negative.

Introduction

As the oligochaetes have great importance for the transformation of energy and the circulation of matter in aquatic ecosystems, the problem of evaluating the function of these annelids, and the structures of their communities, is fundamental not only in selected environments or zones but also in the ecosystem as a whole. Pressure on the land surrounding shallow and eutrophic small water reservoirs in Poland, escalating seasonally, is connected with the high production and abundance of the fauna, particularly of many species of invertebrates. This pressure is manifested by the occurrence of insects that become troublesome for man and livestock (Diptera: some Culicidae, Tabanidae). Small water reservoirs and running waters teeming with invertebrates (also oligochaetes) are often a feeding spot for many land species (chiefly birds).

Kasprzak (1980) and Banaszak & Kasprzak (1980) reported the results of research on the occurrence and density of aquatic oligochaetes and on the role played by these annelids in small aquatic ecosystems in agricultural areas, conducted as a part of studies of energy transformation and matter circulation in the whole agricultural landscape. The aim of this paper is to summarize the most important data on the diversity, abundance, production and respiration of the oligochaete fauna in the small (109 ha), shallow, eutrophic and polymictic Lake Zbechy and the melioration canal discharging into the Obra Canals located in a typical lowland landscape of intensive agriculture in W Poland (Wielkopolska Region).

Methods

The evaluation of abundance, biomass and representativeness of quantitative samples were performed according to Banaszak & Kasprzak (1980). The dry weight of oligochaetes was calculated as 15% of the fresh weight and 1 g dry weight was assumed to correspond to 20.935 kJ. Annual production was calculated following the Bojsen-Jensen method.

The annual respiration was calculated according to the equation proposed by Kamlyuk (1974):

$$R = 0.105 \; W^{0.75}$$

Hydrobiologia 115, 171–174 (1984).

where R (respiration) is expressed as $mg\,O_2 \cdot ind^{-1} \cdot h^{-1}$ and W is the fresh-weight in grams. $Q_{10} = 2$ was a further assumption. Calculations between dates were made on the basis of literature data and of original data of the author on the respiration rates of the following species: *Stylaria lacustris* (Naididae), *Limnodrilus udekemianus, Isochaetides newaensis, Tubifex tubifex, Potamothrix hammoniensis, Psammoryctides barbatus* (Tubificidae), *Lumbriculus variegatus* (Lumbriculidae) and *Lumbricillus rivalis* (Enchytraeidae). $1\,dm^3\,O_2$ consumed was assumed to be equivalent to 20.112 kJ.

The Shannon-Weaver diversity index was calculated according to the formula:

$$H' = - \sum_{i=1}^{S} p_i \cdot \ln p_i$$

where p_i is, for the i-th species, the ratio: average specific numerical abundance (or biomass):total abundance (or biomass). S is the number of species present.

Abundance (numbers and biomass)

The average annual biomass of oligochaetes in the lake amounted to 145.4 mg dry weight $\cdot m^{-2}$ (768.6 ind $\cdot m^{-2}$). In the lake four species of Tubificidae (*Potamothrix hammoniensis, Limnodrilus hoffmeisteri, Psammoryctides barbatus, P. albicola*) made up the greatest number (220.8 ind $\cdot m^{-2}$) and biomass (132.3 mg dry weight $\cdot m^{-2}$) of oligochaetes. *P. hammoniensis* was the most abundant species (90.8 mg dry weight $\cdot m^{-2}$). The greatest density of oligochaetes and greatest number of species was situated in the littoral, in the depth zone 0.5–1.5 m (annual biomass from 85.4 to 22.1 mg dry weight $\cdot m^{-2}$). The annual average biomass of oligochaetes in the canal (628.1 mg dry weight $\cdot m^{-2}$) was more than four times greater than the average biomass in the lake. In contrast to the lake, the canal showed a considerable increase in the biomass of *L. hoffmeisteri* (367.8 mg dry weight $\cdot m^{-2}$). The density of mature individuals in the canal showed seasonal differences. In the winter-spring period and in autumn *L. hoffmeisteri* was dominant, while in the summer period *P. hammoniensis* predominated. In the littoral benthos, smaller forms (average individual fresh weight: 0.20 mg) often dominated, while in the eulittoral, large forms (11.4 mg fresh weight) frequently prevailed.

The highest average density of Tubificidae in the lake and the canal was observed in spring and autumn, less frequently also in summer. This regularity is connected with properties of the reproduction biology of these animals. The regularity with which this process occurs in the Tubificidae is the result of the successive stages of their life history, particularly of the development, resorption and regeneration of the reproductive apparatus. Continuous reproduction for a considerable length of time is possible in the Tubificidae; among others, in *P. hammoniensis* and *L. hoffmeisteri,* i.e. the dominant species in the lake and canal. Changes in the time of development and resorption of the reproductive system occur according to the age of individuals, the life-span being 2–3 years on average (Timm, 1974). In the second year of their life, reproduction occurs twice; in the first and third year only once (Archipova, 1976; Poddubnaya, 1980). In exceptional cases *P. hammoniensis* can reproduce for five years (Jónasson & Thorhauge, 1972).

Species composition and diversity

The species structure of the oligochaete community in a small lake in an agricultural landscape (Banaszak & Kasprzak, 1980) is the characteristic one of European eutrophic lowland lakes. This is indicated by a lack of certain representatives of the Tubificidae (*Rhyacodrilus coccineus, Limnodrilus profundicola, Peloscolex ferox*) and Lumbriculidae (*Stylodrilus heringianus*), associated with oligotrophic and mesotrophic lakes, by the preponderance of *P. hammoniensis, L. hoffmeisteri, P. barbatus* and *P. albicola* in the total numbers and biomass of the oligochaete communities of the bottom of the littoral and profundal and by the low number of species of oligochaetes. The occurrence of the species under discussion, particularly the dominance of *P. hammoniensis* in the oligochaete community, is of paramount importance as an evaluation index of the lake trophy. This species shows higher resistance to negative environmental conditions in the lake than other Tubificidae, particularly as regards a low concentration of oxygen (Berg, *et al.,* 1962; Berg & Jónasson, 1965; Timm, 1970), and is associated mainly with eutrophic lowland lakes

(Milbrink, 1973; Jónasson, 1977). These data confirm Bonomi's (1967) observations on the structural variation of the species of Tubificidae in the course of eutrophication in Lago Maggiore, Lang & Lang-Dobler's (1980) investigations on Lake Geneva and Milbrink's (1973) studies of lakes in S Sweden.

An analysis of the diversity of oligochaetes measured in terms of their absolute numbers and as diversity expressed by the Shannon-Weaver index, shows a remarkable difference for the lake and the canal. Hitherto, data have shown that the fauna of the lake is much more diversified than that of the canal where, also in other groups, the number of species observed was markedly lower. Similarly, the greater diversity of the oligochaete fauna is confirmed by the comparison of the species diversity indices H', calculated separately for the numbers (H'_n) and biomasses (H'_b) of oligochaetes in the lake and in the canal (Tables 1 and 2). In the case of both ecosystems under study, the values of H' for the biomass do not correspond to those for the numbers. This is due to the fact that the domination structure of species in the investigated communities changes according to the criteria used, since numerically dominant species do not always predominate as regards biomass. The Shannon-Weaver index can indicate community organization (level of accumulate information) and average entropy:individual in the community (Loucks, 1970).

The position of species in communities can be established from the point of view of both their numbers and biomass (Whittaker, 1966). The relation between H'_n and H'_b is characterized by a high correlation coefficient (r = 0.87; P < 0.05), which suggests that both species diversity evaluation indices are equally valid. The correlation between H'_n and number of species (r = 0.97; P < 0.05) is fairly high and shows that the number of species affects the value of H'_n in communities. The number of species has an even greater influence on diversity than by H'_b. There is a negative correlation between H'_n and the logarithm of density (r = -0.70; P < 0.05) (Fig. 1). Although Odum (1962) locates among indices which characterize structure, in my opinion, this parameter shows, above all, the functional properties of the ecosystem (energy and matter circulation).

Production and respiration

The lake and the canal are characterized by the high production of the oligochaete communities per existing biomass unit (Table 2). This is indicated by the productivity index $P:\bar{B}$ (production: standing crop), that ranges from 64 to 82%. These values are higher than those of the biomass turnover $P:\bar{B}$ calculated on the basis of data (only for the reproductive season) for *L. hoffmeisteri* ($P:\bar{B}$ = 56–67%; Archipova, 1976) and for all the oligo-

Table 1. Mean values of Shannon-Weaver species diversity index (H') for oligochaetes in the lake and in the canal (probability: 0.01 < P < 0.05, according to Poole (1974)).

	Lake	Canal
H'_{number}	1.8610	1.6987
$H'_{biomass}$	1.0446	0.9448

Table 2. Comparison of mean annual biomass ($\bar{B}$), annual production (P), annual respiration (R) of oligochaetes in the lake and the canal. The values are expressed as kJ m^{-2} a^{-1}.

		$\bar{B}$	P	R	$P:\bar{B}$	R:P
Lake	1977	8.16	54.95	63.42	6.7	1.15
	1978	1.88	16.21	18.15	8.2	1.12
Canal	1977	27.63	178.04	290.00	6.4	1.63
	1978	17.90	131.26	641.52	7.3	4.89

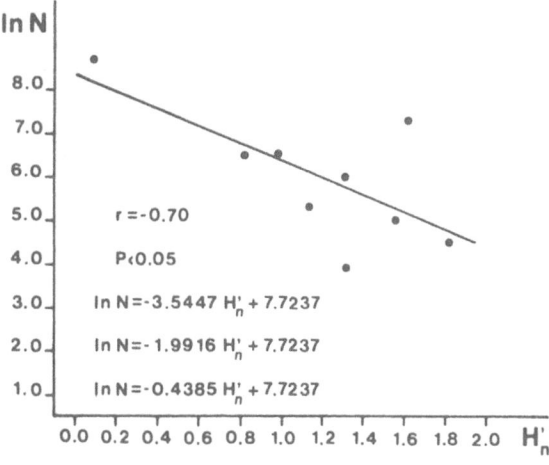

Fig. 1. Correlation between oligochaete abundance, expressed as lnN, and Shannon-Weaver species diversity index (H'_n).

chaetes in the littoral of Mazurian lakes ($P:\bar{B}$ = 40%; Kajak & Dusoge, 1976). For *P. hammoniensis* in the profundal of the eutrophic Lake Esrom, Jónasson & Thorhauge (1976) give a $P:\bar{B} = 1$ and Winberg *et al.* (1972) calculated for the whole oligochaetes community a $P:\bar{B} = 5.4$.

The ratio between respiration in the canal and that in the lake was 4.6 in 1977 and 35.3 in 1978 (Table 2).

Conclusions

1. The trophic status of the lake investigated is indicated by
a) the absence in the lake of species characteristic of oligotrophic and mesotrophic lakes in Poland;
b) a prevalence in numbers and biomass of the whole community of species characteristic mainly of eutrophic lakes; and
c) the comparatively low density of oligochaetes.

2. The frequent occurrence of the larger oligochaetes in the eulittoral shows that processes in the environment cause the circulation of the matter accumulated in the sediments to slow down, whereas in the littoral the prevalence of smaller forms accelerates matter circulation.

3. The relation between species diversity and number of species of oligochaetes is characterized by a high positive correlation, while the relation between species diversity and density of oligochaetes is characterized by a high negative correlation.

4. A comparison of respiration and production shows a surplus of energy expended on respiration by the oligochaetes both in the lake and the canal. In these animals, the expense of energy on respiration is 1.1–4 times higher than on tissue production.

References

Archipova, N. R., 1976. Osobennosti biologii i produkciya Limnodrilus hoffmeisteri Clap. (Oligochaeta, Tubificidae) na serych ilach Rybinskogo vodochranilashcha. In B. A. Vaynsteyn (ed.), Biologiya i sistematika presnovodnych bespozvonochnych. Trudy Inst. Biol. Vnutr. Vod 34 (37): 5–15.

Banaszak, J. & K. Kasprzak, 1980. Evaluation of occurrence and density of Oligochaeta, Mollusca and Chironomidae in bottom deposits of Lake Zbęchy and in melioration channel within agricultural landscape. Pol. ecol. Stud. 6: 221–245.

Berg, K. & P. M. Jónasson, 1965. Oxygen consumption of profundal lake animals at low oxygen content of the water. Hydrobiologia 26: 131–144.

Berg, K., P. M. Jónasson & K. W. Ockelmann, 1962. The respiration of some animals from the profundal zone of a lake. Hydrobiologia 19: 1–39.

Bonomi, G., 1967. L'evoluzione recente del Lago Maggiore rivelata dalle cospicue modificazioni del macrobenton profondo. Mem. Ist. ital. Idrobiol. 21: 197–212.

Jónasson, P. M., 1977. Lake Esrom Research 1867–1977. In C. Hunding (ed.), Danish Limnology Reviews and Perspectives. Folia limnol. scand. 17: 67–89.

Jónasson, P. M. & F. Thorhauge, 1972. Life cycle of Potamothrix hammoniensis (Tubificidae) in the profundal of a eutrophic lake. Oikos 23: 151–158.

Jónasson, P. M. & F. Thorhauge, 1976. Production of Potamothrix hammoniensis (Tubificidae) in the profundal of eutrophic Lake Esrom. Oikos 27: 204–209.

Kajak, Z. & K. Dusoge, 1976. Benthos of Lake Sniardwy as compared to benthos of Mikolajskie Lake and Lake Taltowisko. Ekol. pol. 24: 77–101.

Kamlyuk, L. V., 1974. Energeticheskiy obmen u svobodnozhivushchich ploskich i kolchatych chervey i factory ego opredelyayushche. Zh. obshch. Biol. 35: 874–885.

Kasprzak, K., 1980. Oligochaeta community structure and function in agricultural landscapes. In R. O. Brinkhurst & D. G. Cook (eds.), Aquatic Oligochaete Biology. Plenum Press, New York: 411–431.

Lang, C. & B. Lang-Dobler, 1980. Structure of tubificid and lumbriculid worm communities, and three indices of trophy based upon these communities, as descriptions of eutrophication level of Lake Geneva (Switzerland). In R. O. Brinkhurst & D. G. Cook (eds.), Aquatic Oligochaete Biology. Plenum Press, New York: 457–470.

Loucks, O. L., 1970. Evolution of diversity, efficiency, and community stability. Ann. Zool. 10: 17–25.

Milbrink, G., 1973. On the use of indicator communities of Tubificidae and some Lumbriculidae in the assessment of water pollution in Swedish lakes. Zoon 1: 125–139.

Odum, E. P., 1962. Relationship between structure and function in ecosystems. Jap. J. Ecol. 12: 108–118.

Poddubnaya, T. L., 1980. Life cycles of mass species of Tubificidae (Oligochaeta). In R. O. Brinkhurst & D. G. Cook (eds.), Aquatic Oligochaete Biology. Plenum Press, New York: 175–184.

Poole, R. W., 1974. An introduction to quantitative ecology. McGraw-Hill Book Company, 532 pp.

Timm, T., 1970. On the fauna of the Estonian Oligochaeta. Pedobiologia 10: 52–78.

Timm, T., 1974. O zhiznenych ciklach vodnych oligochet v akvariumach. Gidrobiol. Issled. 6: 97–118.

Whittaker, R. M., 1966. Forest dimensions and production in the Great Smoky Mountains. Ecology 47: 103–121.

Winberg, G. G., V. A. Babitsky, S. I. Gavrilov, G. V. Gladky, I. S. Zakharenkov, R. Z. Kovalevskaya, T. M. Mikheeva, P. S. Nevyadomskaya, A. P. Ostapenya, P. G. Petrovich, J. S. Potaenko & O. F. Yakushko, 1972. Biological productivity of different types of lakes. In Z. Kajak & A. Hilbricht-Ilkowska (eds.), Productivity problems of freshwaters. Polish Scientific Publishers, Warsaw & Krakow: 383–404.

Oligochaeta of the epigean and underground fauna of the alluvial plain of the French upper Rhône (biotypological trial)[1]

J. Juget[2]
Université Claude Bernard, Lyon I, Dpt. Biologie animale et Ecologie 'L.A. C.N.R.S. 367, 43, Boulevard du 11 Novembre 1918, F-69622 Villeurbanne Cédex, France

Keywords: aquatic Oligochaeta, biotypology, French upper Rhône

Abstract

Within a framework of interdisciplinary research, a biotypological trial of the epigean and underground fauna (Oligochaeta and Aphanoneura) of the alluvial plain of the French upper Rhône is proposed, according to the partition of this hydrographic complex into superficial and subterranean functional sets.

Introduction

The French upper Rhône and its alluvial plain represent one of the topics of the interdisciplinary programme of environmental research on 'Ecological management of the water resources' by C.N.R.S. (Roux, 1982).

The idea of a cartographical method for this purpose is based on the arrangement of the fluvial ecosystem into subsystems or functional sets. The latter comprise the principal fluvial axis, the secondary channels still active and the 'lônes' or oxbows more or less isolated and warped, abandoned by the river. The functional sets are distributed perpendicular to the principal channel, therefore the oldest are situated on the margin of the main bed. In fact, this relates to developmental stages which are sensitive to the geomorphological environment and the impact of human activities (Bravard, 1982; Amoros *et al.,* 1982; Pautou & Bravard, 1982; Richardot-Coulet *et al.,* 1982). They can be defined and classified schematically as follows, excluding the tributaries which are not considered in this study:

– The *eupotamon* serves to designate the main channel and the secondary arms supplied by the same kind of water and affected by the same hydrological factors.

– The *parapotamon* designates the old channels which are isolated upstream by the alluvial deposits, but still connected downsteam. The supply is mixed (fluvial waters from downstream, underground waters from upstream).

– The *plesiopotamon* specifies the isolated arms without a permanent contact with the current but periodically flooded by the river (in order of 0 to 5 days/year).

– The *paleopotamon* serves to indicate the old meandering arms which have lost all direct contact with the river.

The interest of studying the aquatic phyto- and zoocoenoses, especially the communities of Oligochaeta, derives from the fact that they integrate the conditions which define such chronosequences.

This study is restricted to the sector of 'Jons-Villette d'Anthon' (Fig. 1). The evolution of the channels of the river in this sector during the last two centuries was reconstructed by Dorgelo (1973), and the geomorphological, hydrological and sedimentological studies of the river system are the subject of various recent publications (B.R.G.M., 1969, 1973; Gibert *et al.,* 1977; Reygrobellet *et al.,* 1981; Reygrobellet & Dole, 1982; Seyed

[1] Structure and dynamics of the French upper Rhône – XXXVI.
[2] With the collaboration of C. Amoros, A. Berly, M. J. Dole, M. Provot, J. L. Reygrobellet and H. Tachet.

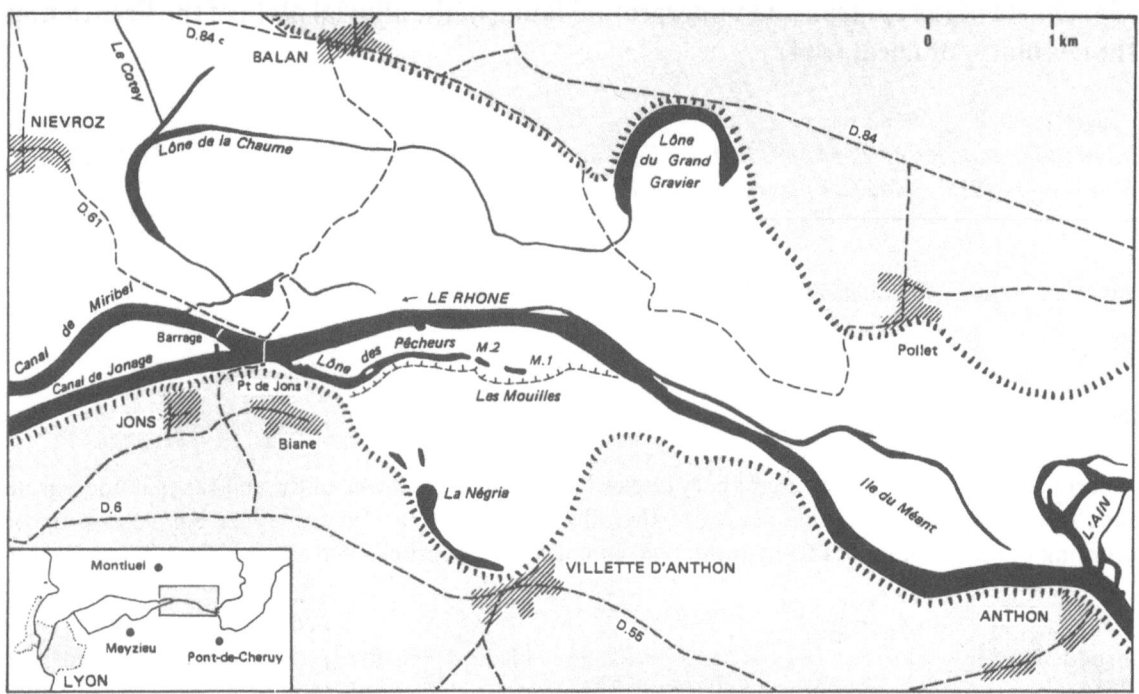

Fig. 1. Map of the upper Rhône in the sector of Jons-Villette d'Anthon, showing some relict appendices.

Reihaini *et al.,* 1982; Bournaud & Cellot, 1981; Bravard, 1982; Lafont & Juget, 1976; Juget, 1980; Juget *et al.,* 1976, 1979; Juget & Roux, 1982.

Sampling methods

A triangular dredge was used for sampling in the main bed (eupotamon) near the mouth of the 'Lône des Pêcheurs' and an Ekman grab in the other functional sets ('Lône des Pêcheurs', 'Les Mouilles', 'Lône du Grand Gravier'), excluding the zooperiphyton growing on the aquatic macrophytes. Two subsets or functional units were chosen at the 'Lône des Pêcheurs': the zone of confluence with the active Rhône and the upstream zone. The Ballinger & McKee (1971) index–organic sediment index, OSI – defined by the product C (% organic carbon) × N (% Kjeldahl nitrogen) and measured from the superficial layers and the fine sediment fraction, served to determine these two subsets on the ecological plan.

Sampling of underground interstitial fauna was carried out by the method of Bou & Rouch (1967;

Bou, 1974) by pumping out the interstitial water:

– within banks of sand, gravel and pebbles situated on the canalised section of the active current of the Rhône, the Miribel canal, nearly 2 km downstream of the mouth of the 'Lône des Pêcheurs';

– on the upstream extremity of the 'Lône des Pêcheurs', 1 m from the water margin in a pebble bank influenced by the underground outflow of the Rhône;

– on the left bank of the 'Lône du Grand Gravier' in the median part of this old meander, a few metres from the water margin in sediments comprising 5% of fine particles.

Sampling of interstitial water was carried out close to the piezometric level which varied between 30 and 90 cm from the surface of the sediment (average depth 50 cm). Complementary sampling was carried out at depths of 1.5 m, on the canalised section of the Rhône (eupotamon) and between 1 m and 3 m at the 'Lône du Grand Gravier' (paleopotamon).

The concept of a modulated frequency index and its evaluation according to biotypological researches

The study of the characteristic associations of Oligochaeta permits the evaluation of a minimal frequency which is dependent on the choice of descriptors of the species or families. This frequency is calculated for each sampling area (functional set or functional unit) from the modulated frequency index: $\text{MFI} = (P_i\,C_i)\frac{1}{2}$, where O_i is the mean relative abundance of the taxon of rank i, expressed as a percentage of the total of individuals of the same systematic rank (according to the family or the species as the case requires), and C_i is the number of the samples in which the taxon of rank i is present, expressed as a percentage of all samples where other taxa of the same systematic rank are present (belonging to the other families or species of the same family according to the rank considered) (at least 8 samples taken in each of four seasons for two consecutive years).

We consider as the threshold value 0.1 of the maximum value of the index, which varies from 0 to 100. This method of calculation entails the elimination of the subresident species which have a relative abundance <1%.

Before calculation at the species level, the index is applied according to the same conventions to the families of Oligochaeta and Aphanoneura; the Aeolosomatidae and Potamodrilidae being combined in the latter group.

The only species or families taken into account are those belonging to a series of samples which have modulated frequency indices superior or equal to 10.

Such a method of calculation provides the following advantages:

a) The moderation of the relative abundance value by the relative sample frequency takes into account seasonal variations of the populations, some of which are only found at particular periods of the year.

b) The second stage of the calculation, which is based on the relative abundance and the relative sample frequency of the species within a family (and not with respect to all individuals belonging to the different families), reduces the numerical distortions caused by the different recruitment and production models within each family.

The results of the investigations into the superficial and the underground fauna of various func-

tional sets, according to Fig. 1, are shown in Table 1. Each family and each species within each family is listed, with respect to the modulated frequency index where this is ≥10.

Results and discussion

The typical associations of the epigean fluvial communities

Among the four representative families tested by means of the frequency index, the Tubificidae and the Naididae predominate in terms of species richness and numerical density; the Naididae are replaced in the first rank by the Tubificidae within the old arms of the Rhône.

Figure 2 compares the distribution of Tubificidae

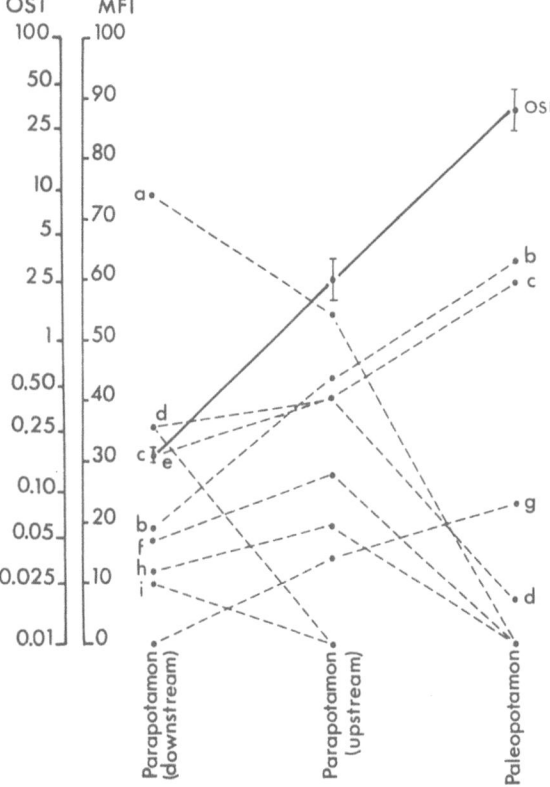

Fig. 2. Relation between organic sediment index (OSI) and modulated frequency index (MFI) of Tubificidae in two function sets of the Rhône river. (a = *Ilyodrilus templetoni*, b = *Potamothrix heuscheri*, c = *Potamothrix hammoniensis*, d = *Limnodrilus hoffmeisteri*, e = *Potamothrix moldaviensis*, f = *Limnodrilus claparedeanus*, g = *Aulodrilus pluriseta*, h = *Tubifex tubifex*, i = *Branchiura sowerbyi*.)

Table 1. Comparison of the hierarchic order of dominance of the families and species of Annelida tested by the modulated frequency index, from superficial and subterranean functional sets of the French upper Rhône (Zoological nomenclature according to Brinkhurst, 1981).

Superficial functional sets

Eupotamon	Parapotamon (Lône des Pêcheurs) downstream (OSI = 0.17 ± 0.02)	Parapotamon (Lône des Pêcheurs) upstream (OSI = 2.59 ± 0.85)	Plesiopotamon (Les Mouilles)	Paleopotamon (Lône du Grand Gravier)
Naididae	Tubificidae	Tubificidae	Tubificidae	Tubificidae
Nais bretscheri	Ilyodrilus templetoni	Ilyodrilus templetoni	Potamothrix heuscheri	Potamothrix heuscheri
Nais elinguis	Limnodrilus hoffmeisteri	Potamothrix heuscheri	Limnodrilus hoffmeisteri	Potamothrix hammoniensis
Nais alpina	Potamothrix moldaviensis	Limnodrilus hoffmeisteri	Potamothrix hammoniensis	Aulodrilus pluriseta
Nais communis	Potamothrix hammoniensis	Potamothrix hammoniensis	Tubifex tubifex	Naididae
Nais barbata	Potamothrix heuscheri	Limnodrilus claparedeanus	Limnodrilus claparedeanus	Stylaria lacustris
Stylaria lacustris	Limnodrilus claparedeanus	Tubifex tubifex	Ilyodrilus templetoni	Slavina appendiculata
Chaetogaster setosus	Tubifex tubifex	Aulodrilus pluriseta	Enchytraeidae	Lumbriculidae
Nais behningi	Branchiura sowerbyi	Naididae	Marionina riparia	Stylodrilus parvus
Lumbriculidae	Naididae	Nais communis	Marionina argentea	Bythonomus lemani
Stylodrilus heringianus	Specaria josinae	Dero digitata	Naididae	
Bythonomus lemani	Vejdovskyella intermedia	Stylaria lacustris	Nais communis	
Enchytraeidae	Amphichaeta leydigii	Ophidonais serpentina	Chaetogaster diastrophus	
Propappus volki	Nais pardalis	Specaria josinae	Lumbriculidae	
Enchytraeus buchholzi	Stylaria lacustris	Dero obtusa	Stylodrilus parvus	
Marionina argentea	Ophidonais serpentina	Nais variabilis		
Cernosvitoviella carpatica	Nais communis	Nais pardalis		
Marionina riparia	Paranais frici	Slavina appendiculata		
Tubificidae		Enchytraeidae		
Limnodrilus hoffmeisteri		Cognettia glandulosa		
		Marionina riparia		

Subterranean functional sets

Underflow	Underflow	Phreatic sheet
Tubificidae	Lumbriculidae	Lumbriculidae
Rhyacodrilus balmensis	Trichodrilus leruthi	Trichodrilus leruthi
Limnodrilus hoffmeisteri	Stylodrilus parvus	Stylodrilus parvus
Rhyacodrilus phreaticola	Stylodrilus heringianus	Tubificidae
Potamothrix vejdovskyi	Enchytraeidae	Rhyacodrilus amphigenus
Rhyacodrilus amphigenus	Marionina argentea	Rhyacodrilus phreaticola
Spirosperma velutinus	Cernosvitoviella carpatica	Rhyacodrilus balmensis
Enchytraeidae	Enchytraeus buchholzi	Enchytraeidae
Marionina argentea	Cognettia glandulosa	Marionina argentea
Propappus volki	Marionina riparia	Cernosvitoviella carpatica
Cernosvitoviella carpatica	Tubificidae	Enchytraeus buchholzi
Enchytraeus buchholzi	Potamothrix heuscheri	Naididae
Marionina riparia	Rhyacodrilus falciformis	Pristina idrensis
Lumbriculidae	Limnodrilus hoffmeisteri	Pristina foreli
Stylodrilus heringianus	Naididae	

Table I. (Continued).

Eupotamon	Parapotamon (Lône des Pêcheurs) downstream (OSI = 0.17 ± 0.02)	upstream (OSI = 2.59 ± 0.85)	Plesiopotamon (Les Mouilles)	Paleopotamon (Lône du Grand Gravier)
Underflow (Continued)	Underflow (Continued)	Underflow (Continued)		Phreatic sheet (Continued)
Trichodrilus leruthi		*Nais communis*		Dorydriidae
Naididae		*Pristina foreli*		*Dorydrilus michaelseni*
Vejdovskyella intermedia		*Nais pardalis*		Enchytraeidae
Pristina foreli				*Achaeta sp.*
Nais elinguis				*Marionina argentea*
Nais communis				*Fridericia (? regularis)*
Pristina idrensis				Tubificidae
Chaetogaster diastrophus				*Rhyacodrilus phreaticola*
Dorydriidae				*Rhyacodrilus balmensis*
Dorydrilus michaelseni				*Haber turquini*
Aeolosomatidae				Naididae
Aeolosoma spp.				*Pristina idrensis*
Potamodrilidae				Aeolosomatidae
Potamodrilus fluviatilis				*Aeolosoma spp.*
Tubificidae				
Rhyacodrilus amphigenus				
Rhyacodrilus phreaticola				
Haber turquini				
Rhyacodrilus balmensis				
Enchytraeidae				
Marionina argentea				
Achaeta sp.				
Cernosvitoviella carpatica				
Fridericia (? regularis)				
Dorydriidae				
Dorydrilus michaelseni				
Aeolosomatidae				
Aeolosoma spp.				
Lumbriculidae				
Stylodrilus heringianus				
Naididae				
Chaetogaster setosus				
Nais elinguis				
Vejdovskyella intermedia				
Chaetogaster diastrophus				
Nais communis				

≥1 m

Subterranean functional sets

and the trophic index of the sediments (OSI) in two of the functional sets. One of the retained sectors corresponds to the downstream sector of the 'Lône des Pêcheurs', which acts as a decantation basin for the fine mineral suspensions transported by the Rhône at the moment of overflow with a consequent three-dimensional dilution of the loading of organic carbon and chlorophyll pigments of the sediments. The second section deals with the upstream zone of the 'Lône des Pêcheurs' where the numerous hydrophytes and helophytes are linked in the narrowing part of the channel facilitating the sedimentation of organic matter. The third sector corresponds to the 'Lône du Grand Gravier', of which the banks are colonized by the hydrophyte *Cladium mariscus* and form a peat-bog.

The positive correlation between the trophic index and the frequency index which characterizes *Potamothrix heuscheri* and *Potamothrix hammoniensis* is reversed for *Potamothrix moldaviensis* and *Ilyodrilus templetoni*. The rarefaction of the genus *Limnodrilus* in the sediments of the 'Lône du Grand Gravier' is consistent with the absence of this taxon noted in certain high mountain lakes, connected with peat-bogs (Juget & Giani, 1974). The antagonism of *Branchiura sowerbyi* and *Aulodrilus pluriseta* must be correlated with the thermal preferenda of these two species, the thermal inertia of the waters from upstream of the 'Lône des Pêcheurs' and the 'Lône du Grand Gravier', due to the inflow of groundwater being more propitious to *Aulodrilus pluriseta*.

In the study of Naididae, their feeding behaviour appears to interfere with the assessment of the species of this family. The microphytes and particularly the diatoms represent in the prospected area a determining element of the primary productivity level, resulting in a positive correlation between the concentration of organic carbon and chlorophyll pigments and the numerical density of the frustules of diatoms with respect to the other particles of the fine sediment fraction.

The positive or negative selection of diatoms by the Naididae (algivorous or limivorous species) can be estimated from the selectivity index (SI) which combines the numerical density of the frustules ingested and the ratio between the maximal lengths of the frustules and other ingested sediment particles (Provot, 1982).

$SI = (\bar{a}\,\bar{n}\,\bar{b}^{-1})\frac{1}{2}$, where $\bar{n}$ is the mean number of diatoms ingested, $\bar{a}$ the mean length of the longest ingested diatoms, $\bar{b}$ the mean length of the other longest ingested particles in a group of N individuals of a given species, all containing diatoms in their gut. This index is more useful than the two components (i.e. numerical density and size of the ingested diatoms) taken separately to define the algivorous or limivorous behaviour of the Naididae. This is confirmed by the highly significant coefficient of correlation ($r = 0.93***$) between the selectivity index and the percentage of occurrence defined as ratio between the number of individuals with diatoms in their gut and the total number of individuals investigated belonging to the species tested in the study area (Fig. 3).

The ethological separation into three categories of the 16 species tested on the basis of their gut contents, as defined in Fig. 3, is in general agreement with their ecological partitioning in the prospected hydrological complex.

Those taxa classified in the category of 'limivorous s. stricto' include species which perform passive negative selection of the large diatoms, only the smallest, as a rule empty, frustules being eaten with the other fine mineral particles. The taxa belong to three rheophilic species, characteristic of the eupotamon, *Nais bretscheri, Nais alpina* and *Nais elinguis*. They inhabit an unstable environment which is unfavourable to the primary productivity of plankton and benthos and to sedimentation of the microphytes.

The category of 'limivorous s. lato' includes taxa which operate relatively little selection between large and small diatoms and between microphytes and sediment particles. These taxa belong either to the species characteristic of the zone of confluence of the parapotamon and the eupotamon (*Vejdovskyella intermedia, Paranais frici*) or to the species with the highest frequency index in this zone (*Specaria josinae*) and again to ubiquitous species like *Stylaria lacustris*.

To the algivorous class belong taxa which actively select large algae in preference to small particles. This class includes in particular taxa which from preference inhabit the upstream zone of the parapotamon, the plesiopotamon or the paleopotamon (*Nais communis, Slavina appendiculata, Chaetogaster diastrophus*).

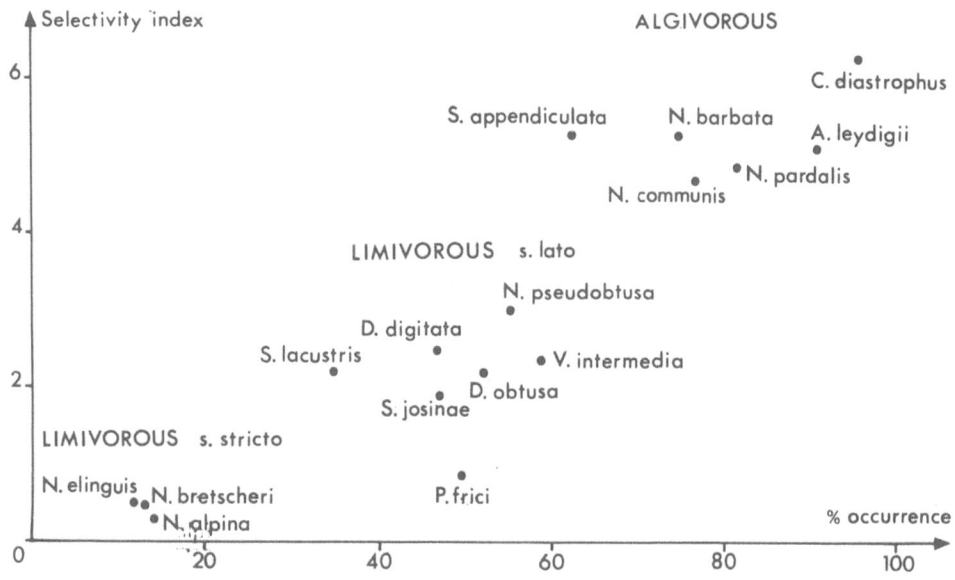

Fig. 3. Relation between the selectivity index (SI) and the percentage of occurrence of zooids feeding on diatoms for 16 species of Naididae inhabiting the hydrographic complex studied (from Provot 1982).

The typical associations of the underground communities

The structure and the form of the communities characteristic of the underflow and the phreatic sheet can be summarised by the following features:

a) The transition from the superficial to the underground field causes either a reversal of the hierarchical order of dominance of the families present (eupotamon) or more generally a substitution of the dominant groups in which certain forms (Potamodrilidae, Dorydrilidae, Rhyacodrilinae) are peculiar to the underground interstitial waters.

b) The occurrence in the deep interstitial horizons (–1.5 m) resulting from the downward movement of such epigean species as *Stylodrilus heringianus* (whose cocoons were observed at this horizon) as well as *Vejdovskyella intermedia* is counter-balanced by the presence in the epigean communities of the plesio- and the paleopotamon of typical taxa of the underground field, such as *Marionina argentea* and *Stylodrilus parvus*. The presence of the last two species in the surface waters reveals a border situation and an effect of source related to the exchanges between the phreatic and the superficial waters and according to the intricate structure of hypogean and epigean species communities.

c) Comparison of the associations of species belonging to the families Tubificidae, Enchytraeidae and Dorydrilidae, which dominate the eupotamic and paleopotamic fields, beyond a depth of 1 m reveals a strong analogy. This is indicative of the existence on the right bank of the Rhône of an alluvial plain with a saturated sheet thickness of 30 m supported by a supply coming from the Dombes and running in direction of the Rhône in a northeast–southwest orientation (B.R.G.M., 1973).

The simultaneous presence in underflow of the Rhône, within the local phreatic sheet and within the regional karstic networks, of a vast complex of taxa, especially *Dorydrilus michaelseni*, *Stylodrilus parvus*, *Rhyacodrilus balmensis*, *Rhyacodrilus phreaticola*, *Haber turquini*, *Marionina argentea*, *Cernosvitoviella carpatica* and *Aeolosoma gineti*, testifies to the functional unity of the underground field which underlies the superficial aquatic spaces.

References

Amoros, C., M. Richardot-Coulet & G. Pautou, 1982. Les 'ensembles fonctionnels': des entités écologiques qui traduisent l'évolution de l'hydrosystème en intégrant la géomorphologie et l'anthropisation (exemple du Haut-Rhône français). Revue Géogr. Lyon 57: 51–62.

182

Ballinger, D. G. & G. D. McKee, 1971. Chemical characterization of bottom sediments. J. Wat. Pollut. Cont. Fed. 43: 216–227.

Bou, C., 1974. Recherches sur les eaux souterraines - 25 - Méthodes de récolte dans les eaux souterraines interstitielles. Ann. Spéléol. 29: 611–619.

Bou, C. & R. Rouch, 1967. Un nouveau champ de recherche sur la faune aquatique souterraine. C.r. Acad. Sci., Paris 265: 369–370.

Bournaud, M. & B. Cellot, 1981. Methodologie du prélèvement dans un fleuve en relation avec les mouvements de la faune benthique. Conv. Rech. 78–123. Univ. Lyon 1, Minis. Envir., 75 pp.

Bravard, J. P., 1982. A propos de quelques formes fluviales de la vallée du Haut-Rhône français. Revue Géogr. Lyon 57: 39–48.

B.R.G.M., 1969. Connaissances de l'hydrologie de la plaine de Lyon - Rapp. 5, 69 S.G.N. 16 JAL.

B.R.G.M., 1973. Synthèse des connaissances sur le système aquifère de l'est lyonnais. Rapp. 73, S.G.N. 199 JAL.

Brinkhurst, R. O., 1981. A contribution to the taxonomy of the Tubificinae (Oligochaeta: Tubificidae). Proc. biol. Soc. Wash. 94: 1048–1067.

Dorgelo, J., 1973. Etude de la végétation dans les anciens lits du Rhône et des moustiques qui lui sont liés, de Lyon au confluent de l'Ain. D.E.S. Univ. Lyon 1, 97 pp.

Gibert, J., R. Ginet, J. Mathieu, J. L. Reygrobellet & A. Seyed-Reihani, 1977. Structure et fonctionnement du Haut-Rhône français, 4. Le peuplement des eaux phréatiques; premiers résultats. Ann. Limnol. 13: 83–97.

Juget, J., 1980. Aquatic Oligochaeta of the Rhône-Alpes area. Current research priorities. In R. O. Brinkhurst & D. G. Cook (eds.), Aquatic Oligochaete Biology. Plenum Press, N.Y.: 241–252.

Juget, J. & N. Giani, 1974. Répartition des Oligochètes lacustres du massif de Neouvieille (Hautes-Pyrénées) avec la description de Peloscolex pyrenaicus n. sp. Ann. Limnol. 10: 33–53.

Juget, J. & A. L. Roux, 1982. Un lône du Rhône, zone humide en position de lisière dans l'espace et dans le temps. Bull. Ecol. 13: 109–124.

Juget, J., C. Amoros, D. Gamulin, J. L. Reygrobellet, M. Richardot, P. Richoux & C. Roux, 1976. Structure et fonctionnement du Haut-Rhône français. 2 - Etude hydrologique et écologique de quelques bras morts. Premiers résultats. Bull. Ecol. 7: 479–492.

Juget, J., B. J. Yi, C. Roux, P. Richoux, M. Richardot-Coulet, J. L. Reygrobellet & C. Amoros, 1979. Structure et fonctionnement du Haut-Rhône français, 7. Le complexe hydrographique de la 'Lône des Pêcheurs' (ancien méandre du Rhône). Schweiz. Z. Hydrol. 41: 395–417.

Lafont, M. & J. Juget, 1976. Les Oligochètes du Rhône - Relevés faunistiques généraux. Ann. Limnol. 12: 253–268.

Pautou, G. & J. P. Bravard, 1982. L'incidence des activités humaines sur la dynamique de l'eau et l'évolution de la végétation dans la vallée du Haut-Rhône français. Revue Géogr. Lyon 57: 63–79.

Provot, M., 1982. Aspects écophysiologiques des relations entre le phytobenthos (diatomées) et les Naididae (Oligochètes). D.E.A. Univ. Lyon 1, 46 pp.

Reygrobellet, J. L. & M. J. Dole, 1982. Structure et fonctionnement des écosystèmes du Haut-Rhône français, 19 - Connaissance des milieux interstitiels régionaux (2): extension à la Lône du Grand Gravier. Pol. Arch. Hydrobiol. 29: 485–500.

Reygrobellet, J. L., J. Mathieu, R. Ginet & J. Gibert, 1981. Structure et fonctionnement du Haut-Rhône français, 8. Hydrologie de deux stations phréatiques dont l'eau alimente des bras morts. Int. J. Speleol. 11: 129–139.

Richardot-Coulet, M., C. Amoros, J. L. Reygrobellet & A. L. Roux, 1982. Diagnose des ensembles fonctionnels aquatiques définis sur le Haut-Rhône français. Application à une cartographie écologique d'un système fluvial. Eau Québec 15: 146–153.

Roux, A. L., 1982. Cartographie polythématique appliquée à la gestion écologique des eaux. Etude d'un hydrosystème fluvial: le Haut-Rhône français. Edit. C.N.R.S. Centre Region. Public. Lyon, 116 pp.

Seyed-Reihani, A., J. Gibert & R. Ginet, 1982. Structure et fonctionnement des écosystèmes du Haut-Rhône français, 23. Etude écologique comparée de 3 stations interstitielles en amont de Lyon. Pol. Arch. Hydrobiol. 29: 501–511.

Comparison of different methods of water quality evaluation by means of oligochaetes

T. D. Slepukhina
Institute of Lake Research of the Academy of Sciences of the U.S.S.R., Sevastjanova 9, 196199 Leningrad, U.S.S.R.

Keywords: aquatic Oligochaeta, water quality, bioindication

Abstract

Using as examples water bodies in different geographical zones (Lake Ladoga, the River Sukhona, shallow-water ponds of the North Caucasus) the known methods of water quality evaluation by means of oligochaetes are considered. There is no unique universal method of pollution bioindication in this way. In water bodies of all types the mass development of oligochaetes and the reduction of their species diversity are indications of: (1) large quantities of organic matter; (2) favourable oxygen regime; (3) absence or insignificant quantity of heavy metal solids, petroleum substances and agricultural chemicals; (4) intensive self-purification of the water body.

Introduction

The oligochaetes are the group of invertebrates most often enlisted for bioindication purposes, because their mass development is enhanced in places of strong organic contamination. They give information on the average status of a water body during a long period preceding the time of observation, which may be of primary concern. It is known also that oligochaetes cannot serve as indices of single or intermittent pollution (excluding the cases of catastrophic death due to excessive stress) because of the extended duration of their life cycles.

In the present report the term 'water quality' implies only one of water characteristics, namely a degree of organic pollution of the water body.

There are several methods of water quality assessment which use oligochaetes. Some of these methods are, in my opinion, suitable only for the water bodies for which they have been proposed. Others can be used in everyday work on water bodies of different types. But is there any method which can be considered as universal? Let us try to apply some of these methods to different types of water bodies.

Study area and methods

I obtained material from water bodies in different geographical zones:
1) Lake Ladoga is the largest oligotrophic water body in Europe in which local centres of organic pollution are registred. Samples of benthos were collected during 1975–1979 at 281 stations.
2) Shallow warm ponds of the North Caucasus Steppe, on the shores of which are cattle-breeding farms. Samples of benthos were obtained at 870 stations during 1962–1963 and 1969–1970 on 147 ponds.
3) We also have material from 17 polluted stations on the Sukhona River (Vologda District).

All samples were collected with an Ekman dredge (replicate samples at each station) and sieved through mesh no. 38.

Results and discussion

A number of investigators working on American Great Lakes (Wright, 1955; Carr & Hiltunen, 1965; Howmiller & Beeton, 1971) consider a high density

Hydrobiologia 115, 183–186 (1984).

of oligochaetes as an index of organic pollution of the water body. Clearly in water bodies of all types mass development of oligochaetes accompanied by sharp reduction of benthos species diversity indicates the presence of excessive amounts of organic matter.

In Lake Ladoga, on some stations subjected to human influence, accumulation of oligochaetes, particularly *Tubifex tubifex* (O. F. Müller), *Limnodrilus hoffmeisteri* Claparède or *Potamothrix hammoniensis* (Michaelsen), at total population densities of 4 000 ind. m^{-2}, has been observed.

In the River Sukhona the density of *Limnodrilus udekemianus* Claparède and *T. tubifex* exceeds 5 000 ind. m^{-2}, reaching a maximum of 22 000 and a biomass of 475 g m^{-2}. *L. hoffmeisteri* was present in smaller amounts.

In the most contaminated (of 147 studied) pools of the North Caucasus, near the inflow of runoff from farms, the number of oligochaetes on an average fluctuated between 6 000 and 8 000 m^{-2}, with a maximum value of 31 800. Only the following species are present: *L. hoffmeisteri, Limnodrilus helveticus* Piguet, *T. tubifex, P. hammoniensis* and *Psammoryctides albicola* (Michaelsen).

It is important that in investigated water bodies of different types and situated in different geographical zones the same species of oligochaetes are recorded (Table 1). The similarity of species composition between heavily polluted sections of water bodies of different type is explained by the fact that the whole complex of conditions which results from pollution is the dominant environmental factor. The action of other factors is weaker.

Unfortunately, in some cases in polluted places the oligochaetes are absent or poorly developed due to unfavourable oxygen regime, lack of suitable sediment, presence of heavy metals and other causes. Therefore some authors (Finogenova & Alimov, 1976) consider the absolute number of oligochaetes as an unreliable criterion for pollution assessment.

The analytical method of Goodnight & Whitley (1961), which utilises the percentage of oligochaetes in a sample is, in my opinion, more restricted. According to the data of these authors, if there are 60% of oligochaetes, then the river is in good condition, if there are 60–80%, contamination may be suspected; above 80% there is no doubt that pollution is present. Our data on the River Sukhona confirm

this statement (Alexandrova, *et al.*, 1975). In Lake Ladoga this criterion is not usable because here oligochaetes are the leading group both on pure sands of varying coarseness, in the south part of the lake, and on the muds of its central and north parts, including the maximal depth of 200–230 m (Slepukhina & Alekseeva, 1982).

In pools of the North Caucasus, in heavily polluted areas with oligochaete densities of 6 000–8 000 m^{-2}, their percentage in samples in the majority of cases was about 80%; however in a number of polluted pools the oligochaetes content was below 30% by weight where there was abundant development of *Chironomus plumosus* L. (Slepukhina, 1969).

The index of King & Ball (1964), i.e. the ratio of insect weight to that of oligochaetes, is far from being universal due to the same reasons as in the case of the method of Goodnight & Whitley. Besides, at the same level of contamination, a small admixture of insecticide could change this relationship in favour of oligochaetes (Aston, 1973) while and admixture of heavy metal ions may favour insects (Brinkhurst, 1966; Liperovskaya, 1970). Seasonal variation and emergence of chironomids would cause the largest fluctuations in the value of this index (Finogenova & Alimov, 1976).

Zahner's tables (Zahner, 1965) proposed for the Bodensee where the class of water purity depends on relationships between the number of *T. tubifex* and *Limnodrilus*, have not found confirmation in the data obtained by me and apparently are important only for the Bodensee and lakes of similar type.

The present data on three types of water bodies have shown that the diagrams of Milbrink (1973) can serve as a good illustration of the distribution of excessive organic matter in sediments. However, in different water bodies the species composition of group-indices will be somewhat different.

Parele & Astapenok (1975) suggest for the rivers of Latvia the index $D = T/O$ where T is the number of tubificids and O is the total number of oligochaetes in the sample. If $D = 0.3$, waters are relatively pure; at $D = 0.3-0.55$, contamination is weak; if $D = 0.55-0.8$ rivers are polluted; if $D = 0.8-1.0$, heavy pollution is present.

As it is seen from Table 1, there are only Tubificidae in the polluted part of River Sukhona, so D equals 1. In the ponds of North Caucasus, Tubificidae dominate among oligochaetes practically every-

Table 1. Oligochaete species present in different areas. A = not polluted; B = organically polluted areas.

Oligochaete species	Presence in				
	Lake Ladoga		Ponds North Caucasus		Polluted part of Sukhona
	A	B	A	B	
Stylaria lacustris (L.)	+		+		
Arcteonais lomondi (Martin)	+				
Ripistes parasita (Schmidt)	+				
Dero digitata (Müller)	+				
D. dorsalis Ferronière	+				
D. obtusa d'Udekem	+				
D. sp.	+		+		
Nais barbata Müller	+				
N. simplex Piguet	+				
N. behningi Michaelsen	+			•	
N. communis Piguet	+		+		
N. elinguis Müller			+		
N. variabilis Piguet	+				
N. bretscheri Michaelsen	+				
N. sp.			+		
Ophidonais serpentina (Müller)	+		+		
Uncinais uncinata (Ørsted)	+				
Homochaeta naidina Bretscher	+				
Aulodrilus limnobius Bretscher	+				
Isochaetides newaensis (Michaelsen)	+				
Limnodrilus udekemianus (Claparède)	+	+	+	+	+
L. helveticus Piguet	+	+	+	+	
L. hoffmeisteri Claparède	+	+	+	+	+
L. claparedeianus Ratzel	+		+		
Ilyodrilus bedoti (Piguet)			+		
I. bavaricus Öschmann			+		
I. moldaviensis Vejdovsky & Mrazek	+				
Psammoryctides albicola (Michaelsen)	+		+	+	
P. barbatus (Grube)	+		+		
P. sp.			+		
Tubifex tubifex (Müller)	+	+	+	+	+
Potamothrix hammoniensis (Michaelsen)	+	+	+	+	
Peloscolex ferox (Eisen)	+				
P. velutinus (Grube)	+				
Alexandrovia onegensis Hrabĕ	+				
Tubificidae gen. sp.	+		+		
Enchytraeidae gen. sp.	+				
Lumbriculus variegatus Müller	+				
Lamprodrilus isoporus Michaelsen	+				
Stylodrilus heringianus Claparède	+				
Rhynchelmis limosella Hoffmeister	+				

where, irrespective of pollution degree, and D is near 1 (Naididae are rare and connected with vegetation). In Lake Ladoga D oscillates from 0 to 1 and may serve apparently as a pollution indicator in oligosaprobic and polysaprobic zones, but in mesosaprobic regions this index is not reliable (Table 2). In this table, there are data of index D at 10 stations in each zone (but in the polysaprobic zone we have only 5 stations).

General analysis of the species composition of all associations of oligochaetes does not give hopeful results (Brinkhurst, 1966). We are also sure that many species of oligochaetes are exceedingly plastic (Finogenova, 1976).

Table 2. Index D in different zones in Lake Ladoga (June 1979).

Number of station	Depth (m)	Index D
Oligosaprobic		
69	51	0
67	120	0
66	130	0
64	91	0
59	53	0
58	30	0.1
51	90	0
44	63	0.1
7	198	0.2
5	80	0
β-mesosaprobic		
65	25	0.1
61	70	0.3
57	20	0.5
56	26	0.5
53	4	0.9
35	66	0.4
34	41	0.3
31	24	1.0
20	23	1.0
α-mesosaprobic		
70	136	1.0
60	18	0.4
52	12	1.0
44	25	0
32	45	1.0
22	35	1.0
9	10	1.0
6	30	0
4	27	0
Polysaprobic		
13	9	1.0
12	11	1.0
11	12	1.0
3	5	1.0
1	24	1.0

Thus, only mass development of a few oligochaete species is undoubtedly evidence of large amounts of easily oxidized organic matter in many habitat. Moreover, additional conclusions on the status of water body or biotope can be made at once:
1) Sharp deficits of oxygen are not typical.
2) There is no high concentration of heavy metal salts and poisons.
3) Mass development of oligochaetes is responsible for exceedingly intensive self-purification of a water body. Daily, oligochaetes ingest a quantity of sediment nine times greater than their body weight (Alsterberg, 1922; Ztvetkova, 1969).

So in the River Sukhona, on the stations where oligochaetes were maximal and their biomass ac-counted for 475 g m^{-2} in summer, 4 275 g of sediment per day, or 128 kg per month were ingested.

References

Alexandrova, D. N., T. D. Slepukhina, N. J. Senatskaya, A. V. Kulishov, L. F. Zhekhnovskaya, A. N. Okhlopkova, M. F. Veselova & A. A. Kurochkina, 1977. Present status and self-purification intensity of the River Sukhona head. Hydrobiological Journal 13: 89–91. (in Russian)

Alsterberg, G., 1922. Die respiratorischen Mechanismen der Tubificiden. Lunds Univ. Årsskr. (N.F.) 18: 1–176.

Aston, R. I., 1973. Tubificids and water quality: a review. Envir. Pollut. 5: 1–10.

Brinkhurst, R. O., 1966. The Tubificidae (Oligochaeta) of polluted waters. Verh. int. Ver. Limnol. 16: 854–859.

Carr, I. F. & I. K. Hiltunen, 1965. Changes in the bottom fauna of western Lake Erie from 1939–1961. Limnol. Oceanogr. 10: 551–569.

Finogenova, N. P., 1976. Importance of oligochaetes as indicators of water pollution. In: G. G. Winberg (ed.), Hydrobiological principles of water self-purification: 51–60 (in Russian).

Finogenova, N. P. & A. F. Alimov, 1976. Evaluation of water pollution degree according to bottom animals composition. In: G. G. Winberg (ed.), Methods of Biological Analysis of Fresh Waters: 96–106 (in Russian).

Goodnight, C. I. & T. S. Whitley, 1961. Oligochaetes as indicators of pollution. Proc. 15th Industr. Waste Conf. Purdue Univ. Ext. Eng. 106: 139–142.

Howmiller, R. P. & A. M. Beeton, 1971. Biological evaluation of environmental quality, Green Bay, Lake Michigan. J. Wat. Pollut. Cont. Fed. 42: 123–133.

King, D. Z. & R. C. I. Ball, 1964. A quantitative biological measure of stream pollution. J. Wat. Pollut. Cont. Fed. 36: 650–653.

Liperovskaya, E. S., 1970. Some features of oligochaetes distribution versus pollution of the River Moscow muds. In: Biological Processes in Sea and Continental Waters. Abstracts of communications of 2nd All-Union Congress of the Hydrobiological Society. Kishinew: 221–222 (in Russian).

Milbrink, G., 1973. Communities of Oligochaeta as indicators of the water quality in Lake Hjälmaren. Zoon 1: 77–88.

Parele, E. A. & E. B. Astapenok, 1975. Tubificids (Oligochaeta, Tubificidae) – indicators of water body quality. Latv. PSR Zin at. Akad. Vest. 9: 44–46. (in Russian)

Slepukhina, T. D. & N. A. Alekseeva, 1982. Bottom invertebrates. In: N. A. Petrova (ed.), Anthropogenic Eutrophication of Ladoga. Leningrad, Nauka: 181–191 (in Russian).

Slepukhina, T. D., 1969. Zoobenthos of ponds of the Steppe area of North Caucasus. Proceedings Laboratory of Limnology, Leningrad 23: 167–176 (in Russian).

Tzvetkova, L. I., 1969. On the role of saprobic oligochaetes in oxygen balance of water bodies. M.Sc. thesis, Leningrad. 20 pp. (in Russian)

Wright, S., 1955. Limnological survey of western Lake Erie U.S. Fish and Wildlife Serv. Spec. Sci., Rep. Fisheries 139: 341.

Zahner, R., 1965. Organismen als Indicatoren für den gewasserzustand. Arch. Hygiene und Bacteriologie 149(3/4): 243–256.

On aquatic Oligochaeta Lumbricomorpha in Europe

Pietro Omodeo

Istituto di Biologia Animale, Università di Padova, Via Loredan 10, I-35100 Padova, Italy
Present address: Dipartimento di Biologia, II Università di Roma – Tor Vergata Via Raimondo, I-00173
Roma, Italy

Keywords: aquatic Oligochaeta, Lumbricomorpha, biology of aquatic earthworms, breeding of aquatic earthworms

Abstract

Twelve species of aquatic earthworms have been found in European inland waters where they occupy different environments. Two species live exclusively in the sapropel, that is in the black fetid asphyctic mud, while four species can be found either in the sapropel or in the more oxygenated gyttja. The remaining species prefer this type of soil, but *Eisenia spelaea* in Italy lives only within submerged litter in mountain streams. The biology of these earthworms is discussed and the possibility of their utilization by man is considered.

Introduction

Aquatic oligochaetes are an interesting subject for those naturalists and biologists who are engaged in the study of nutrition of aquatic vertebrates, water pollution, colonization of great reservoirs, and in many other fields.

Thus many researchers in the systematics and faunology of this taxon have been resumed, which were initiated and developed by eminent zoologists: Frank Beddard, Daniele Rosa, Wilhelm Michaelsen, Leo Černosvitov and Sergěj Hrabě. The researchers in physiology, biology and cytology of these annelids, in which František Vejdovsky and John Stephenson led the field, have received a new impulse, and the exploitation of quite new areas of research has begun: ultrastructure, population dynamics, toxicology, auto- and synecology, and so on.

One important result therefore follows quickly on the heels of another, so quickly that sometimes it is difficult to keep pace. For this reason, among many others, we are very grateful to Giuliano Bonomi who has offered us the opportunity to meet in such a beautiful place and to hear and speak of our common work.

My contribution to this work is an account of some Sparganophilidae, Criodrilidae and Lumbricidae which live in European waters. These worms have received little attention up to now, perhaps because the specialists have little opportunity of collecting them, and the collectors are often unable to classify them (Casellato & Manea, 1981).

My paper will be introductory and not exhaustive, my intention being to stimulate further researchers on these animals.

Geographic distribution and ecology

In Europe twelve species of earthworms have been found that live and reproduce under water or in submerged mud, and thus must be considered as truly aquatic. Three species belonging respectively to the Sparganophilidae and Criodrilidae appear to have always been aquatic, and nine species belonging to the terrestrial Lumbricidae seem to have come back independently to aquatic life during their evolutionary history.

These earthworm species occupy different benthic niches in which they are common and often quite abundant. The glossoscolecid *Sparganophi-*

lus tamesis owes its name to the marsh weed *Sparganium* and to the Thames river; it lives in England, France, and in the whole of Northern America. Some specialists claim that it was introduced into Europe by man, but the evidence is not definitive. According to Černosvitov (1945) and Brinkhurst & Jamieson (1971), *S. tamesis* dwells in the muddy bottom of ponds, rivers and minor watercourses, and tolerates the asphyctic sapropel.

Criodrilus lacuum Hoffmeister, 1845, is known from Europe, the Middle East and the Eastern North America. A similar, little known species, *Criodrilus ochridensis* Georgevitch, 1950, is endemic to the Ohrida lake (Macedonia). *C. lacuum* lives in the mud of rivers, canals, ponds, lakes; in Italy it prefers mud rich in organic detritus (gyttja), but it also lives in the fetid sapropel. It is often associated with *Eiseniella neapolitana* and *Haplotaxis gordioides* but is less tolerant of brackish water than *Haplotaxis* and disappears where the sea water has some influence.

Criodrilus is related to the genera *Alma* and *Drilocrius* which are however more specialized for aquatic life, having two 'claspers' which are utilized during copulation and, sometimes, branchial diverticula in the caudal segments.

The lumbricid genus *Eisenia* resembles *Criodrilus* in ovarian structure, chromosome number and in some reproductive peculiarities. It includes two species with wide distribution (*foetida* and *nordenskioldi*) and three species with restricted distribution (*lucens, submontana* and *spelaea*).

Eisenia spelaea Rosa, 1901, in the Italian Alps and the northern Apennines is tied to two peculiar biotopes; the submerged litter in mountain streams, and the guano mixed with swept foliage in caves. It seems to prefer calcareous waters since in the Alps it has been found only where limestone abounds. According to some authors, the Carpatho-Balkanic *E. submontana* (Vejdovsky, 1875) lives either in water or under the bark of fallen trees. Perhaps it has been mistaken (Zicsi, 1966) for *Eisenia lucens* (Waga, 1857), a more northern species characterized by its luminescence (as is well known, no luminescent organism inhabits fresh waters).

Eisenia spelaea is tetraploid, the basic chromosome number being 17, *Eisenia submontana* is hexaploid with the same basic chromosome number (Omodeo, 1962), while *E. foetida* is diploid with 22 chromosomes as is *Criodrilus lacuum* (specimens from Italy).

Among the Lumbricidae the genus *Eiseniella* is specially adapted for living in inland waters. *E. tetraedra* is semiaquatic and its cocoons may develop under water. *E. neapolitana* (Oerley, 1885), *E. lacustris* (Černosvitov, 1931) and *E. ochridana* (Černosvitov, 1931) are truly aquatic. *E. neapolitana* has the same ecology as *C. lacuum* with which it is often collected; it is smaller, but has large, protruding setae. Quite common in continental Italy, this species inhabits Switzerland, the Balkan peninsula, Turkey and the Middle East; probably it is present in Ireland, and was introduced into California.

About *E. ochridana* and *E. lacustris*, which are found in the very ancient lake Ohrida, little is known but the depth at which they live.

The genus *Allolobophora* includes, together with many strictly terricolous species, at least three species well adapted to life in submerged soils: *A. dubiosa, A. molleri* and *A. limicola*. An excellent paper by Zicsi (1963) deals with *A. dubiosa* (Oerley, 1881) which is confined to the lower Danubian basin and lives in rather oxygenated mud, rich in organic matter.

A. molleri Rosa, 1889 (=*A. moebii* Michaelsen, 1895), is known from the Iberian peninsula, Algeria, Morocco, Madeira, the Canary islands and Mexico. This worm is bound to the sapropel (though it can be found in the gyttja) and in some places exists in very crowded communities. It is indifferent to the granulometry of the soil and to its content in vegetable detritus: in Northern Africa it is common even in streams which flow through quite barren terrain.

A. limicola Michaelsen, 1890, occupies a large area between those of the two congeneric aquatic species: it is known from Germany, Switzerland, Belgium, Southern Scandinavia and Massachusetts. The biology and ecology of this species is little known; the few available data suggest that it lives in the same way as *A. dubiosa* (Černosvitov, 1935; Graff, 1953).

Two aquatic species are classified in the genus *Helodrilus*. *H. oculatus* Hoffmeister, 1845, is known from Great Britain, the Netherlands, Belgium, Germany, Switzerland, France and the western half of the Italian peninsula; it lives in submerged clayey mud, either oxygenated or asphyctic, and in black reducing soils (Bouché, 1971), and seems to be indifferent to the granulometry of the soil.

Helodrilus patriarchalis (Rosa, 1893) inhabits

northeastern Italy, Yugoslavia, Greece, Crete, the Middle East and Iran. It lives in putrid submerged soils (sapropel), often together with *Eiseniella neapolitana* and *Criodrilus lacuum*; its density can be very high (Casellato & Manea, 1981). Sometimes, after flooding or a similar event, this species can be collected in aerated soils, but in such cases it lives at considerable depth, and apparently does not reproduce.

The record of aquatic earthworms would be incomplete without mention of an exotic species introduced by man into Europe and probably elsewhere: *Ocnerodrilus occidentalis*, Eisen, 1878. I found this worm, typical of inundated rice fields, in western Sardinia, living in brackish waters together with the polychaetes *Mercierella enigmatica* and *Nereis* sp. (Omodeo, 1984).

Biology

The biology of aquatic earthworms shows some uncommon characteristics related to their habitat. The most evident feature is the abundance of blood and haemoglobin: the smaller specimens with transparent integuments may appear bright red just like *Tubifex* and *Haplotaxis*. This abundance is obviously connected to the low oxygen tension of their habitat.

Another unusual character shown by *Criodrilus, Sparganophilus, H. patriarcalis* and *E. neapolitana* (and also by some exotic aquatic species) is the presence in the young individuals of a caudal blasteme of growth and the dorsal displacement of the anus: in terrestrial earthworms a caudal blasteme either of growth or of regeneration can be observed only during the diapause, when the worms have emptied the gut.

A third characteristic is that of the seasonal reproduction restricted to few months, in contrast to the uninterrupted reproductive activity of terrestrial species, provided that humidity and food are sufficient.

C. lacuum, H. patriarcalis, H. oculatus and *A. dubiosa* reproduce from the late spring to the summer (from May to August); *E. neapolitana* has a longer reproductive period, from April to September.

Throughout Algeria and Morocco *A. molleri* was found to be immature in February and March, but in the last week of March some specimens began to show the *tubercula pubertatis*, suggesting that this species also reproduces in the spring and summer. In Spain *A. molleri* is mature in July.

The cocoons of these species develop in fresh water, where the cocoons of terrestrial earthworms swell and then degenerate. This indicates different imbibition properties of their extra-ovular yolk.

Earthworms which live in the sapropel derive their nourishment mainly from micro-organisms which abound in the ingested mud or sand (Brinkhurst & Jamieson, 1971). These worms evacuate their faeces on the soil surface: when the sapropel is black and is covered by light-coloured sand, these faeces are conspicuous, being dark against the light background.

Comparative morphology

As suggested above, *Sparganophilus* and *Criodrilus* have aquatic ancestors.

Sparganophilus is remarkable for the thinness and the simple structure of the gut, devoid of Morren's glands, gizzard and typhlosole. Yet its clitellar apparatus is similar to that of terrestrial Glossoscolecidae: the clitellum is short, begins anteriorly (segm. 15–22) and is provided with *tubercula pubertatis*.

The gut of *Criodrilus* also is simple, lacking both Morren's glands and gizzard, but is endowed with a typhlosole. The clitellum of this worm begins anteriorly, but is very long, occupying segments 15–45. Correspondingly the cocoons are long and spindle-shaped. Spermathecae and *tubercula pubertatis* are lacking: this seems to be a very primitive condition. The pigmentation of the epidermis may be greenish or even a deep glossy black (Casellato & Manea, 1981).

The aquatic lumbricid species have unmodified gut and clitellar structure, but *A. molleri* has an unusually long clitellum. This species is remarkable for its green colour. The above-mentioned species and all the species of *Eiseniella* have a body with quadrangular section and the dorsal side gutter-shaped; the incidence of these anatomical features may be connected with the water and mud flux and the low oxygen concentration.

Other anatomical modifications are rare in Lumbricidae which have reverted to aquatic life; I can mention only the unusual length and position of the clitellum of *A. molleri* (segm. 37–55).

The farming of aquatic oligochaetes

In recent years the practice of farming earthworms to transform organic waste has spread in many countries. The worms consume manure, droppings and kitchen waste and convert them into castings which are very useful in horticulture; the worms themselves are utilized as bait or to feed fish and domestic animals.

The species commonly bred are: *Eisenia foetida* in Northern America and Europe, *Eudrilus eugeniae* in Africa and Northern America, and *Pheretima* spp. in southern Asia. These species, which are terrestrial, are farmed in beds of soil mixed with waste, water being sprayed to control humidity. This method and these animals are unfortunately unfit for the conversion of materials in which there is abundance of water and little oxygen, as is the case with the asphyctic muds of sewage treatment plants.

Some of the earthworm species considered here may well be suitable for this purpose because of their way of feeding and their endurance, but the rigidity of their reproductive cycle makes their 'domestication' difficult.

Acknowledgements

This work was supported by the Italian National Research Council, C.N.R. grant no. CT.81.00298.04.

References

Bouché, M. B., 1972. Lombriciens de France. Écologie et systématique. Inst. natn. Rech. agron., Paris. 671 pp.

Brinkhurst, R. O. & B. G. M. Jamieson, 1971. Aquatic Oligochaeta of the world. Oliver & Boyd, Edinb. 860 pp.

Casellato, S. & M. R. Manea, 1981. Oligochaeta lumbricomorpha in waters of northeastern Italy. Boll. Mus. civ. stor. nat. Verona 7: 593–600.

Černosvitov, L., 1931. Zur Kenntnis der Oligochaeten-fauna des Balkans 2. Die wasser-bewohnenden Lumbriciden aus dem Ochridasee. Zool. Anz. 95: 96–103.

Černosvitov, L., 1935. Monographie des tschechoslovakischen Lumbriciden. Arch. přír. výzk. Čech. 19 (1): 1–86.

Černosvitov, L., 1945. Oligochaeta from Windermere and the Lake district. Proc. zool. Soc. Lond. 114: 523–548.

Graff, O., 1953. Die Regenwürmer Deutschlands. SchrReihe ForschAnst. Landw. Braunschw.-Volkenrode 7: 1–81.

Omodeo, P., 1962. Oligochètes des Alpes. Mem. Mus. civ. Stor. nat. Verona 10: 71–96.

Omodeo, P., 1984. The earthworm fauna of Sardinia. Revue Ecol. Biol. Sol. 21: 123–134.

Zicsi, A., 1963. Beobachtungen über die Lebensweise des Regenwurmes Allolobophora dubiosa (Örley) 1880. Acta zool. hung. 9: 219–263.

Zicsi, A., 1966. Beiträge zur Kenntnis der ungarischen Lumbricidenfauna. 4. Opusc. zool., Bpest 6: 187–190.

Diversity and zoogeography of marine Tubificidae (Annelida, Oligochaeta) with notes on variation in widespread species

H. R. Baker
Department of Biology, University of Victoria, P.O. Box 1700, Victoria, British Columbia, Canada V8W 2Y2

Keywords: aquatic Oligochaeta, Tubificidae, zoogeography

Abstract

The specific and generic diversities of the marine Tubificidae (Annelida, Oligochaeta) of the NE Pacific are compared to those of the NE and NW Atlantic as well as to those of Heron Island, Australia. Diversity in the NE Pacific is relatively high when compared to that of the NE Atlantic. The Tubificidae of the NW Atlantic (limited to the eastern coast of the USA) show two distinct zoogeographic regions: Florida to Cape Hatteras; Cape Hatteras to Massachusetts. Diversity, both in terms of the number of species and number of genera, is approximately the same in these two regions, and is similar to that of both the NE Pacific and Heron Island. Evidence suggests that the widespread marine species, in particular *Tubificoides pseudogaster*, have a range of morphotypes across their distributions. The apparent wide distributions of these species may be due to a taxonomy unable to resolve the differences between the morphotypes. The tubificid oligochaete fauna of the NE Atlantic appears impoverished compared to the other regions examined. The NE Pacific, NW Atlantic, and Heron Island regions are not dominated by one group of species while the NE Atlantic fauna is dominated by *Tubificoides benedeni* and *Clitellio arenarius*.

Introduction

In recent years it has become apparent that a large number of marine oligochaetes exist in the worlds oceans, mainly of the families Tubificidae and Enchytraeidae; however, the zoogeography of these groups is practically unknown. The marine tubificid faunas of three widely separated areas are compared here to the fauna of the NE Pacific. This comparison includes only those species that have been found in the intertidal zone or shallow water. This restriction was necessary as most of the investigations in the NE Pacific have been limited to the intertidal zone.

Recent work in the NE Pacific (Pt. Conception, California, USA, to Dixon Entrance, northern British Columbia, Canada), a cold temperate region (Briggs, 1974), has revealed a rich littoral marine tubificid fauna (Brinkhurst & Baker, 1979; Baker,

1981a, 1982, 1983a, 1983b, 1983c; Baker & Brinkhurst, 1981; Baker & Erséus, 1982; Erséus, 1980a; Strehlow, 1982). Most of the collections in this region have been made in British Columbia. Twelve genera with 28 species occur in B.C.; only two additional species can be added from the rest of the region.

Collections in the NE Atlantic (Ireland, N. Ireland, England, Scotland, Sweden, and Norway), also a cold temperate region (Briggs, 1974), combined with a review of the literature (including data for W. Germany) show 11 genera with 24 species (littoral and sublittoral) (Erséus, 1975a, 1975b, 1978a, 1978b, 1979a, 1979b, 1980a, 1980b, 1980c, 1982a; Erséus & Kossmagk-Stephan, 1982).

A review of the available literature for the NW Atlantic (eastern coast of the USA), a warm temperate area (Briggs, 1974), showed that 2 distinct regions could be distinguished based on the distribution of tubificid species: a region along the west coast of Florida to Cape Hatteras, North Carolina (Florida region), and a region between Cape Hatteras and Massachusetts (Cape Cod region) (Baker, 1981b; Baker & Erséus, 1979; Baker & Brinkhurst, 1981; Brinkhurst & Baker, 1979; Erséus, 1978a, 1979a, 1979b, 1979c, 1979d, 1979e, 1980a, 1981a, 1981b, 1982a, 1982b, 1982c, 1983a, 1983b; Erséus & Loden, 1981). The boundary between these two areas lies at Cape Hatteras, an area long recognized as a significant zoogeographic boundary (Briggs, 1974; Hayden & Dolan, 1976). In the Florida region there are 15 genera with 30 species; in the Cape Cod region, 13 genera with 39 species.

Heron Island, Australia, which lies in a tropical region (Briggs, 1974) at the southern edge of the Great Barrier Reef, has 11 genera with 32 species (Erséus, 1979d, 1980a, 1981a, 1981c, 1982a, 1983a; Erséus & Jamieson, 1981; Jamieson, 1977). Sampling at Heron Island has covered only the Heron and Wistari reefs (Erséus, pers. commun.); the very high diversity here is amazing for such a small area.

Comparison of the regions

The number of both genera and species of marine Tubificidae in the NE Pacific is higher than that of the NE Atlantic (Tables 1, 2). Three species are common to these areas (*Tubificoides pseudogaster*, *Monopylephorus rubroniveus*, *M. parvus*). The NE Pacific fauna lacks 3 genera found in the NE Atlantic (*Spiridion*, *Clitellio*, *Adelodrilus*) but does include 5 genera (*Tectidrilus*, *Rhizodrilus*, *Nootkadrilus*, *Discordiprostatus*, and *Vadicola*) not found in the NE Atlantic; the latter three genera are found only in the NE Pacific. The NE Atlantic fauna lacks endemic genera entirely; only 3 species occur with

any regularity (*Clitellio arenarius*, *Tubificoides benedeni*, and *Tubifex costatus*). These species were found in every type of habitat throughout the intertidal range. This is in distinct contrast to the NE

Table 1. Species of the northeast Pacific and northeast Atlantic regions, asterisked species are those shared with the northeast Pacific region.

Northeast Pacific	Northeast Atlantic
PHALLODRILINAE	PHALLODRILINAE
Aktedrilus locyi	*Adelodrilus cooki*
oregonensis	*pusillus*
n.sp. 1	*Aktedrilus curvipenis*
Bacescuella labeosa	*monospermathecus*
Bathydrilus n.sp. 1	*sphaeropenis*
n.sp. 2	*Bacescuella arctica*
Discordiprostatus longisetosus	*parvithecata*
Nootkadrilus compressus	*Bathydrilus rarisetis*
frigidus	*Phallodrilus parthenopaeus*
gracilisetosus	*postspermathecatus*
grandisetosus	*prostatus*
hamatus	*rectisetosus*
verutus	*Spiridion insigne*
Phallodrilus tempestatis	
RHYACODRILINAE	RHYACODRILINAE
Monopylephorus cuticulatus	*Monopylephorus parvus**
parvus	*rubroniveus**
rubroniveus	
Rhizodrilus pacificus	
Vadicola aprostatus	
LIMNODRILOIDINAE	LIMNODRILOIDINAE
Limnodriloides monothecus	*Limnodriloides agnes*
victoriensis	*scandinavicus*
Tectidrilus diversus	
verrucosus	
TUBIFICINAE	TUBIFICINAE
Tubificoides apectinatus	*Clitellio arenarius*
coatesae	*Tubificoides amplivasatus*
nerthoides	*benedeni*
pseudogaster	*heterochaetus*
n.sp. 1	*pseudogaster**
n.sp. 2	*Tubifex costatus*
n.sp. 3	*litoralis*

Table 2. Number of species and genera per subfamily for each region (genera:species).

Subfamily	Region				
	Northeast Pacific	Northeast Atlantic	Cape Cod	Florida	Heron Island
Phallodrilinae	6:14	6:13	6:11	7:12	5:13
Rhyacodrilinae	3: 5	1: 2	2: 8	4: 8	4: 9
Limnodriloidinae	2: 4	1: 2	3: 8	4:10	2:10
Tubificinae	1: 7	3: 7	2:12	0: 0	0: 0
Total	12:30	11:24	13:39	15:30	11:32

Pacific where species ranges are often quite narrow and well defined (Baker, unpubl. observ.). The composition of the fauna in terms of the number of genera per subfamily is similar between these two regions (Tables 2, 3).

The Cape Cod fauna is quite diverse (Tables 2 and 4) and has more genera and species than the NE Pacific; 5 species are shared (Table 3). The Cape Cod region shares 8 species with the NE Atlantic (compare Tables 1 and 4); thus the northern regions of the Atlantic do show a slightly greater similarity to each other than do either to the NE Pacific. It is interesting to note that of the three common NE Atlantic species only *C. arenarius* and *T. benedeni* are found in the Cape Cod region; *T. costatus* is not.

There is a fundamental difference between the Cape Cod and Florida regions in that the latter is dominated by genera belonging to the Limnodriloidinae, Rhyacodrilinae, and Phallodrilinae (Table 2) whereas the Tubificinae are an important faunal element of the Cape Cod region (12 of 39 species). There are no known members of the Tubificinae in the Florida region, although Shirley and Loden (1982) found *Tubificoides hererochaetus* and described *Tubificoides denouxi* from the Calcasieu estuary in Louisiana (Gulf of Mexico). Other species of *Tubificoides* remain to be described from the Gulf of Mexico (Baker, unpubl. observ.). Future sampling will undoubtedly show the existence of tubificine species in the Florida region. The only species common to Florida and the NE Pacific is *Limnodriloides monothecus* (Erséus, 1982a); to Florida and the NE Atlantic *Phallodrilus rectisetosus* (two different subspecies; Erséus, 1981c).

Heron Island (Table 5) is totally distinct from the NE Pacific in terms of species but shares some of the cosmopolitan genera (see below).

Table 3. Number of species and genera shared by the northeast Pacific and the other regions relative to the total number of species or genera of the two regions being compared (shared species or genera:total species or genera).

	Northeast Atlantic	Cape Cod	Florida	Heron Island
Number of shared:total species	3:50	5:64	1:59	0:62
Number of shared:total genera	7:16	6:19	5:22	4:19

Table 4. Species and genera of the Florida and Cape Cod regions. Asterisked species are those shared with the northeast Pacific region.

Florida	Cape Cod
PHALLODRILINAE	**PHALLODRILINAE**
Adelodrilus acochlearis	*Adelodrilus anisosetosus*
magnithecatus	*cristatus*
Aktedrilus floridensis	*magnithecatus*
Bathydrilus adriaticus	*multispinosus*
Inanidrilus bulbosus	*Aktedrilus monospermathecus*
Peosidrilus biprostatus	*Bathydrilus longus*
Phallodrilus caudatus	*Peosidrilus biprostatus*
extremus	*Phallodrilus coeloprostatus*
rectisetosus	*obscurus*
sabulosus	*parviatriatus*
tenuissimus	*Uniporodrilus granulothecatus*
Uniporodrilus granulothecatus	
RHYACODRILINAE	**RHYACODRILINAE**
Heterodrilus bulbiporus	*Heterodrilus bulbiporus*
minisetosus	*minisetosus*
occidentalis	*occidentalis*
pentcheffi	*pentcheffi*
Kaketio ineri	*Monopylephorus evertus*
Marcusaedrilus hummelincki	*irroratus*
luteolus	*parvus**
Parakaketio longiprostatus	*rubroniveus**
LIMNODRILOIDINAE	**LIMNODRILOIDINAE**
Limnodriloides baculatus	*Limnodriloides agnes*
barnardi	*barnardi*
hastatus	*medioporus*
*monothecus**	*monothecus**
rubicundus	*rubicundus*
vespertinus	*Smithsonidrilus marinus*
Smithsonidrilus marinus	*Thalassodrilides belli*
Tectidrilus bori	*milleri*
squalidus	
Thalassodrilides gurwitschi	
TUBIFICINAE	**TUBIFICINAE**
	Clitellio arenarius
	*Tubificoides apectinatus**
	benedeni
	brownae
	diazi
	dukei
	heterochaetus
	intermedius
	longipenis
	maueri
	*pseudogaster**
	wasselli

The NE Pacific shares very few species with the other regions discussed here (Table 3) but does show more affinities to the NE Atlantic and Cape Cod regions than to Heron Island and the Florida region.

As can be seen from Table 2 the NE Pacific has

Table 5. Species and genera of Heron Island.

PHALLODRILINAE

Aktedrilus parviprostatus
Bathydrilus rohdei
     superiovasatus
Coralliodrilus atriobifidus
     avisceralis
     oviatriatus
     parvigenitalis
Jamiesoniella athecata
Phallodrilus albidus
     clavatus
     filitheeatus
     geniculatus
     rectisetosus

RHYACODRILINAE

Gieredrilus inermis
Heterodrilus claviatriatus
     jamiesoni
     keenani
     queenslandicus
     scitus
Heronidrilus bihamis
     fastigatus
Macquaridriloides heronae

LIMNODRILOIDINAE

Limnodriloides armatus
     australis
     tenuiductus
     uniampullatus
Marcusaedrilus capricornae
     irregularis
     grandiculus
     minusculus
     sacculatus
     tuber

approximately the same number of species and genera as the Cape Cod, Florida, and Heron Island regions. All of these show a higher number of species and all except Heron Island have a higher number of genera than the NE Atlantic (Table 2); these regions are all similar in terms of latitudinal range except for Heron Island (see above).

There are only four genera common to all 5 regions; *Phallodrilus, Aktedrilus, Bathydrilus,* and *Limnodriloides.* The subfamilies Phallodrilinae, Rhyacodrilinae, and Limnodriloidinae are well represented and appear to be cosmopolitan (Table 2). The Tubificinae are best represented in the temperate latitudes but are unknown to date in Florida (see above discussion) and Heron Island.

Variation in widespread species

Most species of marine Tubificidae have quite restricted distributions. There are only 5 species that appear to be at all widespread (*Tubificoides pseudogaster, Monopylephorus rubroniveus, M. parvus, Limnodriloides monothecus,* and *Phallodrilus rectisetosus*). One common characteristic of these 5 species is that the genital systems are all relatively simple as compared to the other members of their respective genera. These forms may, therefore, be classified as single species over their geographic range simply because they lack the suite of complex characters necessary to adequately express variation.

M. rubroniveus has been regarded as being composed of up to 10 synonyms two of which have now been separated as distinct taxa (*M. limosus, M. kermadecensis*; Baker & Brinkhurst, 1981). All of the remaining 8 synonyms differ slightly in various characters. *M. parvus* has also been described as varying in some characters between material from different localities (see Marcus, 1965; Baker & Brinkhurst, 1981). Erséus (1982a) noted several differences between material of *L. monothecus* from different areas but declined to erect new taxa based on these differences. Erséus (1979d, 1981c) has erected three subspecies of *P. rectisetosus* from different areas (Italy and France; Florida; Heron Island) based on minor differences in the genitalia. *T. pseudogaster,* redescribed from type material by Baker (1980), also displays variation in material from different geographical areas.

I have recently examined additional specimens of *T. pseudogaster* from Sweden (Tjärnö), Germany (Schlei Fjord), England (Hull River estuary), Canada (Frobisher Bay, Northwest Territories), and USA (Friday Harbour, Washington). Penis sheath lengths of material from the above localities are shown in Fig. 1. The new material examined showed significant variation with regard to penis sheath lengths ($P \leqslant 0.05$) of the above material belonging to the same population; Baker, unpubl. data) and appears to be different than the material from the type locality of *pseudogaster* (Kysing Fjord, Denmark; Dahl, 1960; Baker, 1980). There is a definite cline in the penis sheath lengths of the new material (Fig. 1). Length and width of the atria and vasa deferentia also show significant variations between the above populations. However, while these differ-

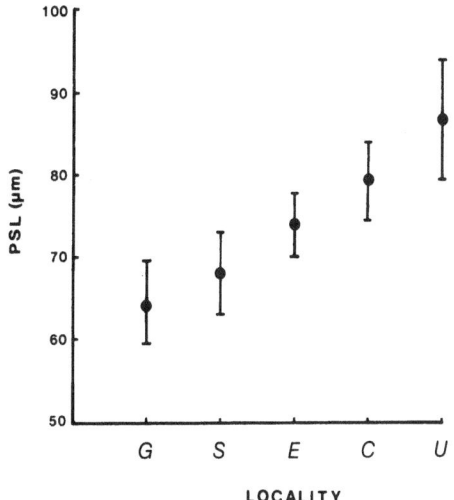

Fig. 1. 95% confidence intervals about the means of penis sheath lengths (PSL) of *T. pseudogaster* material from Schlei Fjord, Germany (G); Tjärnö, Sweden (S); Hull River estuary, England (E); Frobisher Bay, Canada (C); and San Juan Island, USA (U). Localities arranged in order of increasing geographical distance from the type locality of *T. pseudogaster*.

ences do exist, it is not known if they are genotypic or phenotypic in nature; the differences are not pronounced enough to warrant specific or subspecific separation of the various populations.

This problem is one of the main failings of the primarily morphological taxonomy in use today; in simple forms, with few distinctive characters, it becomes almost impossible to distinguish phenotypic from genotypic variation. This situation is aggravated by the lack of modern studies on intra-specific variation in more complex, easily distinguishable species. Until the scope of variation within distinct species in a limited geographical range is known, it will be impossible to determine if the variation in widespread species represents intra- or inter-specific variation. Thus, given the very high number of endemic marine species, these so-called widespread species may in fact reflect the results of an inadequate taxonomy rather than truly cosmopolitan species.

Conclusions

It would appear that there is a fundamental difference between the tubificid faunas of the NE Atlantic and the other regions studied. In all 4 of the other regions the littoral tubificid fauna is rich and in the NE Pacific (Baker, unpubl. observ.) and NW Atlantic (Florida and Cape Cod regions) there is a distinct change in species composition with latitudinal change. In the NE Atlantic the littoral tubificid fauna is not as rich and is dominated by *Tubificoides benedeni* and *Clitellio arenarius* (Baker, pers. observ.) over the whole region.

The NE Pacific, Cape Cod, Florida, and Heron Island regions have littoral tubificid faunas with high diversity, definite species ranges, and no dominant group of species. The NE Atlantic fauna usually has a relatively low diversity with broad species ranges, and is dominated by *Tubificoides benedeni* and *Clitellio arenarius*.

The very few widespread marine tubificid species are all simple forms; their apparent wide distributions may be a taxonomical artifact.

Acknowledgements

My thanks to Dr. R. O. Brinkhurst, Dr. C. Erséus, Ms. K. Coates and Ms. B. Day for their criticisms of this manuscript. I also thank Dr. C. Erséus, Dr. O. Pfannkuche, Dr. B. Barnett, and Dr. J. Wacasey for the loan of *T. pseudogaster* material. Dr. C. Erséus very generously allowed me access to unpublished information. This work was supported by a Natural Sciences and Engineering Research Council of Canada post-graduate award.

References

Baker, H. R., 1980. A redescription of Tubificoides pseudogaster (Dahl) (Oligochaeta: Tubificidae). Trans. am. microsc. Soc. 99: 337–342.

Baker, H. R., 1981a. Phallodrilus tempestatis n.sp., a new marine tubificid (Annelida: Oligochaeta) from British Columbia. Can. J. Zool. 59: 1475–1478.

Baker, H. R., 1981b. A redescription of Tubificoides heterochaetus (Michaelsen) (Oligochaeta: Tubificidae). Proc. biol. Soc. Wash. 94: 564–568.

Baker, H. R., 1982. Two new Phallodrilinae genera of marine Oligochaeta (Annelida: Tubificidae) from the Pacific northeast. Can. J. Zool. 60: 2487–2500.

Baker, H. R., 1983a. Vadicola aprostatus nov.gen., nov.sp., a marine oligochaete (Tubificidae: Rhyacodrilinae) from British Columbia. Can. J. Zool. 60 (1982): 3232–3236.

Baker, H. R., 1983b. New species of Bathydrilus Cook (Oligochaeta; Tubificidae) from British Columbia. Can. J. Zool. 61: 2162–2167.

196

Baker, H. R., 1983c. New species of Tubificoides Lastochkin (Oligochaeta; Tubificidae) from the Pacific Northeast and the Arctic. Can. J. Zool. 61: 1270–1283.

Baker, H. R. & R. O. Brinkhurst, 1981. A revision of the genus Monopylephorus and redefinition of the subfamilies Rhyacodrilinae and Branchiurinae (Tubificidae: Oligochaeta). Can. J. Zool. 59: 939–965.

Baker, H. R. & C. Erséus, 1979. Peosidrilus biprostatus n.g., n.sp., a marine tubificid (Oligochaeta) from the eastern United States. Proc. biol. Soc. Wash. 92: 505–509.

Baker, H. R. & C. Erséus, 1982. A new species of Bacescuella Hrabĕ (Oligochaeta, Tubificidae) from the Pacific coast of Canada. Can. J. Zool. 60: 1951–1954.

Briggs, J. C., 1974. Marine zoogeography. McGraw-Hill Book Co., S. Francisco, 475 pp.

Brinkhurst, R. O. & H. R. Baker, 1979. A review of the marine Tubificidae (Oligochaeta) of North America. Can. J. Zool. 57: 1553–1569.

Dahl, I. O., 1960. The Oligochaete fauna of 3 Danish brackish water areas. Meddr. danm. Fisk. Havunders. 2: 1–20.

Erséus, C., 1975a. Peloscolex amplivasatus sp.n. and Macroseta rarisetis gen. et sp.n. (Oligochaeta, Tubificidae) from the west coast of Sweden. Sarsia 58: 1–8.

Erséus, C., 1975b. On the systematic position of Rhyacodrilus prostatus Knöllner (Oligochaeta, Tubificidae). Zool. Scr. 4: 33–35.

Erséus, C., 1978a. New species of Adelodrilus and a revision of the genera Adelodrilus and Adelodriloides (Oligochaeta, Tubificidae). Sarsia 63: 135–144.

Erséus, C., 1978b. Two new species of the little-known genus Bacescuella Hrabĕ (Oligochaeta, Tubificidae) from the North Atlantic. Zool. Scr. 7: 263–267.

Erséus, C., 1979a. Bermudrilus peniatus n.g., n.sp., (Oligochaeta, Tubificidae) and two new species of Adelodrilus from the northwest Atlantic. Trans. am. micros. Soc. 98: 418–427.

Erséus, C., 1979b. Taxonomic revision of the marine genera Bathydrilus Cook and Macroseta Erséus (Oligochaeta, Tubificidae), with descriptions of six new species and subspecies. Zool. Scr. 8: 139–151.

Erséus, C., 1979c. Inanidrilus bulbosus gen. et sp.n., a marine tubificid (Oligochaeta) from Florida, USA. Zool. Scr. 8: 209–210.

Erséus, C., 1979d. Taxonomic revision of the marine genus Phallodrilus Pierantoni (Oligochaeta, Tubificidae), with descriptions of thirteen new species. Zool. Scr. 8: 187–208.

Erséus, C., 1979e. Uniporodrilus granulothecatus n.g., n.sp., a marine tubificid (Oligochaeta) from estern United States. Trans. am. micros. Soc. 98: 414–418.

Erséus, C., 1980a. Taxonomic studies on the marine genera Aktedrilus Knöllner and Bacescuella Hrabĕ (Oligochaeta, Tubificidae), with descriptions of seven new species. Zool. Scr. 9: 97–111.

Erséus, C., 1980b. Redescriptions of Phallodrilus parthenopaeus Pierantoni and P. obscurus Cook (Oligochaeta, Tubificidae). Zool. Scr. 9: 93–96.

Erséus, C., 1980c. New species of Phallodrilus (Oligochaeta, Tubificidae) from the Arctic deep sea and Norwegian fjords. Sarsia 65: 57–60.

Erséus, C., 1980d. Two new records of the Caribbean marine tubificid Kaketio ineri Righi and Kanner (Oligochaeta) Proc. biol. Soc. Wash. 93: 1220–1222.

Erséus, C., 1981a. Taxonomic revision of the marine genus Heterodrilus Pierantoni (Oligochaeta, Tubificidae). Zool. Scr. 10: 111–132.

Erséus, C., 1981b. Taxonomy of the marine genus Thalassodrilides (Oligochaeta: Tubificidae). Trans. am. micros. Soc. 100: 333–344.

Erséus, C., 1981c. Taxonomic studies of Phallodrilinae (Oligochaeta, Tubificidae) from the Great Barrier Reef and the Comoro Islands with descriptions of ten new species and one new genus. Zool. Scr. 10: 15–31.

Erséus, C., 1982a. Taxonomic revision of the marine genus Limnodriloides (Oligochaeta, Tubificidae). Verh. naturwiss. Ver. Hamburg, NF 25: 207–277.

Erséus, C., 1982b. Revision of the marine genus Smithsonidrilus Brinkhurst (Oligochaeta, Tubificidae). Sarsia 67: 47–54.

Erséus, C., 1982c. Parakaketio longiprostatus gen. et sp.n., a marine tubificid (Oligochaeta) from Florida, USA. Zool. Scr. 11: 195–197.

Erséus, C., 1983a. Taxonomic studies of the marine genus Marcusaedrilus Righi & Kanner (Oligochaeta, Tubificidae), with descriptions of seven new species from the Caribbean area and Australia. Zool. Scr. 12: 25–36.

Erséus, C., 1983b. New records of Adelodrilus (Oligochaeta, Tubificidae), with descriptions of two new species from the North-west Atlantic. Hydrobiologia 106: 73–83.

Erséus, C. & B. G. M. Jamieson, 1981. Two new genera of marine Tubificidae (Oligochaeta) from Australia's Great Barrier Reef. Zool. Scr. 10: 105–110.

Erséus, C. & K. J. Kossmagk-Stephan, 1982. A new species of Aktedrilus (Oligochaeta, Tubificidae) from the North Sea coast of the Federal Republic of Germany. Zool. Anz. 209: 91–96.

Erséus, C. & M. S. Loden, 1981. Phallodrilinae (Oligochaeta: Tubificidae) from the east coast of Florida, with descriptions of a new species of Adelodrilus. Proc. biol. Soc. Wash. 94: 819–825.

Hayden, B. P. & R. Dolan, 1976. Coastal marine fauna and marine climates of the Americas. J. Biogeogr. 3: 71–81.

Jamieson, B. G. M., 1977. Marine meiobenthic Oligochaeta from Heron and Wistari Reefs (Great Barrier Reef) of the genera Clitellio, Limnodriloides and Phallodrilus (Tubificidae) and Grania (Enchytraeidae). Zool. J. linn. Soc. 61: 329–349.

Marcus, E., 1965. Naidomorpha aus brasilianischen Brackwasser. Beitr. neotrop. Fauna 4: 61–83.

Shirley, T. C. & M. S. Loden, 1982. The Tubificidae (Annelida, Oligochaeta) of a Louisiana estuary: ecology and systematics with the description of a new species. Estuaries 5: 47–56.

Strehlow, D. R., 1982. Aktedrilus locyi Erséus, 1980 and Aktedrilus oregonensis n.sp. (Oligochaeta, Tubificidae) from Coos Bay, Oregon, with notes on distribution with tidal height and sediment type. Can. J. Zool. 60: 593–596.

Distribution and habitat characteristics of Naididae and Tubificidae in the inland waters of Israel and the Sinai Peninsula

C. Pascar-Gluzman[1] & C. Dimentman[2]

[1] *Department of Biology, Ben Gurion University of the Negev, Beer Sheva 84120, Israel*
Present address: Instituto de Embriología, Biología e Histología, Facultad de Ciencias Médicas, Universidad Nacional de La Plata, La Plata 1900, Argentina
[2] *Department of Zoology, The Hebrew University of Jerusalem, 91904 Jerusalem, Israel*

Keywords: aquatic Oligochaeta, Naididae, Tubificidae, Israel & Sinai distribution, habitat characteristics

Abstract

Nineteen species of Naididae and fourteen species of Tubificidae were found in collections from inland waters in Israel and Sinai. The local distribution patterns of the majority of these species were determined. The ranges of several physical and chemical variables characterizing the habitats of each species were defined. A correlation was found between the distribution patterns of some species and the following variables: salinity, temperature, dissolved oxygen, current velocities and stability of the habitats.

Introduction

Several studies have been carried out on the aquatic Oligochaeta of Israel (Rosa, 1893; Stephenson, 1913; Černosvitov, 1938; Gitay, 1965, 1968; Por, 1968, Por & Masry, 1968; Brinkhurst, 1981; Pascar-Gluzman, 1981). However, most of these studies are taxonomic works based on limited material, collected principally from Lake Kinneret and the Huleh swamps. Unfortunately, they do not provide sufficient information to determine the habitat characteristics and regional distribution patterns of oligochaete species in Israel. The present study is directed towards a geographical and ecological description of the local distribution of Naididae and Tubificidae within Israel and some adjacent areas in the Sinai Peninsula (Fig. 1).

Methods

Collections from the Inland Water Ecological Service (I.E.S.) and the Zoological Museum of the Hebrew University of Jerusalem were examined. Whole mounts were identified on the basis of setae

and genital features of mature specimens (Brinkhurst, 1971). Water samples for chemical analysis were simultaneously collected with many of the oligochaete samples and analyzed according to standard methods. These analyses included measurements of conductivity, biological oxygen demand (BOD), chemical oxygen demand (COD), nitrates, ammonia en chlorides. Field measurements included determination of temperature, current velocity and dissolved oxygen concentration.

Results

Thirty-three taxa of aquatic Oligochaeta were identified from a total of 645 samples. The number of locations and samples in which each species was collected is listed in Table 1. Twenty-two of these species, fifteen naidids and seven tubificids, have not been recorded from the area under consideration by previous authors. In contrast, two naidids (*Pristina bilobata* (Bretscher) and *Pristina aequiseta* Bourne) and one tubificid (*Potamothrix heuscheri* (Bretscher)) that were identified by Černosvitov (1938), Gitay (1968) and Brinkhurst & Jamie-

Hydrobiologia 115, 197–205 (1984).
© Dr W. Junk Publishers, Dordrecht.

Table 1. Frequency of occurrence of Naididae and Tubificidae species collected in Israel and adjacent areas, 1977–1982.

Species	Code in Tables 2–4	Number of localities	Number of samples
NAIDIDAE			
Chaetogaster diaphanus (Gruithuisen)*	CD	1	2
Dero digitata (Müller)	DDI	11	12
D. dorsalis Ferroniere*	DDO	3	3
D. nivea Aiyer	DN	2	4
D. obtusa D'Udekem*	DO	19	24
D. (Aulophorus) furcatus (Müller)*	DF	7	8
Homochaeta naidina Bretscher*	HN	1	1
Nais bretscheri Michaelsen*	NB	5	8
N. communis Piguet*	NC	11	13
N. elinguis Müller*	NE	34	45
N. pardalis Piguet	NP	30	40
N. simplex Piguet*	NS	7	7
Paranais litoralis (Müller)*	PL	2	2
Pristina idrensis Sperber*	PRI	40	52
P. longiseta Ehrenberg*	PRL	8	13
Slavina sp.*	SL	1	1
Specaria sp.*	SP	4	4
Stephensoniana trivandrana (Aiyer)	ST	14	24
Stylaria lacustris (Linnaeus)*	STL	4	5
TUBIFICIDAE			
Aulodrilus limnobius Bretscher*	AL	4	5
A. pluriseta (Piguet)*	AP	20	24
Limnodrilus hoffmeisteri Claparede	LH	31	47
L. udekemianus Claparede*	LU	11	12
Peloscolex (Embolocephalus) kurenkovi Sokolskaya*	PK	3	4
Phallodrilus sp.*	PH	1	1
Potamothrix bavaricus (Oschmann)	POB	66	81
P. hammoniensis (Michaelsen)	POH	15	15
Psammoryctides albicola Michaelsen	PA	56	104
Rhyacodrilus sodalis (Eisen)*	RS	5	5
Tubifex ignotus (Stolc)	TI	9	9
T. nerthus Michaelsen, Brinkhurst*	TN	3	3
T. tubifex (Müller)	TT	37	50
Varichaeta israelis Brinkhurst	VI	13	17

* New records for Israel.

son (1971) were not found in the present study.

The presence of Naididae and Tubificidae species in the various geographic and climatic regions studied are indicated in Table 2. The local distribution of each species is illustrated in Figs. 2–29. The results from the distribution maps and Table 2 reveal that the numbers of Naididae and Tubificidae species found in the humid regions are much higher than in the arid regions; thirty-two species compared to nineteen.

Fourteen species are restricted to the humid regions, whereas only one rare species (*Phallodrilus* sp.) was found in the arid zone and was not collected in the humid regions. The Upper Jordan Basin has more Naididae and Tubificidae species than any of the other regions. Similarly, the Lower Jordan Basin is the richest of all the arid regions of the area.

Three species (*Pristina idrensis, Potamothrix bavaricus* and *Tubifex tubifex*) were found almost all over the country, while six other species are restricted to the Rift Valley. One of these (*Varichaeta israelis*) inhabits this area from the Upper Jordan in the north to East Sinai in the south.

A second species (*Aulodrilus limnobius*) was collected along the River Jordan. Four species were found each at a single location only; three of these

Table 2. Geographical and climatic distribution of Naididae and Tubificidae in the areas studied.

Species	Humid regions				Arid regions			
	Golan Heights	Coastal Plain	Galilee, Samaria, Judea	Upper Jordan Basin	Lower Jordan Basin	Dead Sea	Negev	Sinai
NAIDIDAE								
CD				+				
DDI		+		+				
DDO	+							
DN		+	+					
DO	+	+	+	+				
DF		+			+			+
HN		+						
NB				+				
NC		+		+	+			
NE		+	+	+	+			
NP	+	+		+	+			
NS	+	+		+				
PL		+			+			
PRI	+	+	+	+	+	+	+	+
PRL		+		+		+		
SL				+				
SP	+		+	+				
ST		+	+	+	+			
STL		+		+				
TUBIFICIDAE								
AL				+	+			
AP	+	+	+	+				
LH		+	+	+	+			
LU		+	+		+			
PK			+	+				
PH								+
POB	+	+	+	+	+	+	+	+
POH	+	+	+	+	+			
PA	+	+	+	+	+			
RS				+	+		+	
TI			+	+	+	+	+	
TN		+	+	+				
TT		+	+	+	+	+	+	+
VI				+	+	+		+
Total number of species	10	22	16	26	17	6	5	6

(*Chaetogaster diaphanus, Nais bretscheri* and *Slavina* sp.) in the Upper Jordan, a fourth species (*Phallodrilus* sp.) in a well on the Red Sea coast.

Species richness of the habitats. The occurrence of species of Naididae and Tubificidae in a wide range of aquatic habitats is presented in Table 3.

In general, the highest species richness was found in running waters, i.e. Jordan River, streams and springs. In contrast, the number of species in stagnant waters was relatively low; e.g. Lake Kinneret (Sea of Galilee), reservoirs, semi-permanent ponds, wells and cisterns. The species richness of temporary rainpools was particularly low. An exception to the general pattern in stagnant waters was found in permanent ponds, which showed high species richness.

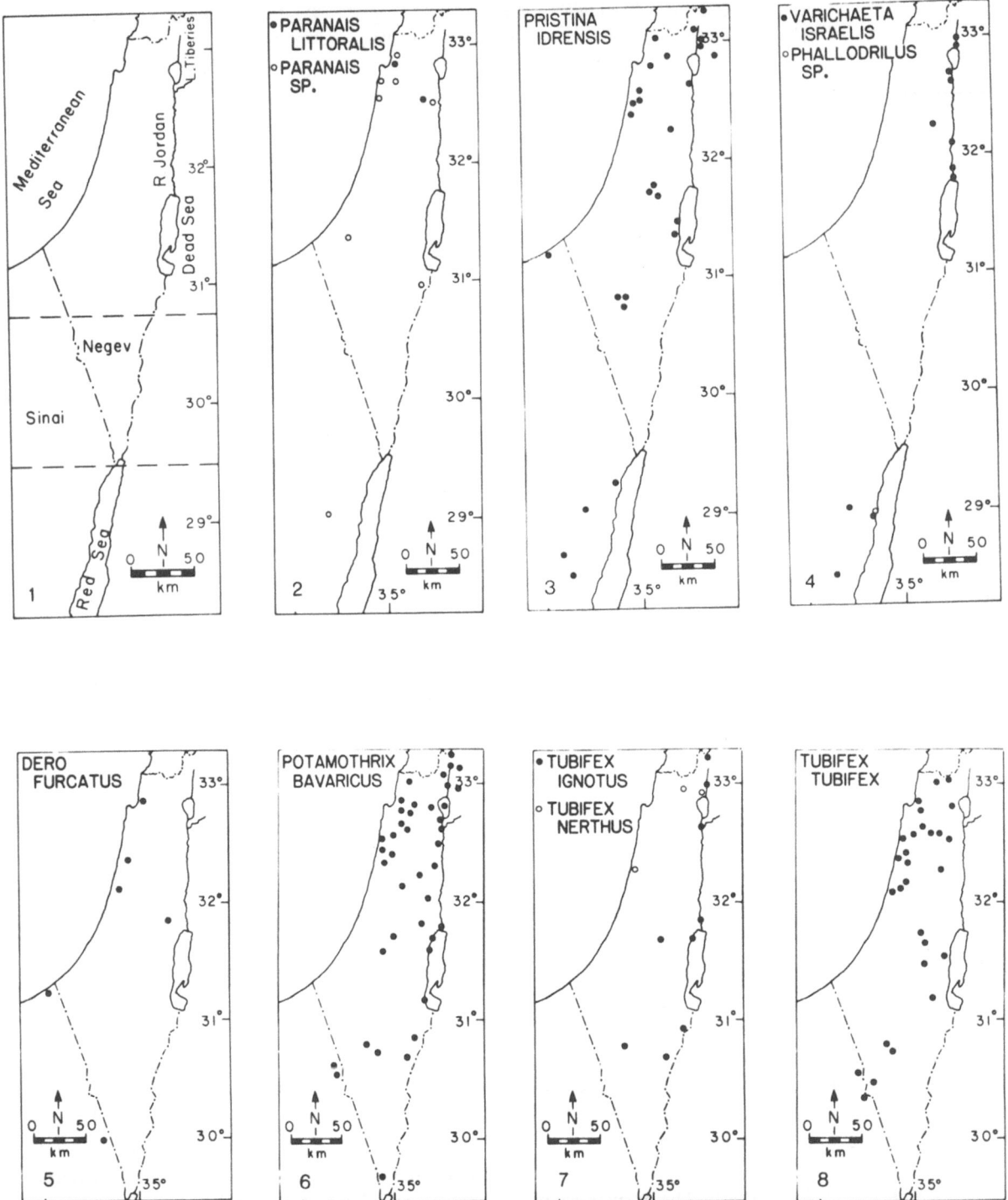

Figs. 1–8. 1. Study area. *2–8.* Distribution patterns of Naididae and Tubificidae distributed throughout the study area.

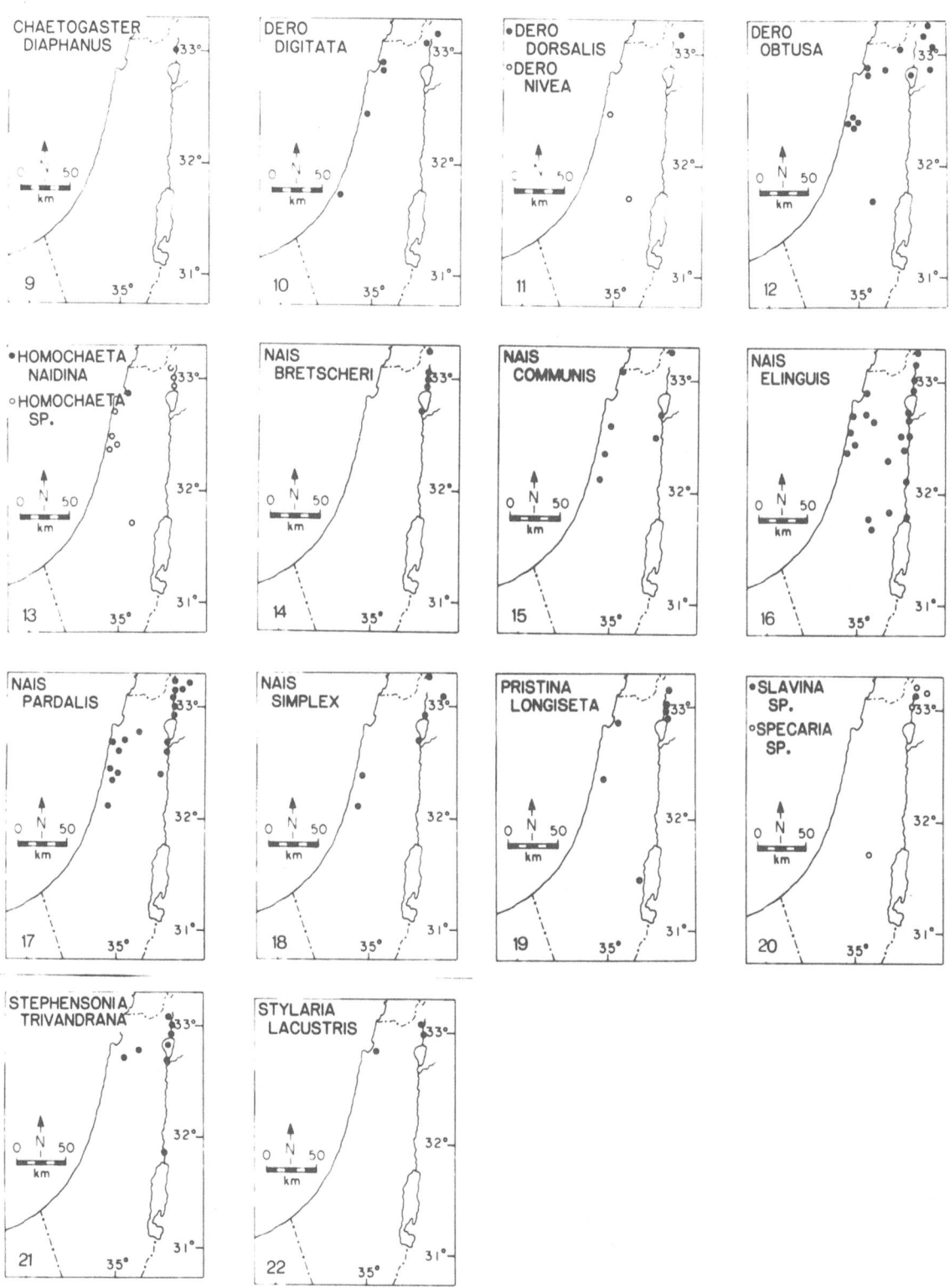

Figs. 9–22. Distribution patterns of Naididae restricted to northern and central parts of the study area.

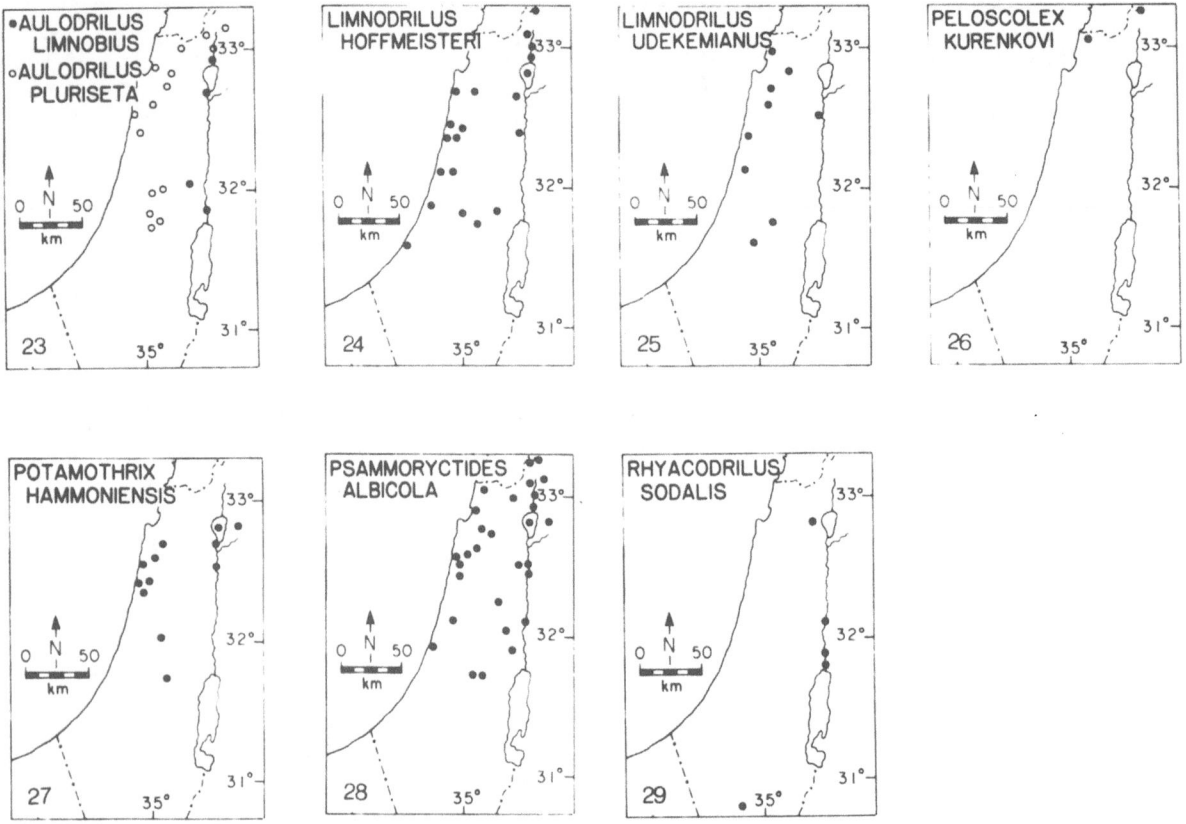

Figs. 23–29. Distribution patterns of Tubificidae restricted to northern and central parts of the study area.

Physical and chemical variables. The observed range of current velocity, temperature, pH, dissolved oxygen, BOD, COD, nitrate, ammonia, chlorides and conductivity of the sites in which each species was collected is presented in Table 4. Specific relationships between species occurrence and these variables are indicated below.

Current velocity. Pristina idrensis, Varichaeta israelis and *Psammoryctides albicola* were collected from a wide range of current velocities (0–150 cm⁻¹). In contrast, *Nais bretscheri* was restricted to running waters (> 20 cm⁻¹), while *Tubifex nerthus* was found only in standing waters.

Temperature. Pristina idrensis and *Potamothrix bavaricus* were found within a wide temperature range (8–32 °C and 13–35 °C respectively), while *Dero digitata, Chaetogaster diaphanus, Paranais*

litoralis and *Peloscolex kurenkovi* were not collected from temperature above 20 °C.

Dissolved oxygen. Dero furcatus was collected at the lowest measured dissolved oxygen levels (1.3 mg l⁻¹). *Peloscolex kurenkovi* was restricted to habitats characterized by high dissolved oxygen (lowest value 9.7 mg l⁻¹).

BOD and COD. Limnodrilus hoffmeisteri was found at the highest BOD and COD values recorded in this study and exhibited the greatest tolerance range (BOD: 0–81 mg l⁻¹; COD: 0–260 mg l⁻¹).

Nitrates and ammonia. Potamothrix hammoniensis was the only species found at the highest recorded nitrate values and exhibited the greatest tolerance range (0–100 mg l⁻¹). *Limnodrilus hoffmeisteri* was collected from the greatest range of ammonia values (0–45 mg l⁻¹).

Table 3. Occurrence of Naididae and Tubificidae species in various types of water bodies in Israel and adjacent areas. Abbreviations: Riv = river; str. = streams; spr. = springs; res. = reservoirs; perm. = permanent; temp. = temporary; cist. = cisterns.

Species	Habitats							
	Running waters			Standing waters				
	Riv.	Str.	Spr.	Lake & res.	Perm. ponds	Semi-perm. ponds	Temp. pools	Wells & cist.
NAIDIDAE								
CD	+							
DDI		+	+		+	+		
DDO			+		+			
DN		+			+			
DO		+		+	+	+		+
DF		+	+		+			+
HN		+						
NB	+							
NC	+	+	+					
NE	+	+	+		+			
NP	+	+	+	+	+	+		
NS	+	+	+		+			
PL		+						
PRI	+	+	+		+	+		+
PRL	+	+						
SL		+						
SP	+	+	+			+		
ST	+	+		+		+		
STL	+	+				+		
TUBIFICIDAE								
AL	+	+						
AP	+	+	+		+			
LH	+	+	+	+	+	+		
LU		+	+		+			
PK	+		+		+			
PH								+
POB	+	+	+	+	+		+	
POH	+	+	+	+				
PA	+	+	+	+	+	+		
RS	+		+		+			
TI	+	+	+		+			
TN						+		
TT		+	+		+	+	+	+
VI	+	+	+					+
Total number of species	21	26	20	7	18	11	2	6

Conductivity and chlorinity. Brackish and saline waters proved to be inhabited by two species: *Paranais litoralis* and *Tubifex tubifex*. The first species was restricted to brackish and saline waters (conductivity: 3.0–14.4 mmho/cm^{-2}; chlorinity: 413–5144 mg l^{-1}), while the second species was also collected from fresh water habitats (conductivity: 0.3–11.6 mmho cm^{-2}; chlorinity: 5–4800 mg l^{-1}).

Discussion

The relatively high number of species that were found in the present study compared with the results of previous works in this area reflects the great diversity of habitats examined. The absence of three species in this study that are known from Israel may simply be due to low sampling in the

Table 4. Ranges of physical and chemical variables from Nadidae and Tubificidae habitats.

Species	Veloc. (cm s⁻¹)	Temp. (°C)	pH	DO		BOD (mg l⁻¹)	COD (mg l⁻¹)	NO₃ (mg l⁻¹)	NH₄ (mg l⁻¹)	Cl (mg l⁻¹)	Cond. (mS cm⁻²)
				(%)	(mg l⁻¹)						
NAIDIDAE											
CD	107	17–18	7.6	112–116	–	0	0	9	0	33	0.4
DDI	10	14–16	–	–	2.5–7.5	0	8–52	–	–	917	3.74
DN	–	2.5–30	8.4	–	6.8	–	–	–	–	161–948	1.1–3.7
DO	13–53		7.0–9.0	20–176	5.6–15.2	2	12	1–24	0–3	–	2.1
DF	6	19–27	7.0–8.5	16–129	1.3–10.3	0–4	12–16	1–26	0–1	145–281	0.6–3.47
HN	2	15	8.5	100	–	8	16	8	0	413	3.7
NB	21–90	21–30	7.4–8.0	96–112	–	0–8	0–48	5–8	0–1	22–61	0.3–0.5
NC	0–84	16–28	6.8–8.5	22–140	4.2–9.7	0–3	0–128	0–12	0–1	25–1739	0.3–6.5
NE	0–100	13–32	7.0–9.0	34–150	2.8–14.8	0–56	0–180	0–37	0–17	30–2552	0.4–8.1
NP	0–91	11–31	7.3–9.0	30–150	9.6–13.5	0–13	0–56	0–37	0–6	17–2085	0.3–6.7
NS	0–84	10–25	7.3–8.3	79–176	15.2	0–3	0–12	0–8	0–1	25–339	0.3–1.5
PL	2	15–20	8.1–8.5	100	8.5	0–8	16–44	8–9	0	413–5144	3.0–14.4
PRI	0–150	8–32	6.7–9.0	56–132	5.2–12.0	0–20	0–56	0–82	0–5	20–1686	0.3–5.9
PRL	1–75	17–32	7.3–8.3	40–150	–	0–60	0–36	2–14.0	0–1	17–824	0.4–3.1
SL	–	31	8.3	150	–	–	–	8	0	36	0.4
SP	21–107	17–22	8.1	112–115	–	0	0	9–11	0	33–71	0.3–0.4
ST	5–90	17–30	7.4–9.0	79–112	–	0–13	0–40	3–37	0–6	17–1339	0.4–4.8
STL	1–71	17–25	7.3–7.9	40–132	–	0	0–12	2–7	0–1	17–824	0.5–3.1
TUBIFICIDAE											
AL	2–55	23–29	6.8–8.1	96–104	7.2	0	0–12	1–7	0–5	60–3193	0.4–9.0
AP	0–66	15–30	7.1–9.0	20–90	2.8–8.6	0–15	0–28	0–15	0–2	30–1008	0.4–3.7
LH	0–69	9–30	7.1–9.0	20–200	6.5–16.9	0–81	0–260	0–70	0–45	33–994	0.3–3.6
LU	1–51	15–28	7.2–9.0	21–116	2.1	0–62	0–184	0–37	0–37	60–1580	0.6–5.8
PK	1–90	15–20	7.8–8.1	90–116	9.7–10.2	0	0–4	3–8	0–3	14–36	0.3–0.7
PH	0	24	7.1	–	7.5	–	–	10	0	824	3.5
POB	13–35	13–35	6.8–9.5	20–200	2.5–16.0	0–40	4–128	0–70	0–8	3–3207	0.3–11.6
POH	20–62	16–32	7.2–9.0	56–116	7.3–11.3	0–12	0–72	0–100	0–8	4–2897	0.8–8.3
PA	0–150	15–30	6.8–9.0	22–188	3.5–10.2	0–32	0–180	0–69	0–5	3–2985	0.3–6.7
RS	0–100	13–29	7.1–7.4	96–124	–	0	4–24	1–3	3–5	867–1570	3.4–6.0
TI	0–57	18–29	7.4–8.2	83–104	–	0–2	0–44	0–6	0–2	107–2647	0.4–6.8
TN	0	20–28	6.5–7.5	85	–	13	24	4	0	78	0.6
TT	0–50	12–32	7.1–9.5	25–200	3.7–16.8	0–28	0–132	0–37	0–4	5–4800	0.3–11.6
VI	0–150	18–30	7.1–8.2	87–124	7.5–7.7	0–31	0–62	0–10	0–5	34–3207	0.4–9.4

Huleh reserve and Lake Kinneret. However, the absence of two of these species which were found by Černosvitov (1938) in the Huleh swamps may indicate their disappearance from our area due to the drainage of these swamps.

The high species richness of Naididae and Tubificidae in the humid regions compared to that of arid areas apparently results from the high diversity of aquatic habitats in the humid regions. Arid regions are especially poor in permanent stable freshwater habitats. The majority of freshwater habitats in this area are either temporary rainpools or springs subjected to catastrophic flooding.

Distribution patterns: three species distributed throughout the country: *Pristina idrensis, Potomothrix bavaricus* and *Tubifex tubifex* (Figs. 3, 6, 8) are euryoeic, being found in a wide range of environmental conditions (Table 4). *Tubifex tubifex* is equally capable of colonizing freshwater as well as oligohaline habitats (Table 4). The present study also indicates that *Tubifex tubifex* can inhabit temporary rainpools which are extremely common in arid regions. *Potamothrix bavaricus*, an eurythermal and euryhaline species (Table 4) has also been found in temporary rainpools (Table 3). *Pristina idrensis*, an eurythermal species (Table 4) is probably distributed throughout the country because of its ability to inhabit small and unstable springs.

The distribution of the majority of the species is limited by a complex of unknown ecological factors. However, for at least some of these species is it possible to ascertain the probable limiting factors from the physical and chemical data (Table 4). For example, the distribution of *Paranais litoralis* (Fig. 2) positively correlates with the distribution of saline habitats in Israel and Sinai. This species is known to inhabit similar environments in other parts of the world (Brinkhurst & Jamieson, 1971; Grigelis, 1980). Marine-brackish water habitats are also known to be inhabited by various species of *Phallodrilus* (Brinkhurst & Jamieson, 1971); one of these species was found in a brackish water body in Sinai. High temperature and/or low levels of dissolved oxygen apparently limit the distribution of *Peloscolex kurenkovi* in Israel (Table 4). This species, which is known from Kamchatka (U.S.S.R.) was collected in several springs in northern Israel (Fig. 26) which were characterized by relatively low temperatures and high levels of dissolved oxygen (Table 4). *Nais bretscheri* was found in habitats characterized by running water (> 20 cm s^{-1}) and low salinity (< 0.50 mmho cm^{-2}). It was collected from the Upper Jordan River but not from the Lower Jordan River, which is characterized by higher salinities.

Aquatic habitats suffering from heavy organic pollution as indicated by high levels of BOD, COD, nitrates and ammonia are inhabited by *Limnodrilus hoffmeisteri* (Table 4).

The physical and chemical factors in Table 4 enable us to explain the distribution patterns of several aquatic oligochaetes in the areas studied. However, at the present time we lack sufficient data on the nature of the substrates in the various habitats which may be the major factor determining the distribution of the majority of oligochaetes. Future investigation of such data will enable us to understand better the distribution patterns of many oligochaete species in this country.

Acknowledgements

We would like to thank Drs. M. Ladle and G. J. Bird for their generous assistance in identification of the material. We are grateful to Prof. F. D. Por, for his continuous encouragement of this study. We also would like to thank the staff of the I.E.S. laboratory and especially Mr. R. Ortal, Dr. H. Bromley-Schnur, Ms. S. Greenberg and Dr. G. Herbst, for their invaluable help.

References

Brinkhurst, R. O., 1971. A guide for the identification of British Aquatic Oligochaeta. Freshwat. Biol. Ass. Sci. Publ. (2nd Edn) 22: 55 pp.
Brinkhurst, R. O., 1981. A contribution to the taxonomy of the Tubificidae (Oligochaeta: Tubificidae). Proc. biol. Soc. Wash. 94: 1048–1067.
Brinkhurst, R. O. & B. G. M. Jamieson, 1971. Aquatic Oligochaeta of the World. Oliver & Boyd, Edingburgh, 860 pp.
Černosvitov, L., 1938. The Oligochaeta. In R. Washbourne & R. F. Jones (eds.), Report on the Percy Sladen Expedition to Lake Huleh. Ann. Mag. nat. Hist. 11: 535–549.
Gitay, A., 1965. The benthic fauna of Lake Tiberias. M.Sc. Thesis, The Hebrew University, 39 pp. (English summary).
Gitay, A., 1968. Preliminary data on the ecology of the level-bottom fauna of Lake Tiberias. Israel J. Zool. 17: 81–96.
Grigelis, A., 1980. Ecological studies of aquatic oligochaetes in the USSR. In R. O. Brinkhurst & D. G. Cook (eds.), Aquatic Oligochaete Biology. Plenum Press, New York: 225–240.
Pascar-Gluzman, C., 1981. A preliminary list of aquatic Oligochaeta from Israel: Naididae and Tubificidae. Israel J. Zool. 30: 230–232.
Por, F. D., 1968. The invertebrate zoobenthos of Lake Tiberias: 1. Qualitative aspects. Israel J. Zool. 17: 51–79.
Por, F. D. & D. Masry, 1968. Surival of a nematode and an oligochaete species in the anaerobic benthal of Lake Tiberias. Oikos 19: 388–391.
Rosa, D., 1893. Viaggio del Dr. E. Festa in Palestina, nel Libano e regioni vicine. II. Lumbricidi. Boll. Musei. Zool. Anat. comp. R. Univ. Torino 8(160): 1–14.
Stephenson, J., 1913. Aquatic Oligochaeta from Tiberias. J. Proc. asiat. Soc. Beng. 9: 53–56.

A taxonomic and faunistic survey of the marine Tubificidae and Enchytraeidae (Oligochaeta) of Italy. Introduction and preliminary results

Giuliano Bonomi[1] & Christer Erséus[2]

[1]C.N.R.-Instituto Italiano di Idrobiologia, Largo Vittorio Tonolli, 50/52, I-28048 Pallanza NO, Italy
[2] Swedish Museum of Natural History, Stockholm, and (postal address:) Department of Zoology, University of Gothenburg, Box 25059, S-400 31 Göteborg, Sweden

Keywords: aquatic Oligochaeta, Tubificidae, Enchytraeidae, zoogeography

Abstract

On the basis of material collected along the Ligurian, Tyrrhenian, Ionic and Adriatic coasts of Italy, a tentative list of 26 species (18 tubificids, 8 enchytraeids) of marine Oligochaeta is presented. Most of the species are new to science, and it can therefore be concluded that there is a very high diversity of oligochaetes in the Mediterranean Sea.

Introduction

During 1901–1917, Prof. U. Pierantoni, at Stazione Zoologica di Napoli, described seven oligochaete species (families Enchytraeidae and Tubificidae) from coarse sublittoral sand (so-called 'Amphioxus' sand) in the Bay of Naples. The publications by Pierantoni indeed represent pioneering work in the field of oligochaete research, as they are the first accounts of truly marine Oligochaeta. Despite this, it was almost 50 years after Pierantoni before oligochaete workers around the world began to report on marine representatives of this group to any notable extent.

At present, we know more than 200 species of marine Oligochaeta, and the number is very rapidly increasing. To date, however, very little is known about the Mediterranean members of the group. Hrabě (1971, 1975) has studied the marine Tubificidae of the Adriatic coast of Yugoslavia, but otherwise the literature contains only scattered records of Mediterranean Tubificidae and Enchytraeidae (Čejka, 1913; Hrabě, 1973; Giere, 1974; Erséus & Lasserre, 1976; Erséus, 1979a, b, c, 1980, 1981a, 1982).

To improve the knowledge of the Mediterranean Tubificidae and Enchytraeidae, we are undertaking a survey of the Italian representatives of these families. A sampling programme along the Ligurian, Tyrrhenian (including the Strait of Messina), Ionic and Adriatic coasts of the country was carried out during 1981. The material collected will be taxonomically treated by the junior author elsewhere, and the present paper will, therefore, only provide some preliminary notes on our findings.

Areas studied

As many oligochaetes are known from the subtidal zone of the sea, study areas in which marine research laboratories could provide a boat for subtidal sampling were selected. The facilities of the following institutes were utilized: (1) Acquario Comunale di Livorno, (2) Stazione Zoologica di Napoli, Napoli, (3) Istituto di Idrobiologia di Ganzirri, Universitá di Messina, Messina, (4) Stazione di Biologica Marina, Universitá di Lecce, Porto Cesareo, and (5) Laboratorio di Idrobiologia dello Istituto di Zoologica, Universitá di Bologna, Fano.

In addition to the sampling performed in the areas near these laboratories, some intertidal samples were collected in other parts of Italy (Fig. 1).

Hydrobiologia 115, 207–210 (1984).
© Dr W. Junk Publishers, Dordrecht.

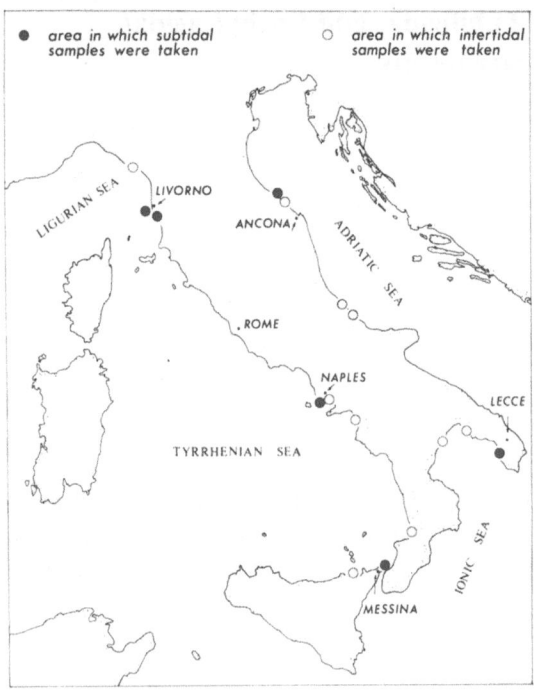

Fig. 1. Map of Italy showing the sampling areas of the present survey.

Methods

More than 50 qualitative samples of sediment were collected. Subtidal samples were taken from boats, using a grab or a dredge, or (at Secca della Meloria, at Livorno) a bucket handled by a diver. Littoral samples were taken digging by hand. The samples were repeatedly washed in sea water, and the organic material suspended in this water was collected in a 0.25 mm sieve. The sieved residues were immediately inspected under a dissection microscope, and all mature specimens of Oligochaeta were fixed in Bouin's fluid. Very few immature specimens were fixed, as taxonomic studies of oligochaetes have to be based upon sexually mature individuals. The material is now being studied at the Department of Zoology in Göteborg, Sweden, using permanent mounts of whole specimens and histological sections.

Preliminary results and discussion

Oligochaetes were found in slightly more than 50% of the samples collected. A list of the species found, as provisionally identified, is provided in Table 1. The high number of species found in the Tyrrhenian Sea reflects the richness of the material collected at Messina, Sicily. As many as 12 of the 18 species of Tubificidae found are new to science and will be described in due time. Among the Enchytraeidae also a majority of the species appear to be new. These enchytraeid species have to be further scrutinized, which will involve comparative studies on material from other geographical areas, before their true number and distinctions can be established.

Two of the new species of *Phallodrilus* lack an alimentary canal and must rely upon some kind of epidermal uptake of nutrients. Several gutless marine Tubificidae have been described from the Caribbean area and the Pacific coast of Australia (Giere, 1979; Erséus, 1979c, d, 1981b; Erséus & Baker, 1982), but up to now none has been known from the Mediterranean Sea.

Heterodrilus subtilis has not been reported since it was originally described by Pierantoni (1917; as *Clitellio subtilis*). The new material will therefore become the basis for a redescription of the species.

Adelodrilus pusillus Erséus, 1978, *Grania ovitheca* Erséus, and *Grania maricola* Southern, 1913 are all known from North-western Europe, but they have not been reported from the Mediterranean to date. *Grania macrochaeta* (Pierantoni, 1901) is most probably one of the four(?) other *Grania* species found, but this particular species is in great need of a more detailed redescription (Coates & Erséus, in prep.).

Some additional species of Italian marine Tubificidae known from the literature are listed in Table 2. At present, there is thus a total of about 35

Table 2. Species of Italian Tubificidae, other than those found in the present survey. They are all reported from the Bay of Naples (Tyrrhenian Sea) only.

Species	Source
Phallodrilus rectisetosus	Erséus (1979c)
Phallodrilus parthenopaeus	Pierantoni (1902)
Aktedrilus magnus	Erséus (1980)
Heterodrilus arenicolus	Pierantoni (1902)
Limnodriloides pierantonii	Erséus (1982)
Limnodriloides agnes	Erséus (1982)
Limnodriloides roseus (sp. dub.)	Pierantoni (1903a)
Limnodriloides pectinatus (sp. dub.)	Pierantoni (1903a)
Thalassodrilides gurwitschi	Erséus (1981a)

Table 1. Provisional list of species found in the present survey.

Species	Presence (+) in			
	Ligurian Sea	Tyrrhenian Sea	Ionic Sea	Adriatic Sea
Family TUBIFICIDAE				
Bathydrilus adriaticus		+		
Bathydrilus (?) sp.n.		+		
Phallodrilus sp.n. A		+		
Phallodrilus sp.n. B		+		
Phallodrilus sp.n. C		+		
Phallodrilus sp.n. D		+		
Adelodrilus pusillus		+		
Bacescuella mediterranea		+		
Aktedrilus monospermathecus				+
Aktedrilus sp.n.		+		+
Coralliodrilus sp.n. A		+		
Coralliodrilus sp.n. B		+		
Coralliodrilus sp.n. C	+	+		
Heterodrilus subtilis		+		
Heterodrilus sp.n. A			+	
Heterodrilus sp.n. B	+			
Limnodriloides appendiculatus		+	+	
Tubificoides sp.n.				+
Family ENCHYTRAEIDAE				
Grania ovitheca		+		
Grania maricola		+		
Grania spp. (4 different?)	+	+	+	
Marionina spp. (2 different?)	+	+	+	

Total number of species: approximately 26.

marine species of Tubificidae and Enchytraeidae known from Italian waters. However, the high number of new species found during the present investigation indicates that we still know only a fraction of the whole marine oligochaete fauna of Italy. It could also be noted that Hrabě (1971, 1973, 1975) has described three additional species of Tubificidae from the Adriatic coast of Yugoslavia (*Limnodriloides monothecus, L. maslinicensis,* and *Spiridion modricensis*); these can be expected to occur on the Italian coast as well.

The present study neither permits detailed conclusions on the ecological significance of marine oligochaetes in Italy, nor does it provide much information on what kind of parameters control the distribution of the different species. Only some general remarks can be made. There is, for instance, no doubt that the highest species diversity is found in coarse subtidal sands of the '*Amphioxus*-type'. In the Strait of Messina in particular (5–20 m depth) we found a very rich fauna of interstitial oligochaetes. Unfortunately, the well-known, old localities in the Bay of Naples where this type of sand is found have been drastically changed by eutrophication and other pollution, and we were unable to find any of Pierantoni's species there. The littoral samples contained very few oligochaetes, if compared with samples from similar habitats in, e.g., Northwestern Europe (cf. Giere, 1975). In particular, the very fine sands of the beaches of the Ionic and Adriatic coasts proved to be very poorly inhabited by oligochaetes.

Conclusions

The present survey, although preliminary, has demonstrated that a rich fauna of Tubificidae and Enchytraeidae is present in the marine waters of Italy. When taxonomically worked up, the materal collected will be an important contribution to the knowledge of the fauna of the Mediterranean Sea as a whole. However, to get a more complete picture of the distribution (in both ecological and geographical terms) of the oligochaetes along the Italian coasts, additional sampling and more detailed studies are needed.

210

Acknowledgements

The 'Consiglio Nazionale delle Ricerche' of Italy provided financial support for this study. We are also deeply indebted to Prof. S. Genovese, Prof. E. Di Domenico, Prof. M. Grasso, Prof. C. Piccinetti, and Prof. E. Sordi, for working facilities at their respective marine laboratories; to Prof. E. Vannini, Universitá di Bologna, for the interest and support he has shown for this project; and to Dr. S. Giacobbe, Mr. P. Martina, Dr. Patrizia Casali, and Dr. Ursula Salghetti-Drioli, for their valuable assistance in the field work.

References

Čejka, B., 1913. Litorea krumbachi n. spec. n. gen. — Ein Beitrag zur Systematik der Enchytraeiden. Zool. Anz. 42: 145–151.

Erséus, C., 1979a. Taxonomic revision of the marine genera Bathydrilus Cook and Macroseta Erséus (Oligochaeta, Tubificidae), with descriptions of six new species and subspecies. Zool. Scr. 8: 139–151.

Erséus, C., 1979b. Re-examination of the marine genus Spiridion Knöllner (Oligochaeta, Tubificidae). Sarsia 64: 183–187.

Erséus, C., 1979c. Taxonomic revision of the marine genus Phallodrilus Pierantoni (Oligochaeta, Tubificidae), with descriptions of thirteen new species. Zool. Scr. 8: 187–208.

Erséus, C., 1979d. Inanidrilus bulbosus gen. et sp.n., a new marine tubificid (Oligochaeta) from Florida (USA). Zool. Scr. 8: 209–210.

Erséus, C., 1980. Taxonomic studies on the marine genera Aktedrilus Knöllner and Bacescuella Hrabĕ (Oligochaeta, Tubificidae), with description of seven new species. Zool Scr. 9: 97–111.

Erséus, C., 1981a. Taxonomy of the marine genus Thalassodrilides (Oligochaeta, Tubificidae). Trans. am. microsc. Soc. 100: 333–344.

Erséus, C., 1981b. Taxonomic studies of Phallodrilinae (Oligochaeta, Tubificidae) from the Great Barrier Reef and the Comoro Islands, with descriptions of ten new species and one new genus. Zool. Scr. 10: 15–31.

Erséus, C., 1982. Taxonomic revision of the marine genus Limnodriloides (Oligochaeta, Tubificidae). Verh. naturwiss. Ver. Hamburg NF 25: 207–277.

Erséus, C. & H. R. Baker, 1982. New species of the gutless marine genus Inanidrilus (Oligochaeta, Tubificidae) from the Gulf of Mexico and Barbados. Can. J. Zool. 60: 3063–3068.

Erséus, C. & P. Lasserre, 1976. Taxonomic status and geographical variation of the marine enchytraeid genus Grania Southern (Oligochaeta). Zool. Scr. 5: 121–132.

Giere, O., 1974. Marionina istriae n.sp., ein neuer Enchytraeidae (Oligochaeta) aus dem mediterranen Hygropsammal. Helgoländer wiss. Meeresunters. 26: 359–369.

Giere, O., 1975. Population structure, food relations and ecological role of marine oligochaetes, with special reference to meiobenthic species. Mar. Biol. 31: 139–156.

Giere, O., 1979. Studies on marine Oligochaeta from Bermuda, with emphasis on new Phallodrilus species (Tubificidae). Cah. Biol. mar. 20: 301–314.

Hrabĕ, S., 1971. On new marine Tubificidae of the Adriatic Sea. Scr. Fac. Sci. nat. Univ. Brno 1: 215–226.

Hrabĕ, S., 1973. On a collection of Oligochaeta from various parts of Yugoslavia. Biol. Vĕst. 21: 39–50.

Hrabĕ, S., 1975. Second contribution to the knowledge of marine Tubificidae (Oligochaeta) from the Adriatic Sea. Vĕst. čsl. Spol. zool. 39: 111–119.

Pierantoni, U., 1901. Sopra una nuova specie di oligochete marino (Enchytraeus macrochaetus n.sp.). Monit. zool. ital. 12: 201–202.

Pierantoni, U., 1902. Due nuovi generi di Oligocheti marini rinvenuti nel Golfo di Napoli. Boll. Soc. Nat. Napoli 16: 113–117.

Pierantoni, U., 1903a. Altri nuovi oligocheti del Golfo di Napoli (Limnodriloides n.gen.). Boll. Soc. Nat. Napoli 17: 185–192.

Pierantoni, U., 1903b. Studii anatomici su Michaelsena macrochaeta Pierant. Mitt. zool. Stat. Neapel 16: 409–444.

Pierantoni, U., 1917. Heterodrilus arenicolus Pierant. e su di una nuova specie del genera Clitellio. Boll. Soc. Nat. Napoli 29: 82–91.

Ecology and importance of Oligochaeta in the biocenosis of zoobenthos in lakes of the National Park of the Lithuanian SSR

Antanas Grigelis
Institute of Zoology and Parasitology, Lithuanian Academy of Sciences, 232021 Vilnius, Lithuanian S.S.R., U.S.S.R.

Keywords: aquatic Oligochaeta, distribution according to sediment type

Abstract

This paper aims to describe the ecology and determine the distribution of Oligochaeta according to sediment type, depth and other environmental factors in 21 lakes of the National Park of the Lithuanian SSR. Diagrams are included showing the distribution of the lakes according to the index of density and biomass of Oligochaeta in the trophic structure of biocenosis of the zoobenthos. Light is thrown on the problem of sediment composition as a significant factor in the distribution and development of aquatic Oligochaeta.

There are 83 lakes with a total area of 3680.7 ha in the territory of the National Park of the Lithuanian SSR. Some of the lakes, such as Gavys, Gavaitis, Vajuonis, Kretuonas and Ūsiai, have been excluded from territory of the National Park as they are more intensively used for recreational purposes.

The lakes of the National Park are located in the hills in a variety of topographical situations. They differ in size, depth, consistency of bottom, length, width and orientation of the principal axis in relation to the direction of the prevailing Western winds (Grigelis *et al.,* 1975). The study of the benthic fauna of these lakes during 1976–1980 showed Oligochaeta to be a leading group of the zoobenthos in sublittoral and profundal zones of some of the lakes. *Potamothrix hammoniensis* formed up to 79.6% of the biomass in certain sections or even in the whole lakes. This species is also widely distributed in the freshwater basins of the Baltic Sea area. The investigations carried out during 1976–1980 established that aquatic Oligochaeta are dominant. At the same time their species composition, distribution, density and biomass were also determined.

The investigated lakes of the National Park have been divided into three groups, according to the index of dominance of aquatic Oligochaeta:

Group 1. Density of the Oligochaeta is from 53 to 70.8% (Fig. 1a), and the biomass from 55.5 to 79.6% (Fig. 1b). The lakes in this group (Lūšiai, Šakarvai, Ūkojas Alksnas) differ in depth and character of the bottom.

Group 2. Density 25 to 47%; biomass 14.7 to 45%.

Group 3. Density 8.7 to 12.1%.

The distribution, species composition and development of aquatic Oligochaeta depend on many factors such as water temperature, chemical and physical properties of the water (Grigelis *et al.,* 1981; Žukaitė, 1980), sediments, bottom microflora and vegetation cover. Verdonschot (1981) showed that in the Delta area of the south-western Netherlands the principal factors governing the distribution of aquatic Oligochaeta were salinity and type of substrate. In the lakes I investigated, there are no differences in the salinity of the water. Apart from bottom substrate, the other factors show little variation and are of minor importance.

The section of Lake Lūšiai which was studied was not extensive but showed considerable variation in sediment type and thermal conditions. In the section starting at 320 m from the shore and extending to a point at a depth of 37 m, I found all types of

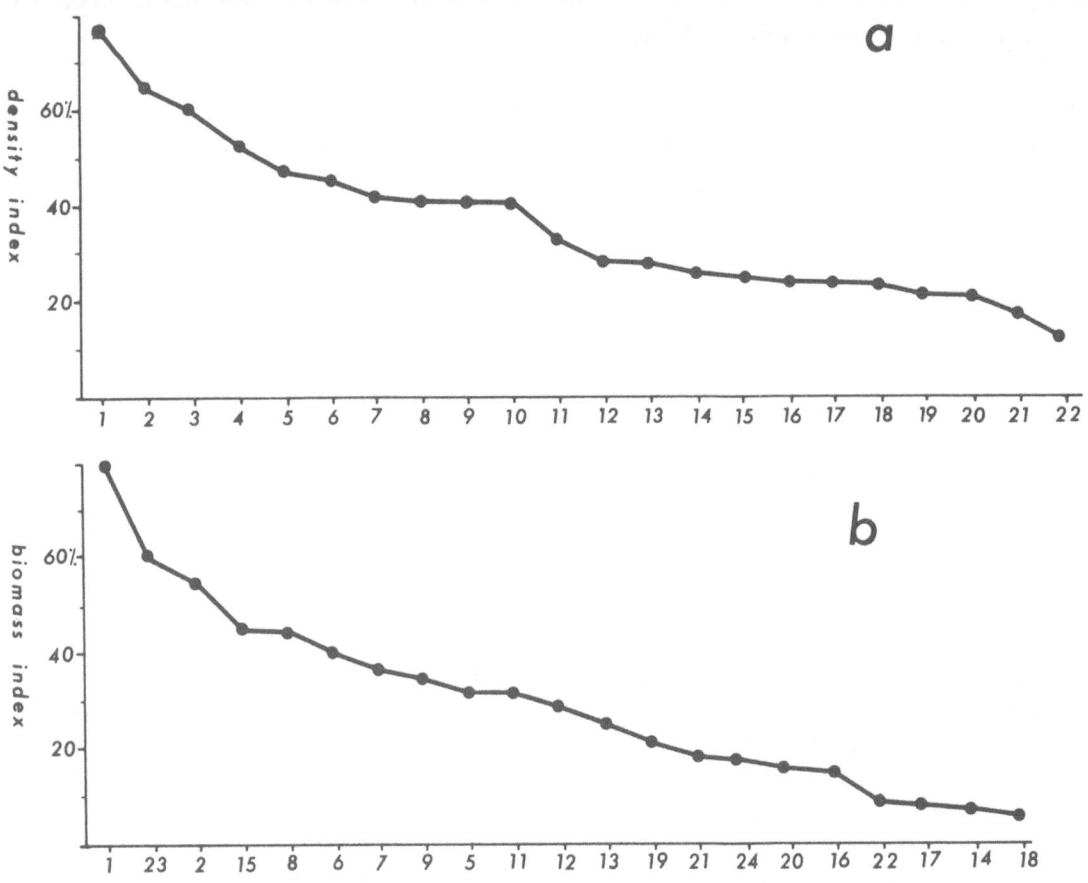

Fig. 1. Distribution of the lakes of the National Park of the Lithuanian SSR, according to the density index of Oligochaeta (a) and to the biomass index (b).
Lakes: 1) Šakarvai, 2) Ūkojas, 3) Alksnas, 4) Lūšiai, 5) Linkmenas, 6) Asalnas, 7) Taramas, 8) Gavys, 9) Lūšykštis, 10) Dringis, 11) Baluošas, 12) Gruodiškis, 13) Usiai, 14) Kiaunas, 15) Dringikštis, 16) Pakasas, 17) Dringis (bay), 18) Almajas, 19) Tauragnas, 20) Žeimenis, 21) Asalnaitis, 22) Utenas, 23) Dringis (south), 24) Vajuonis.

bottom: sand to a depth of 1–1.5 m, grey silt from 1.5–2 to 6 m, grey silt with gravel from 6 to 10 m, light grey silt from 10 to 24 m, carbonate silt from 24 to 27 m, and dark silt at depths greater than 27 m (Fig. 2). Concretions of iron were found at a depth of 15 m in the channel connecting lakes Lūšiai and Asalnas. These concretions were in the form of small balls with a diameter of 2–4 mm.

Along the shore of Lake Lūšiai down to a depth of 1.5 m there is a sand biotope, where the dominant species were *Lumbriculus variegatus* and *Psammoryctides barbatus*. Their density ranged from 160 to 1 280 ind · m^{-2}, with a biomass from 1.72 to 5.42 g m^{-2}; that is, a density of 80–90% and a biomass of 91.5–94.1%. *Stylaria lacustris* and *Ophidonais serpentina* were in second place.

The biotope was more varied at greater depth, in the transitional bottom where sand changes to silt. Here the most numerous species in terms of population density was *P. hammoniensis,* although there were also large numbers of *L. variegatus, S. lacustris* and other species.

At a depth of 6–10 m, in the sub-littoral, there was grey silt, where only *P. hammoniensis* lived.

Grey silt with a large part of carbonate was found at a depth of 10–24 m. There were spots of brown silt and some of blue clay with brown silt.

In the biotope of carbonate silt at a depth of 15 m, *P. barbatus* was the predominant species. The population density was 240 ind · m^{-2}, corresponding to 2.8 g m^{-2} or 51.5% of the total biomass of zoobenthos. The density of *P. hammoniensis* was 160 ind ·

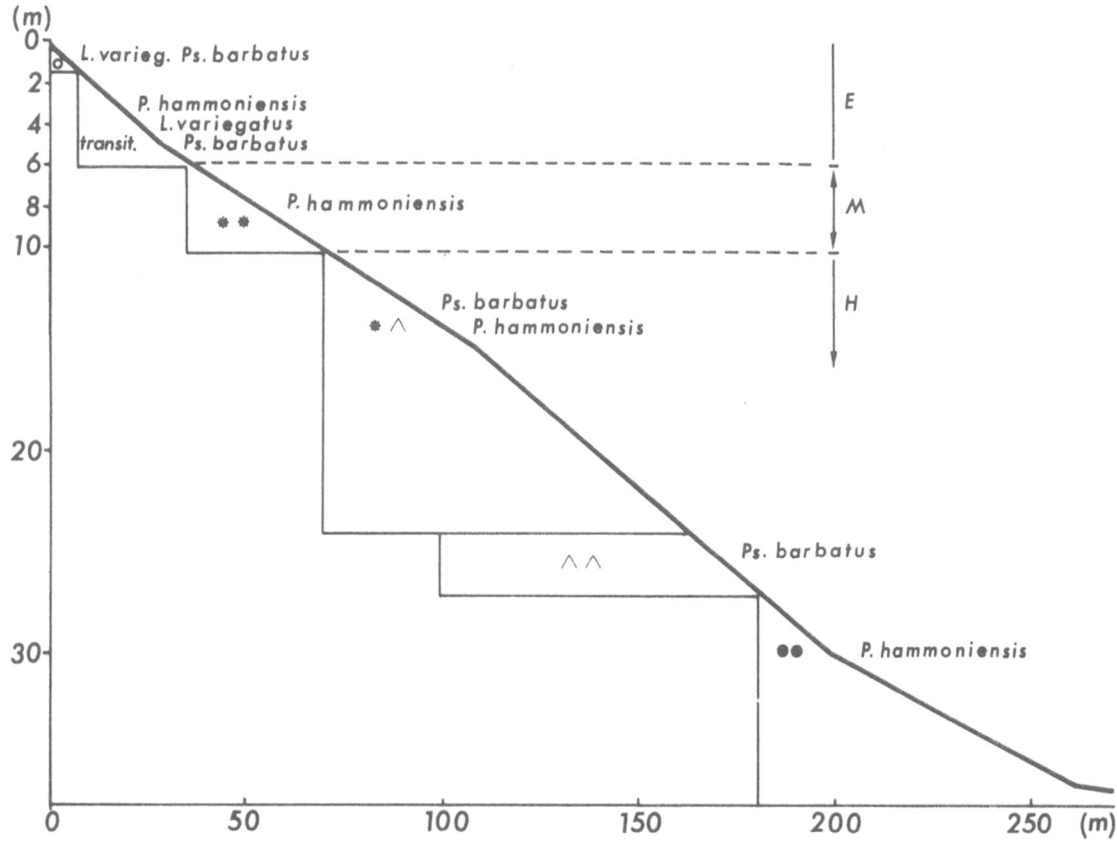

Fig. 2. Distribution of Oligochaeta according to sediment type in Lake Lūšiai.
E – epilimnion, M – metalimnion, H – hypolimnion, O – sand, ✶✶ – gray silt, ✶∧ – light gray silt rich in carbonate, ∧∧ – carbonate silt, ●● – dark silt.

m⁻² (13.8%), corresponding to 0.24 g m⁻² (4.4%).

Oligochaeta were absent in the biotope of brown silt. The biotope of blue clay with brown silt is found at a depth of 18–19 m, where *P. barbatus* was at 280 ind · m⁻² (35% of the total density of zoobenthos), with a biomass of 0.28 g (48%).

P. barbatus (85 ind · m⁻², 1.3 g biomass) and *P. hammoniensis* (140 ind · m⁻², 0.56 g biomass) were encountered deeper in the grey silt, at a depth of 23 m. At depths greater than 27 m there was only *P. hammoniensis* (290 ind · m⁻², biomass 2 g (90%)).

In the hypolimnion of Lake Lūšiai and Lake Šakarvai, the thermal conditions were not so favourable for Oligochaeta to live and reproduce. The temperature of the water near the bottom ranges from 6.4 to 6.6 ° C (Žukaitė, 1980). The data of Archipova (1980) showed that *P. hammoniensis*

begins to breed when the water temperature reaches 6–7 ° C.

P. hammoniensis grew slowly at the lower temperature and did not reach the size of individuals living in the metalimnium zone in lakes Lūšiai and Šakarvai and in the channel of Lake Asalnai.

The maximum density of *P. hammoniensis* (560 ind · m⁻², with 4.92 g biomass) was in the channel of Lake Asalnai during 1976–1980. In the deep zone of Lake Lūšiai, where warm-stenotherm and salmon (Salmonidae) fish were practically absent, *P. hammoniensis* numbered 680 ind · m⁻², with 3.68 g biomass.

214

References

Archipova, N. R., 1980. K biologii Potamothrix hammoniensis (Mich.) (Oligochaeta, Tubificidae) Ribinskoho vodoxranilishcha. Trudȳ Inst. Biol. vnutriennix vod 44: 14–27.

Grigelis, A., G. Balkuvienė, J. Banionienė, O. Nainaitė, A. Orlova, V. Sukackas & E. Žukaitė, 1975. Lietuvos ežerų hidrobiologiniai tyrimai. (Hydrobiological research on Lithuanian lakes.) 302 pp.

Grigelis, A., R. Lenkaitis, O. Nainaitė & E. Žukaitė, 1981.

Peculiarities of distribution of cold-stenotherm hydrobionts in lakes of the National Park of the Lithuanian SSR. Verh. int. Ver. Limnol. 21: 501–503.

Verdonschot, P., 1981. Some notes on the ecology of aquatic oligochaetes in the Delta Region of the Netherlands. Arch. Hydrobiol. 92: 53–70.

Žukaitė, E. I., 1980. Termichieskii rezhim ozior Lūšia i Šakarvai v pieriod lietniei termichiescoi stagnazii v 1978 g. Trudȳ Akad. Nauk Litovskoi SSR (Ser. B) 5: 83–91.

The distribution of aquatic oligochaetes in the fenland area of N.W. Overijssel (The Netherlands)

Piet F. M. Verdonschot

Provincial Water Authority, Department of Watershed Management. P.O. Box 73, NL-8000 AB Zwolle, The Netherlands

Keywords: aquatic Oligochaeta, fenland, cluster analysis, ecological groups

Abstract

A synoptic study of the aquatic ecosystems in the Dutch province of Overijssel revealed the presence of 33 oligochaete taxa in the fendland area of N.W. Overijssel. The material was collected at 101 sampling sites in 1981. Oligochaetes were grouped by means of a normal and inverse cluster analysis. Although most oligochaetes are quite ubiquitous, differences in occurrence and abundance were observed and related to minor differences in environmental parameters.

Introduction

In 1981 the author started a descriptive investigation of surface waters in the province Overijssel (The Netherlands). The primary objective of this research is to provide for an ecological background of a regional classification of aquatic ecosystems in the province of Overijssel, in particular for provincial policy and management purposes. The investigation comprises studies of microphyte, macrophyte and macrofaunal communities. Preliminary results of the distribution of oligochaete worms in a local fenland area (N.W. Overijssel) are presented here.

Study area

The landscape of N.W. Overijssel consists of a large area of undrained fen, situated in a large valley which is intersected by a finely meshed pattern of ditches, canals, peat pits and lakes. This structure of the landscape is the result of human activity. Peat extraction especially, which started in the Middle Ages, has shaped the landscape but this is now a thing of the past. During the last few decades many old peat pits and ditches have gradually been filled in again by hydrosere vegetation, after which some areas have been exploited as reed or hayland or have been used for extensive cattle farming. Other parts have developed into marshy woodlands.

The sampling sites are located in ditches, canals and peat pits (Fig. 1). The lakes will be studied in the near future.

The hydrological situation is mainly defined by water inlet from adjacent polders, the higher diluvial grounds and the Frisian lakes, and water outlet near the village of Vollenhove. The watertable remains fairly constant. The fenland acts like a basin in which the residence time of the water fluctuates between several months and a year. The water is being progressively polluted by a sewage treatment plant near the village of Steenwijk, an intensive water recreation and by the poor quality of the water being let in in dry periods from the Frisian lakes (in origin IJssel/Rhine water quality). The chemical composition of the water can in general be described as hypertrophic. Only some local isolated waters have a better quality.

Hydrobiologia 115, 215–222 (1984).

216

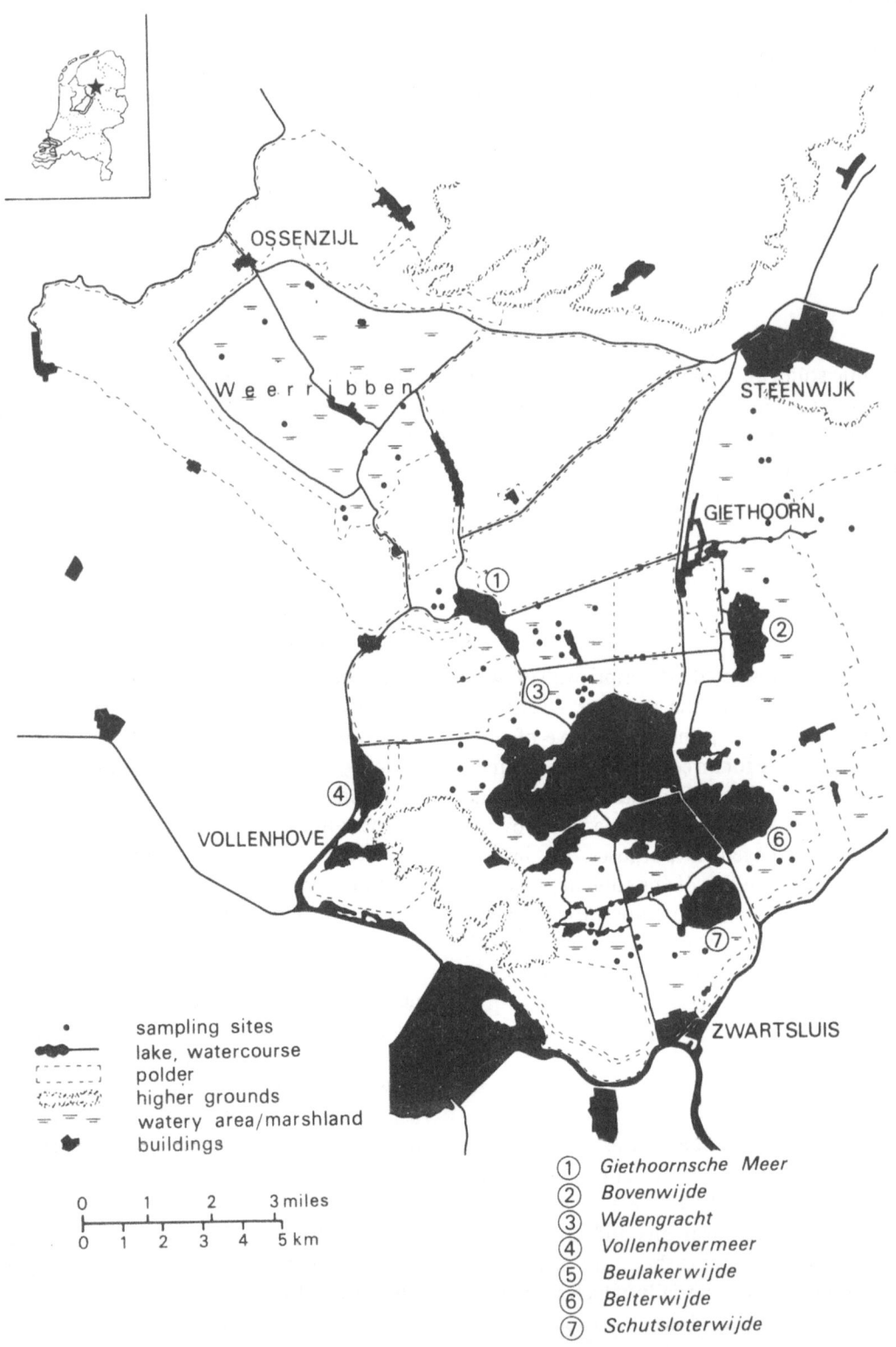

sampling sites
lake, watercourse
polder
higher grounds
watery area/marshland
buildings

0   1   2   3 miles
0 1 2 3 4 5 km

① Giethoornsche Meer
② Bovenwijde
③ Walengracht
④ Vollenhovermeer
⑤ Beulakerwijde
⑥ Belterwijde
⑦ Schutsloterwijde

Fig. 1. The fenland area of N.W. Overijssel. Sampling sites are indicated (·).

Materials and methods

Data presented in this paper were obtained from 101 sites, sampled between May and September 1981. The procedure consisted of sampling the floating and submerged vegetation with a standard net (mesh-width 0.5 mm) over about five metres (distribution over the main micro-habitats), taking separate bottom samples with this standard net (over one metre) near the bank, and taking five samples with an Ekman-Birge dredge at deeper places.

The oligochaetes were separated live from the collected material in the laboratory, conserved in formalin (4%) and cleared (for identification) in a mixture of benzylbenzoate and propanol-2.

Water samples for chemical analyses were taken at the water surface. Bottom types were classified by field observation. A bottom index was calculated by using the following classes: 1 = peat, 2 = silty peat, 3 = peaty silt, 4 = silt or sandy silt, 5 = black silt. At the time of sampling the higher aquatic plants were investigated as well and algal samples were taken. The quantitative data of the subsam-

ples were reduced to nine classes of abundance.

These reduced data were processed by hierarchical agglomerative cluster analysis according to Orloci (1967). Both sampling sites and oligochaete species were classified. These normal and inverse classifications were processed further by nodal analysis. The concepts of constancy, fidelity and average concentration of abundance for the nodes (Boesch, 1977) were used.

Results

In total 33 oligochaete species or higher taxa were identified. The sites at which oligochaetes were found are given in Fig. 1. Cluster analysis was used as a tool to simplify the procedure of shuffling the two-way table.

After studying the dendrograms together with the original data matrix, 'reasonable' groups were determined arbitrarily. The hierarchical relationships for these site and species groups are given in Fig. 2. These species groups are given in Table 1.

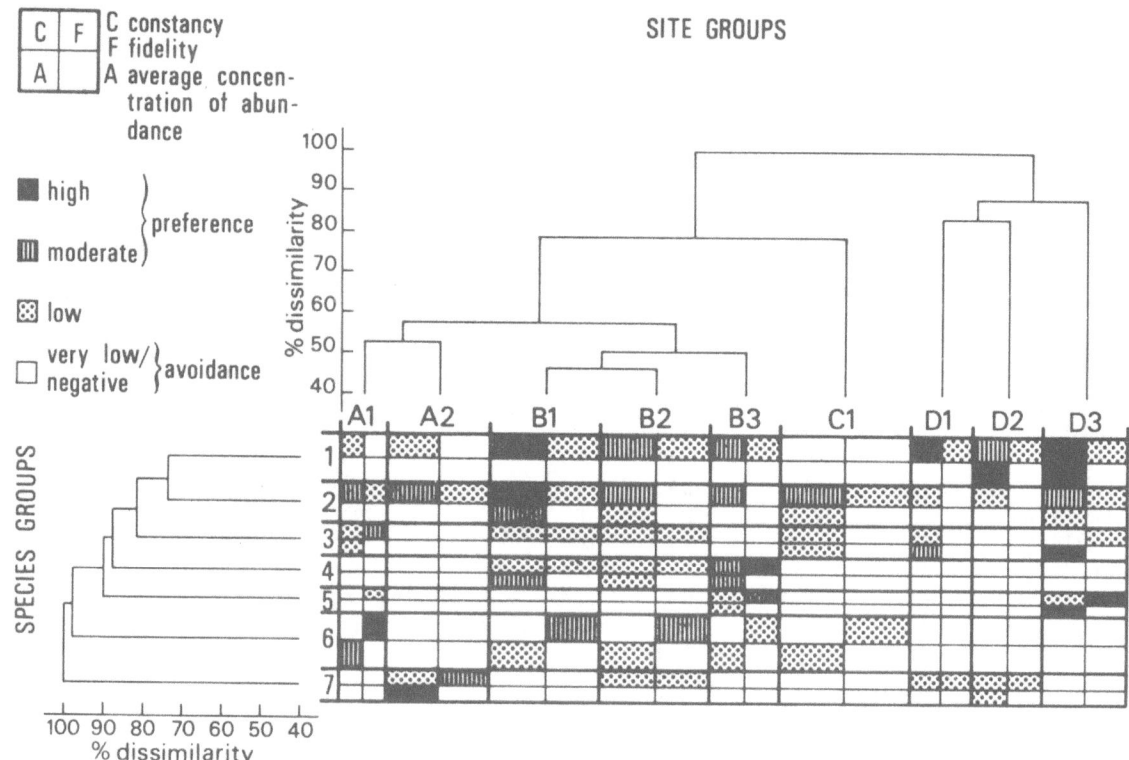

Fig. 2. Nodal constancy, nodal fidelity and average concentration of abundance for the nodes in a two-way table for species groups (explanation, see Table 1) versus site groups (see Table 2). Hierarchical relationships are indicated.

Table 1. Species groups.

Group 1:
Limnodrilus hoffmeisteri
Potamothrix hammoniensis
Psammoryctides barbatus
Limnodrilus profundicola
Psammoryctides albicola
Ilyodrilus templetoni

Group 2:
Stylaria lacustris
Lumbriculus variegatus
Nais variabilis
Aulodrilus pluriseta
Chaetogaster diaphanus

Group 3:
Ophidonais serpentina
Limnodrilus claparedianus
Nais communis
Ripistes parasita

Group 4:
Branchiura sowerbyi
Peloscolex ferox
Nais pardalis
Nais pseudoptusa

Group 5:
Limnodrilus udekemianus
Lumbriculidae
Enchytraeidae

Group 6:
Nais barbata
Chaetogaster diastrophus
Tubifex ignotus
Potamothrix moldaviensis
Uncinais uncinata
Tubifex tubifex
cf. *Rhyacodrilus coccineus*

Group 7:
Dero dorsalis
Dero digitata
Dero sp.
Haemonais waldvogeli

Constancy, fidelity and average concentration of abundance are presented in the nodal diagram (Fig. 2). In the nodal diagram presented the width of the rows and columns is proportional to the number of entities in the respective site and species groups. For an environmental characterization of the site groups the average of the most important parameters was calculated and presented in Table 2.

Discussion

All the oligochaetes collected can occur together (Fig. 2) and thus show overlap in ecological tolerance. Despite this overlap, oligochaetes are classed in different groups due to differences in frequency of occurrence and abundance.

Species group 1 (Table 1) consists of the more common, truly benthic tubificids (mainly found in bottom samples). Especially *Limnodrilus hoffmeisteri* and *Potamothrix hammoniensis* were found throughout the area. The suggestion of Brinkhurst (1964) and Milbrink (1973), that *P. hammoniensis* rather than *Tubifex tubifex* (note its absence) is characteristic of eutrophic water, is supported.

Species group 1 much prefers site groups D2 and

Table 2. Mean values of some environmental parameters of the site groups.

Site group	A1	A2	B1	B2	B3	C1	D1	D2	D3
No. of sampling sites	6	13	14	14	9	17	8	9	11
Depth cm	50	76	86	81	69	52	89	62	72
pH	7.4	7.4	7.6	7.6	7.5	7.3	7.7	7.4	7.5
O_2%	79.7	136.8	148.5	128.8	130.7	129.8	142.5	131.2	144.9
O_2 mg l^{-1}	7.4	13.2	13.4	12.6	12.3	12.1	15.5	12.7	14.7
Ortho-P mg l^{-1}	0.05	0.12	0.04	0.04	0.03	0.08	0.08	0.32	0.03
Tot.-P mg l^{-1}	0.17	0.12	0.26	0.19	0.16	0.27	0.11	0.77	0.14
NH_4^+ mg l^{-1}	0.43	0.89	0.26	0.19	0.70	0.49	0.34	1.62	0.44
NO_3^- mg l^{-1}	0.16	0.14	0.13	0.16	0.26	0.13	0.12	0.09	0.16
T.O.C. mg l^{-1}	14.0	17.6	18.7	17.8	16.8	20.9	14.5	17.0	15.0 "
Conductivity μS	453	403	398	378	436	292	436	386	402
Ca^{2+} mg l^{-1}	104.2	63.9	59.1	60.8	57.2	56.5	75.3	57.3	51.5
Cl^- mg l^{-1}	48.0	47.0	53.4	58.8	51.4	34.4	66.9	49.2	50.4
Fe^{3+} mg l^{-1}	0.49	0.89	0.41	1.07	0.40	0.23	0.17	0.80	0.80
Transparency cm	42	55	36	55	34	42	71	34	46
Algal conc. $\times 10^4$	0.30*	1.37	1.87	2.20	1.83	1.69	1.21	1.76	3.05
Submerged veg. %	20.2	16.3	11.0	9.1	1.6	29.3	1.9	8.3	8.6
Floating veg. %	21.7	40.3	35.4	34.0	16.9	41.1	50.0	15.8	15.1
Bottom index**	1.5	1.5	2.7	2.5	2.7	1.3	2.6	3.8	3.4

* One observation only.

**The bottom index is explained in the text.

Table 3. Significant correlation coefficients and ranges of matching environmental parameters.

Parameter Species	O_2 mg l^{-1}	ortho-P mg l^{-1}	Tot.-P mg l^{-1}	NH_4^+ mg l^{-1}	T.O.C. mg l^{-1}	Fe^{3+} mg l^{-1}	depth cm	transparency cm
Psammoryctides barbatus	0.40 5.5–20.0	*	*	*	*	*	*	*
Limnodrilus profundicola	*	*	*	*	*	*	*	-0.42 5–110
Psammoryctides albicola	*	*	0.75 0.05–1.41	*	*	*	-0.61 55–140	-0.60 20–90
Ilyodrilus templetoni	*	*	-0.80 0.05–0.29	*	*	*	*	*
Stylaria lacustris	-0.23 2.5–20.0	*	0.29 0.02–0.75	*	*	*	*	-0.38 5–140
Aulodrilus pluriseta	*	0.87 0.01–0.33	*	*	*	*	*	*
Chaetogaster diaphanus	*	*	*	*	*	-0.84 0.15–8.40	-0.97 40–140	*
Ophidonais serpentina	*	*	*	*	-0.55 13–20	*	-0.39 40–120	*
Limnodrilus claparedianus	*	*	0.85 0.05–0.59	*	*	*	*	*
Peloscolex ferox	*	*	*	*	*	0.63 0.00–8.40	*	-0.64 10–65
Limnodrilus udekemianus	0.83 5.7–18.0	*	*	*	*	*	*	*
Dero sp.	*	*	*	0.94 0.32–3.10	*	*	*	*

* No significant correlation.

D3 (Fig. 2). Site group D2 is characterized by high average ortho-phosphate, total phosphate and ammonium concentrations (Table 2). The floating vegetation in both site groups is scarce. Site group D3 is further characterized by a high average algal concentration. The bottom in both site groups consists of a silt layer upon a peat or sand subsoil. On the other hand species goup 1 avoids site group C. This site group is characterized by more isolated shallow waters with a dense submerged vegetation, a high average transparency and total organic carbon content, and a low average conductivity. The bottom consists of peat with a low silt content. The frequency of occurrence and abundance of group 1 is directly related to the silt content of the bottom.

A linear correlation coefficient is computed to determine the correlation between an environmental parameter and the occurrence of a single species. This coefficient is always quoted with the restriction of the range of variability of the parameter used. Only the significant correlations (two-sided reliability of 95%) are presented (Table 3).

In this way the unity of species group 1 is supported by the positive correlation of *Psammoryctides albicola* with total phosphate concentration and the negative correlation of this species and of *Limnodrilus profundicola* with transparency. The negative correlation of *Limnodrilus profundicola* with oxygen, found by Dumnicka & Pasternak (1978), supports our data, although Pfannkuche, Jelinek & Hartwig (1973) indicated a high oxygen demand. The preference of *Psammoryctides albicola* for shallow waters, also reported by Chekanovskaya (1962) and Laakso (1969), is in contrast with the ecology of species group 1. The preference of *Psammoryctides barbatus* for well oxygenated waters is supported by earlier findings (Brinkhurst, 1964; Dzwillo, 1966; Särkka, 1969) but does not appear specific for species group 1. This is also true for *Ilyodrilus templetoni*, although Pfannkuche (1977) reported finding this species in eutrophic waters, and Hiltunen (1967) called it saprophilous, together with *Limnodrilus profundicola*.

Species group 2 consists of the more common species. *Stylaria lacustris* and *Lumbriculus variegatus* appeared at almost all sites. Chekanovskaya species. *Stylaria lacustris* and Lumbriculus variegatus appeared at almost all sites. Chekanovskaya (1962) called *L. variegatus* characteristic of marshy waters. Species group 2 prefers site group B1 and is

clearly present at site groups C1 and D3. Site group B1 is characterized by a high average depth and a low average transparency. The site groups C1 and D3, described above, have no average levels of environmental parameters in common. *Stylaria lacustris*, *Lumbriculus variegatus* and *Nais variabilis* feed on algae (Moore, 1979) and occur in waters with fairly well developed vegetation or with an algal bloom. Species group 2 seems to be independent of the trophic state of the water, although *Stylaria lacustris* prefers waters with a high total phosphate content and a low oxygen content and transparency (also reported by Dumnicka & Pasternak, 1978). The preference of *Aulodrilus pluriseta* for higher ortho-phosphate concentrations corresponds to a similar preference in *Stylaria lacustris*. However, Dumnicka & Pasternak (1978) reported a negative correlation of *Aulodrilus pluriseta* with ortho-phosphate concentration. This latter species prefers peaty bottoms (Brinkhurst, 1964). The negative correlation of *Chaetogaster diaphanus*, another member of species group 2, with iron content and depth is based on six observations only.

Species group 3 is rather scattered over several site groups but slightly prefers group A1 and is abundant at some sites of the site groups D1 and D3. Site group A1 is characterized by shallow waters with a dense submerged vegetation. This site group is further characterized by a high average calcium content and transparency and a relatively low average total organic carbon content. The latter feature is found also in site groups D1 and D3. *Ophidonais serpentina*, the most abundant species of this group, prefers shallow waters with low concentrations of total organic carbon. The preference of *Limnodrilus claparedianus*, another member of species group 3, for high total phosphate concentrations (also reported by Pfannkuche, 1977) is in contrast with the ecology of species group 3. *Nais communis* and *Ripistes parasita*, although members of species group 3, do not occur at site group A1 and are therefore probably misclassified, also because of a low number of observations.

Species group 4 prefers site groups B1 and B3 and is clearly present at site group B2. Site group B1 has already been described. Site group B3 is characterized by a low average percentage of submerged and floating vegetation. This group is further characterized by a high average nitrate concentration. This site group consists of wide, eutrophicated canals, subjected to a more or less intensive boating (and other water recreation) activities causing turbulence. The bottom consists of peat with a high silt content. *Branchiura sowerbyi*, especially abundant at site group B3, lives in the silt of standing waters (Chekanovskaya, 1962). The preference of *Peloscolex ferox* for waters of low transparency agrees with its occurrence in species group 4. However Brinkhurst (1964), Dzwillo (1966) and Särkka (1969) reported a preference of *P. ferox* for less productive lakes. A preference of *P. ferox* for silty bottoms is described by Della Croce (1955). *Nais pardalis,* also a member of species group 4, was earlier found in the hypertrophic lake Veluwe. Brinkhurst & Kennedy (1962) reported the occurrence of *Nais pseudoptusa* (species group 4) in silty bottoms, which supports our data.

Species groups 5 and 6 consist of species only found once or twice or contain whole higher taxa. The preference of species group 5 for site group D3 emerges as the result of some scattered findings. The preference of *Limnodrilus udekemianus* (group 5) for water with high oxygen content does not agree with observations by Pfannkuche (1977) and Dzwillo (1966), nor with the occurrence of this species in species group 5 with a preference for site group D3.

Species group 7 consists of oligochaetes (mainly the genus *Dero*) living on the substrate or free swimming. This species group much prefers site group A2 and is clearly present at site group D2. Site group A2 is characterized by relatively deep waters with a fairly well developed submerged vegetation and a highly developed floating vegetation. The bottom consists of peat with a low silt content.

Summary

In 1981, 101 sampling sites in the fenland area of N.W. Overijssel (The Netherlands) were sampled. The chemical composition of the water in these fenlands showed only slight differences in trophic, saprobic and ionic states. Although almost all the oligochaetes collected can occur together, it appeared that there were significant differences in the frequency of occurrence and abundance of the 33 oligochaete taxa.

The differences in occurrence and abundance of oligochaetes were described by normal and inverse cluster analysis. In this way seven oligochaete groups could be distinguished. Five groups out of seven were related to minor differences in environmental parameters.

– Group 1, comprising more common benthic tubificids was more abundant in waters with high phosphate concentrations, a poorly developed floating vegetation and a bottom with a thick silt layer. This group avoided sites with really clean peat bottoms. The frequency of occurrence and abundance of this group is directly related to the silt content of the bottom.
– Group 2, composed of more common species bound to the vegetation or with a more or less planktonic life, seemed to be independent of the trophic state of the water.
– Group 3 was best developed in waters with a low total organic carbon content due to the preference of just one species (*Ophidonais serpentina*) for this environment.
– Group 4 preferred wide, eutrophicated turbulent (due to water recreation) waters without vegetation but with a silty bottom.
– Group 7 was restricted to deep waters with a fairly well developed water vegetation and a relatively clean peat bottom.

Acknowledgements

The author is much indebted to J. Janse, D. Monnikendam and G. Willemsen for assistance in the field and the laboratory. A. Vlasblom (Delta Institute for Hydrobiological Research) and S. Bakker advised on statistical analysis. J. Gardeniers and J. Laseur reviewed the manuscript and provided many constructive suggestions. M. Smies improved the English text. C. Faasen prepared the figures and C. Jansen typed the manuscript.

References

Boesch, D. F., 1977. Application of numerical classification in ecological investigations of water pollution. Spec. scient. Rep. 77, Virginia Inst. mar. Sci., 114 pp.

Brinkhurst, R. O., 1964. Observations on the biology of lake dwelling Tubificidae (Oligochaeta). Arch. Hydrobiol. 60: 385–418.

Brinkhurst, R. O. & C. R. Kennedy, 1962. Some aquatic Oligochaeta from the Isle of Man with special reference to the Silver Burn Estuary. Arch. Hydrobiol. 58: 367–376.

222

Chekanovskaya, O. V., 1962. The aquatic oligochaeta fauna of the U.S.S.R. Opred. Faune SSSR 78: 1-411 (in Russian).

Della Croce, N., 1955. The conditions of sedimentation and their relationship with Oligochaeta populations of Lake Maggiore. Mem. Ist. ital. Idrobiol., Suppl. 8: 39-62.

Dumnicka, E. & K. Pasternak, 1978. The influence of physico-chemical properties of water and bottom sediments in the River Nida on the distribution and numbers of Oligochaeta. Acta hydrobiol., Kraków 20: 215-232.

Dzwillo, M., 1966. Untersuchungen über die Zusammensetzung der Tubificidenfauna im Hamburger Hafen. Abh. Verh. naturw. Ver. Hamburg (NF) 11: 101-116.

Hiltunen, J. K., 1967. Some oligochaetes from Lake Michigan. Trans. am. microsc. Soc. 86: 433-454.

Laakso, M., 1969. Oligochaeta from brackish water near Tvärmine, south-west Finnland. Ann. zool. fenn. 6: 98-111.

Milbrink, G., 1973. On the use of indicator communities of Tubificidae and some Lumbriculidae in the assessment of water pollution in Swedish lakes. Zoon 1: 125-139.

Moore, J. W., 1979. Influence of food availability and other factors on the composition, structure and density on a subarctic population of benthic invertebrates. Hydrobiologia 62: 215-223.

Orloci, L., 1967. An agglomerative method for the classification of plant communities. J. Ecol. 55: 193-206.

Pfannkuche, O., 1977. Ökologische und systematische Untersuchungen an naidomorphen Oligochaeten brackiger und limnischer Biotope. Diss. Univ. Hamburg. 138 pp.

Pfannkuche, O., H. Jelinek & E. Hartwig, 1975. Zur Fauna eines Süsswasserwattes im Elbe-Aestuar. Arch. Hydrobiol. 76: 475-498.

Särkka, J., 1969. The bottom fauna at the mouth of the river Kokemäenjoki, southwestern Finland. Ann. zool. fenn. 6: 275-288.

Oligochaeta from profundal zones of Spanish reservoirs

E. Martinez-Ansemil[1] & N. Prat[2]
[1] Colegio Universitario de Orense, Universidad de Santiago, C. General Franco 35, Orense, Spain
[2] Departamento d'Ecologia, Facultad de Biologia, Universidad de Barcelona, Av. Diagonal 637-645, Barcelona-28, Spain

Keywords: aquatic Oligochaeta, biogeography, ecology

Abstract

The profundal benthic fauna of 63 Spanish reservoirs was studied during 1972–75. Samples were taken in the deepest part of each reservoir with a modified Van Veen grab. The fauna was dominated by tubificid worms and chironomids. Sixteen species of Oligochaeta were found, two of them new for the Spanish fauna (*Potamothrix heuscheri* and *Haber pyrenaicus*). This was the first time *H. pyrenaicus* was found since the species was described in 1974 from the Pyrenees. Four species were frequent in the reservoirs (*Limnodrilus hoffmeisteri, Tubifex tubifex, L. claparedeianus* and *Dero digitata*) but only the first two were abundant as mature individuals. Another interesting fact is the high abundance of immature animals in all samples, accounting for more than 80% of the total individuals examined. The geographical distribution of all species in Spain and their relationship with the typology of Spanish reservoirs are discussed. There seems to be a correlation between the throphic status of the reservoirs and the relative proportion of *L. hoffmeisteri* as compared to *T. tubifex*, the number of individuals of the first species increasing with water eutrophication.

Introduction

During 1972 to 1975 a survey of more than 100 reservoirs all over Spain was conducted by a team from the Department of Ecology of the University of Barcelona. The physical and chemical features of the reservoirs and the composition of their phyto- and zooplankton have been well publicised in a series of papers (Estrada, 1975; Margalef, 1975; Planas, 1975; Margalef *et al.* 1976; Armengol, 1978; Margalef & Mir, 1979). During the general survey, benthic samples were taken in the deepest part of some reservoirs and results concerning the chironomid composition have already been published (Prat, 1980). The data on Oligochaeta which the samples yielded are presented here.

Material and methods

The profundal samples were taken with a modified Van Veen grab covering a surface of 400 cm². Only one sample was taken in each reservoir during each sampling period. The grab was designed to work in different types of substrates as has been previously discussed (Prat, 1980). The mud was then filtered through a 250 μm net, preserved with formol at 4% and sorted in the laboratory at 10 × magnification.

Sixty-three reservoirs were examined with a total of 96 samples and up to 20 000 individuals seen. The reservoirs sampled were distributed all over Spain (Fig. 1); the samples were collected during the spring of 1974 and autumn and winter of the same year. Additional data on the morphological and limnological characteristics of the reservoirs can be obtained in Margalef *et al.* (1976).

Hydrobiologia 115, 223–230 (1984).

Species composition

From the 96 samples examined, 16 species have been identified (Table 1). Two of them are new for the Spanish fauna: *Potamothrix heuscheri* and *Haber pyrenaicus*. Three of the other species had been previously found only very recently (Rodriguez, 1981). These are *Limnodrilus claparedeianus*, *L. profundicola* and *Potamothrix bavaricus*. This demonstrates how poorly the Spanish oligochaete fauna is known; in fact, the first extensive faunistic study has been published only in very recent years (Martinez-Ansemil & Giani, 1980).

Only two other species, *Stylodrilus heringianus* (Prat, 1980) and *Bothrioneurum vejdowskyanum* (Martinez-Ansemil, 1981), had been previously found among the profundal fauna of other Spanish reservoirs. The most interesting discovery is, of course, the presence of *Haber pyrenaicus* in Lake Sanabria, a natural mountain lake sampled together with the reservoirs. This is the first record of this species far from the Pyrenean mountains, in which it seemed to be endemic (Juget & Giani, 1974). Of a

Table 1. Species of Oligochaeta found in the profundal zone of Spanish reservoirs.

	F	% ind.
Uncinais uncinata (Orsted) -i-	3.2	0.035
Nais pseudobtusa Piguet -i-	1.6	0.015
Dero digitata (Müller) -i-	23.8	0.85
Tubifex tubifex (Müller) -m-	46.0	5.8
Limnodrilus claparedeianus Ratzel -m-	31.7	1.9
L. hoffmeisteri Claparède -m-	58.7	7.4
L. profundicola (Verill) -m-	1.6	0.015
L. udekemianus Claparède -m+i-	7.9	0.30
Psammoryctides barbatus (Grube) -m+i-	1.6	0.099
Potamothrix bavaricus (Oschmann) -m-	4.8	0.14
P. hammoniensis (Michaelsen) -m-	1.6	0.71
P. heuscheri (Bretscher) -m-	4.8	0.09
Spirosperma ferox (Eisen) -i-	1.6	0.25
Embolocephalus velutinus (Grube) -i-	3.2	0.13
Haber pyrenaicus (Juget & Giani) -m-	1.6	0.035
? *Telmatodrilus* sp. -i-	1.6	0.005
Branchiura sowerbyi Beddard -i-	4.8	0.40
Limnodrilus -i-	87.3	46.0
Tubificidae (with hair setae) -i-	61.9	35.7
Lumbriculidae -i-	3.2	0.035

F = frequency; % ind. = relative abundance; m = mature; i = immature.

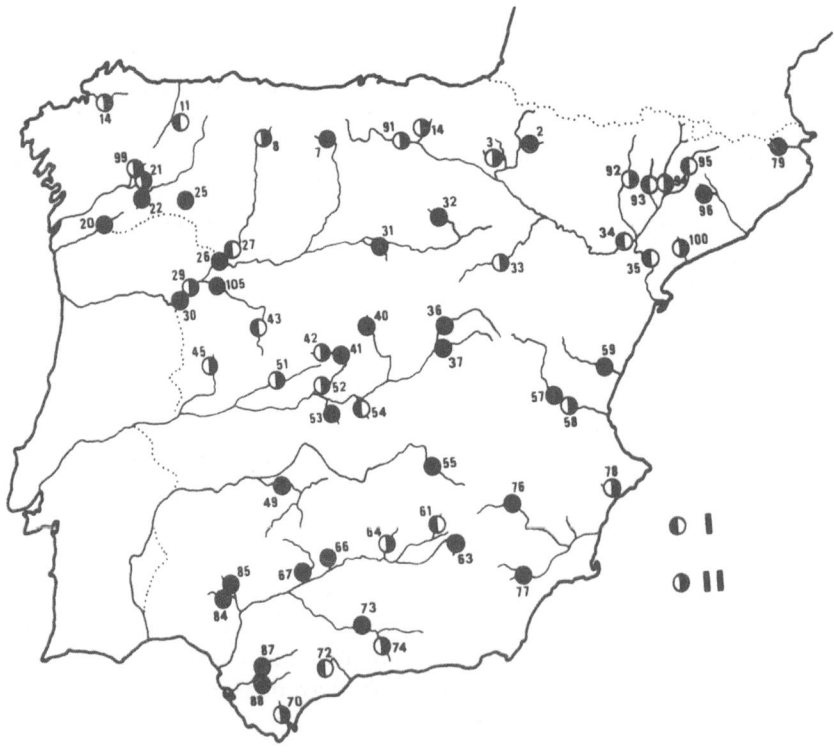

Fig. 1. Sampled reservoirs in the two sampling periods. I: Spring 1974; II: Autumn and Winter 1974.

total of 16 species, only *Limnodrilus hoffmeisteri*, *Tubifex tubifex*, *L. claparedeianus* and *Dero digitata*, present in more than 20% of the reservoirs, are common in our samples, as can be seen in Table 1. A striking feature is the frequency and abundance of immature animals. In fact, out of all the individuals we examined of *Limnodrilus* and tubificids with hair setae, 82% were immature. Only *Limnodrilus hoffmeisteri* and *Tubifex tubifex* are abundant as mature animals.

Patterns of geographical distribution

The most frequent and abundant species are distributed all over Spain. *Limnodrilus hoffmeisteri*, *Tubifex tubifex* and *L. claparedeianus* can be found in the deepest part of many reservoirs (Fig. 2), and the absence of these species from some other reservoirs is probably only a matter of sampling intensity. We think that these species could be found in all our reservoirs if a larger number of samples were taken; differences in relative density are only to be expected as a result of the different trophic status of the reservoirs.

Dero digitata is also a common and well-distributed species as can be seen in Fig. 3, but its presence in the profundal is associated with reservoirs with a high oxygen content in the sediments. The distribution of *Branchiura sowerbyi* is seen to extend to the south, after being previously recorded by De Haro (1964) and Prat (1980) in the north and center of Spain. We believe this is a common species in our reservoirs, especially in the littoral zone (Prat, 1980).

Although very scarce, *Potamothrix bavaricus* and *P. heuscheri* have been recorded in Spain only in the eastern zone where the water is alkaline (Fig. 3) but not in eutrophic reservoirs as has been stated in the literature (see Milbrink, 1980).

Some species, such as *Embolocephalus velutinus*, *Spirosperma ferox* and *Haber pyrenaicus*, seem to be confined to the deepest part of the western Spanish reservoirs (Fig. 4). These reservoirs are oligotrophic, poorly mineralised and with sediment rich in coarse organic material; but no definitive conclusions about the distribution of the above species can be made because an intensive sampling survey is required.

An interesting point is the rareness of *Potamo-*

Fig. 2. Distribution of the most frequent species of Oligochaeta in the profundal zone of Spanish reservoirs (F > 25%).

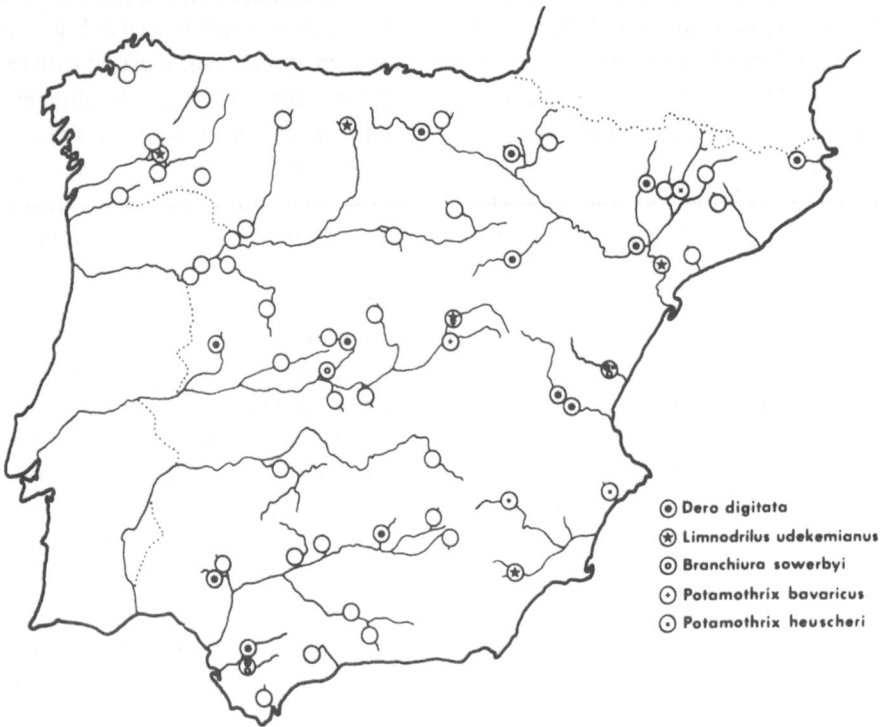

Fig. 3. Distribution of the species of Oligochaeta whose frequency is comprised between 4 and 25% in the profundal zone of Spanish reservoirs.

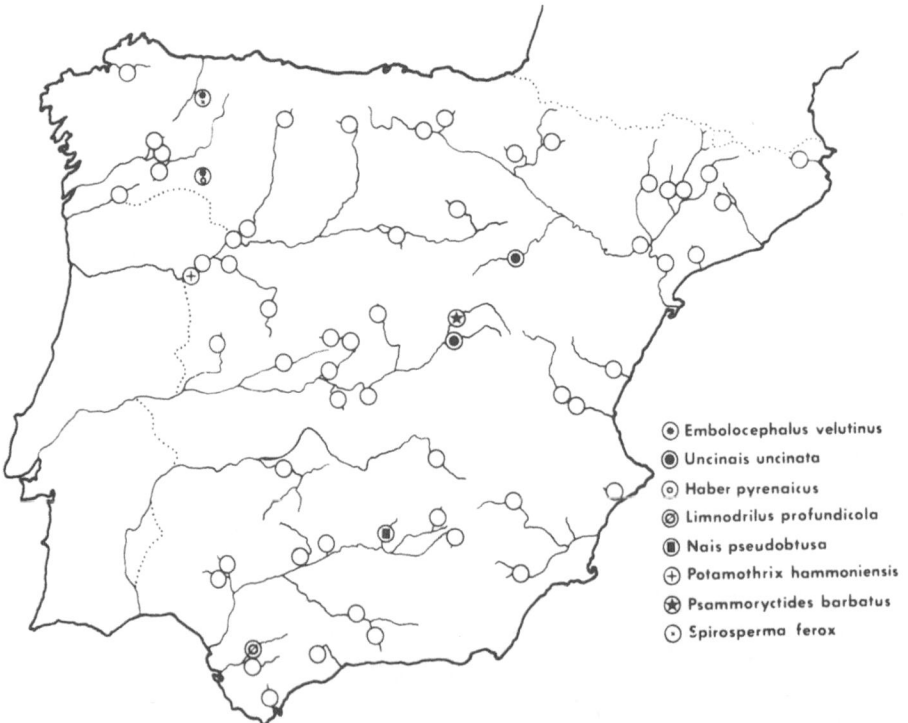

Fig. 4. Distribution of the less frequent species of Oligochaeta in the profundal zone of Spanish reservoirs (F < 4%).

thrix hammoniensis, collected only in two eutrophic reservoirs in western Spain – reservoir 30 in the present survey and reservoir 27 (Prat, 1980) – despite the extreme eutrophic condition of some of the Spanish reservoirs. This species, probably the commonest tubificid in eutrophic lowland lakes in Europe (Milbrink, 1980), seems to be infrequent in Spanish reservoirs. One of the reasons for this rareness could be the almost total absence of lakes in Spain, from where the colonization of the reservoirs by this worm could be expected. This species is also rare in the south-west of France (Giani, pers. commun.).

Another species found in our survey is *Psammoryctides barbatus*, also rare, and present only in the eastern part of Spain (alkaline waters). This is a species very common in alkaline waters and, as Martinez-Ansemil & Giani (1980) have indicated, in the north-west of Spain this species is present only in one river, and that is the most alkaline of all the rivers they sampled.

Depth distribution

The reservoirs sampled were of very different sizes, from very big lakes, more than 100 m deep, to shallow ones with only 3 m of water column. According to the size of the reservoir and the water level at the time of sampling, the samples were taken at very different depths. We were thus provided with information about the depth at which the different species can live.

As can be seen in Fig. 5, many species can be found in both shallow and profundal reservoirs, in particular *Limnodrilus hoffmeisteri* and *Tubifex tubifex*. Depth does not therefore seem to be a critical factor for the majority of the species. Even the naidids, which, in lakes, are normally distributed in the littoral zone (see Learner *et al.*, 1978), can reach considerable depths in some reservoirs, when these are oligotrophic with a good oxygen supply close to the mud. *Dero digitata* has also been found in eutrophic reservoirs, but in this case in shallow ones.

Branchiura sowerbyi was present only in grabs collected in shallow reservoirs. This species colonizes mainly the littoral zone of Spanish reservoirs where in some cases it is very abundant.

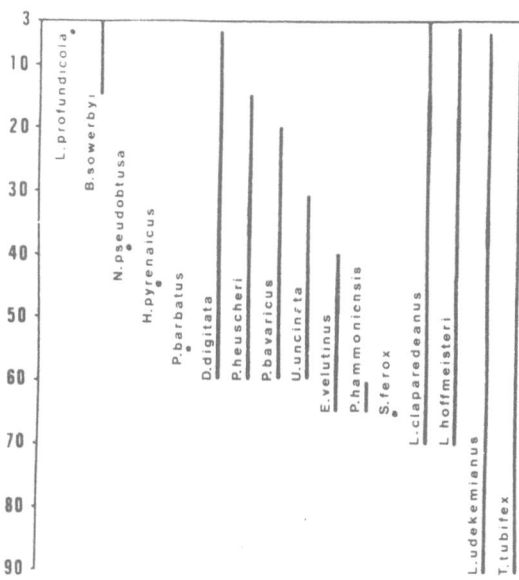

Fig. 5. Bathymetrical distribution of the species of Oligochaeta found in Spanish reservoirs.

The use of Oligochaeta in the typology of a reservoir

One of the aims of geographical studies is the classification of water bodies into different groups. The typology of lakes may be based on physicochemical measurements, indicator species or a combination of both.

The profundal benthos has been extensively used as an indicator of progressive eutrophication of waters (see review by Brinkhurst, 1974), especially the chironomids (Brundin, 1958; Saether, 1975, 1979) and the Oligochaeta (Brinkhurst, 1980; Milbrink, 1980). An attempt at classifying Spanish reservoirs by means of the chironomid fauna has been already made (Prat, 1980), using the indicator species proposed by Brunding (1958) and Saether (1975).

The general survey of Spanish reservoirs has allowed Margalef (1975) and Margalef *et al.* (1976) to classify the reservoirs into six major groups. The key factors for this typology are the mineralization of the waters and the eutrophy defined by phosphorous content, chlorophyll concentration and phytoplankton composition.

Of these six groups (compare Table 2 with Tables XIII and XIV in Margalef *et al.* 1976), the reser-

Table 2. Composition and structure of the Oligochaeta communities in the six groups of reservoirs proposed by Margalef *et al.* (1976).

	I (n=5)		II (n=9)		III (n=6)		IV (n=6)		V (n=17)		VI (n=20)	
	F	D/S	F	D/S	F	D/S	F	D/S	F	D/S	F	D/S
Uncinais uncinata -i-	-	-	-	-	-	-	-	-	-	-	10	0.23
Nais pseudobtusa -i-	-	-	-	-	-	-	-	-	6	0.11	-	-
Dero digitata -i-	-	-	11	0.14	16	0.6	16	2	29	2.2	35	3
Tubifex tubifex -m-	60	60	33	8.1	67	7.5	33	2.1	47	9.7	50	9.60
Limnodrilus claparedeianus -m-	-	-	22	1.4	33	6.4	17	3.5	41	6.8	30	3.5
Limnodrilus hoffmeisteri -m-	60	14.3	67	11.6	100	46.4	50	23.6	71	18.3	45	5.8
Limnodrilus profundicola -m-	-	-	-	-	-	-	-	-	-	-	5	0.4
Limnodrilus udekemianus -m+i-	20	0.7	-	-	-	-	-	-	12	0.5	10	1.4
Psammoryctides barbatus -m+i-	-	-	-	-	-	-	-	-	-	-	5	0.7
Potamothrix bavaricus -m-	-	-	-	-	-	-	-	-	-	-	15	1
Potamothrix hammoniensis -m-	-	-	-	-	17	18	-	-	-	-	-	-
Potamothrix heuscheri -m-	-	-	-	-	-	-	-	-	-	-	10	0.6
Spirosperma ferox -i-	20	7.1	-	-	-	-	-	-	-	-	-	-
Embolocephalus velutinus -i-	40	3.9	-	-	-	-	-	-	-	-	-	-
Haber pyrenaicus -m-	20	1	-	-	-	-	-	-	-	-	-	-
? Telmatodrilus sp. -i-	-	-	-	-	-	-	17	0.4	5.9	0.04	-	-
Branchiura sowerbyi -i-	-	-	11	5.1	-	-	-	-	-	-	5	0.2
Limnodrilus -i-	80	65	67	46	100	257.8	100	173.1	88	100.3	90	67.8
Tubificidae (with hair setae) -i-	80	232.4	56	49.4	50	81	67	18.25	41	16.4	80	121.2
Lumbriculidae -i-	20	0.7	-	-	-	-	-	-	-	-	5	0.07
Total Oligochaeta		385		123		418		223		154		215

n = number of reservoirs in each group; F = frequency; D/S = density per sample; m = mature; i = immature.

voirs of groups I to III are those with poorly mineralized waters and those of groups IV to VI with alkaline waters. Groups I and VI comprise the oligotrophic reservoirs with eutrophy inceasing from group I to III and from group VI to IV. Differences between groups II and III and groups IV and V are mainly of phytoplankton composition (Margalef, 1975).

We have divided the reservoirs we sampled into these six groups, and have noted the frequency and mean density of each oligochaete species per sample in each group (Table 2). As can be seen in Table 2, groups I and VI are those with a more particular species composition. *Spirosperma ferox*, *Embolocephalus velutinus* and *Haber pyrenaicus* were found only in group I (the oligotrophic, poorly mineralized reservoirs) and *Limnodrilus profundicola*, *Psammoryctides barbatus*, *Potamothrix bavaricus* and *P. heuscheri* were present only in group VI (the oligothrophic mineralized ones). These observations must be interpreted rather cautiously because they are only based on a low number of samples, but they provide a hypothesis for future work on the regional distribution of Oligochaeta.

On the other hand, the only example recorded of *Potamothrix hammoniensis* is from one reservoir in group III (the most eutrophic group of reservoirs in the poorly mineralized water zone). This fact is in accordance with the biology of this species, very typical of eutrophic lakes in Europe (Milbrink, 1980), and with the distribution shown by Timm (1980).

The commonest species were *Limnodrilus hoffmeisteri*, *Tubifex tubifex* (present in all the groups) and *L. claparedeianus* (in five groups). The latter species has a low-frequency percentage and low densities in all groups; the frequency and abundance of the first two species seem to be related to the degree of eutrophication of the reservoir.

The lowest densities of *T. tubifex* are in groups III and IV (the most eutrophic) in which *L. hoffmeisteri* are most abundant. Also the density of immature *Limnodrilus* is very high in these groups of reservoirs, while the highest number of immature tubificids with hair setae (a high percentage being *T. tubifex*) was present in groups I and VI (Table 2).

Thus, the density of the most abundant species in our reservoirs (*L. hoffmeisteri* and *T. tubifex*) seems to be related to the eutrophic state, with an increase in abundance of the former species with lake fertility.

References

Armengol, J., 1978. Los crustáceos del planton de los embalses españoles. Oecología aquat. 3: 3–96.

Brinkhurst, R. O., 1974. The Benthos of Lakes. McMillan Press Ltd., London, 190 pp.

Brinkhurst, R. O. 1980. Pollution biology – the North American experience. In: R. O. Brinkhurst & D. G. Cook (eds.), Aquatic Oligochaete Biology. Plenum Press, New York: 471–475.

Brundin, L., 1958. The bottom faunistical lake type system and its application to the southern hemisphere. Moreover a theory of glacial erosion as a factor of productivity in lakes and oceans. Verh. int. Ver. Limnol. 13: 288–297.

De Haro, A., 1964. Sobre la distribución de los Oligoquetos en España. Branchiura sowerbyi Beddard (1892), forma cosmopolita, encontrado en España. Bol. r. Soc. esp. Hist. nat. (Biol.) 62: 137–142.

Estrada, M., 1975. Statistical consideration of some limnological parameters in Spanish reservoirs. Verh. int. Ver. Limnol. 19: 1849–1859.

Juget, J. & N. Giani, 1974. Répartition del Oligochètes lacustres du massif de Néouvielle (Hautes-Pyrénées) avec la description de Peloscolex pyrenaicus, n.sp. Ann. Limnol. 10: 33–53.

Learner, M. A., G. Lochead & B. D. Hugues, 1978. A review of the biology of British Naididae (Oligochaeta) with emphasis on the lotic environment. Freshwat. Biol. 8: 357–375.

Margalef, R., 1975. Typology of reservoirs. Verh. int. Ver. Limnol. 10: 1841–1848.

Margalef, R., D. Planas, J. Armengol, A. Vidal, N. Prat, A. Guiset, J. Toja & M. Estrada, 1976. Limnologia de los embalses españoles. Publicaciones del Ministerio de Obras Públicas, Madrid, no. 123, 454 pp.

Margalef, R. & M. Mir, 1979. Phytoplankton of Spanish reservoirs as dependent from environmental factors and potential indicator of water properties. Atti Convegno sui bacini lacustri artificiali, Sassari, 4–6 Ottobre 1977: 191–205.

Martinez-Ansemil, E., 1981. Estudio taxonómico y ecológico comparativo de los Oligoquetos de los ríos Tambre (Galicia) y Argens (Sur de Francia). Tesis, Univ. Santiago de Compostela, 358 pp.

Martinez-Ansemil, E. & N. Giani, 1980. Premières données sur les Oligochètes aquatiques de la Péninsule Ibérique. Ann. Limnol. 16: 43–54.

Milbrink, G., 1980. Oligochaete communities in pollution biology: the European situation with special reference to lakes in Scandinavia. In: R. O. Brinkhurst & D. G. Cook (eds.), Aquatic Oligochaete Biology. Plenum Press, New York: 433–455.

Planas, D., 1975. Distribution and productivity of the phytoplankton in Spanish reservoirs. Verh. int. Ver. Limnol. 19: 1860–1870.

Prat, N., 1980. Bentos de los embalses españoles. Oecología aquat. 4: 3–43.

Rodriguez, P., 1981. Primeros resultados del estudio de los Oligoquetos acuáticos del País Vasco, 5. Reunión Bienal r. Soc. esp. Hist. nat., Communication P. 130.

Saether, O. A., 1975. Neartic chironomids as indicators of lake typology. Verh. int. Ver. Limnol. 19: 3127–3133.

Saether, O. A., 1979. Chironomid communities as water quality indicators. Holarct. Ecol. 2: 65–74.

Timm, T., 1980. Distribution of aquatic oligochaetes. In: R. O. Brinkhurst & D. G. Cook (eds.), Aquatic Oligochaete Biology. Plenum Press, New York: 55–77.

Index